Lectures on QUANTUM CHROMODYNAMICS

T0331829

Lectures on QUANTUM CHROMODYNAMICS

Andrei Smilga

University of Nantes, France

World Scientific
New Jersey • London • Singapore • Hong Kong

Published by

World Scientific Publishing Co. Pte. Ltd.

P O Box 128, Farrer Road, Singapore 912805

USA office: Suite 1B, 1060 Main Street, River Edge, NJ 07661

UK office: 57 Shelton Street, Covent Garden, London WC2H 9HE

British Library Cataloguing-in-Publication Data
A catalogue record for this book is available from the British Library.

ISBN 981-02-4331-6

Printed in Singapore by Uto-Print

Lectures on QUANTUM CHROMODYNAMICS

To my teachers:

Voldemar Petrovich Smilga
Rudolf Karlovich Bega
Boris Lazarevich Ioffe

Preface

Time to gather stones...

In his well-known popular lectures R. Feynman [1] reflects on the way physical theories are built up and distinguishes two such ways or, rather, two stages in the process of their construction: *(i)* the "Babylonian" stage and *(ii)* the "Greek" stage.

It is not difficult to guess that the term "Babylonian" refers to ancient Babylon and the corresponding physical theory is just geometry. A Babylonian geometer (the words "mathematician" or "physicist" were not yet coined) knew many facts about circles, triangles, and other figures, and his understanding was not purely empirical because he could also *relate* different such facts with each other. He could e.g. heuristically derive a fact A on the basis of some other facts B and C or maybe the fact C on the basis of B and some other fact D, etc. He was actually quite efficient in his main job: the calculation of the areas of land estates. In other words, his theory described the observed experimental facts well and had direct practical applications.

Our Babylonian colleague was lacking, however, a consistent structured system in which a set of basic simple facts are chosen as *axioms* and all others are rigorously derived as *theorems*. Constructing such a system meant the advanced "Greek" stage of understanding geometry. At this stage a physical theory becomes a branch of mathematics where the primary criterion for Truth is not Experiment, but rather Rigor of derivations and internal logical Consistency. Feynman writes that a modern physicist is a

Babylonian rather than a Greek in this respect: he does not care too much about Rigor, and his God and ultimate Judge is Experiment.

Strictly speaking, this is not quite correct (a general statement never is). Some branches of classical and also of quantum physics have now quite reached the Greek stage. As a result, many problems concerning classical and quantum dynamics of Hamiltonian systems are studied now in mathematics departments rather than in physics departments. If you want, rendering a theory understandable to mathematicians may be considered a final goal of research in physics.

Regarding quantum chromodynamics and quantum field theory in general, we are living now in interesting times when we go over from the Babylonian to the Greek stage. A basic concept in quantum field theory is the path integral. Mathematicians still do not understand what it is: they cannot provide a rigorous definition for it. My personal impression, however, is that such a mathematical definition will be given soon.

A transitional period when a physical theory is already essentially constructed and understood at a semi-heuristic level but has not become yet a branch of mathematics, is a proper time to write textbooks. And many such textbooks were written recently. Is there really a need to write another one? It is difficult for me to judge whether this particular book is worth reading, but if you ask me why I have written it, the answer is the following:

There are very good recent monographs [2; 3] on quantum field theory. They are modern and comprehensive but are rather voluminous, and this might be considered a disadvantage. The large size of these books is due to two major choices which their authors make:

- The basics of quantum field theory are described from scratch. No preliminary acquaintance with the subject is required.
- *Many* different quantum field theories are considered there: quantum electrodynamics, quantum chromodynamics, and electroweak theory. The third volume of Weinberg's book is devoted to supersymmetric theories.

Besides general comprehensive monographs, there are also books (cf. [4]–[7]) written in a different style and specially devoted to Yang–Mills theory and to QCD. All these books are "recommended reading" and many pages in my own book were actually written under their influence. That especially concerns Faddeev's and Slavnov's book which was one of my own primary sources when learning QCD.

All these books concentrate, however, on the *perturbative* aspects of QCD, whereas my primary goal is to convey to the reader what is currently known about *nonperturbative* QCD dynamics. In particular, I discuss in some details the instanton solution, the nonperturbative (*viz.* lattice) definition of the path integral including also the "Ginsparg–Wilson revolution" — the modern way to define fermionic path integrals. I further discuss the θ vacuum picture, theoretical and phenomenological aspects of quark–hadron duality, the QCD phase transitions brought about by heat or by squeeze, and, last but not least, confinement. To make the book reasonably self-contained, the perturbative dynamics of QCD is also considered, however. We outline the derivation of the Feynman rules, dwell upon the ultraviolet regularization and renormalization issues, and also review the structure of infrared and collinear singularities in QCD.

Only quantum chromodynamics is discussed* and we assume the reader to be already familiar with QED, the book [8] being our main reference point. That means the acquaintance with the standard operator approach to quantization, derivation of the Feynman rules using Wick's theorem, calculation of Feynman integrals, etc. We describe, however, path integral quantization in field theory and in quantum mechanics — a material not covered by the book [8], which can, of course, be found in almost *any* modern textbook on field theory.

We put emphasis on the principal foundational aspects of QCD and provide only a limited discussion of QCD phenomenology. Actually, one of our assumptions is that the reader has a rough idea about hadron phenomenology, knows about flavor $SU(3)$ symmetry (still, the most basic facts about unitary groups are given in the Appendix) and its implications for the spectrum of mesons and baryons, has heard about confinement, etc.

All these limitations together with a concise manner of presentation make the book relatively thin (which was one of my goals). Being influenced by the style of the Landau course, I have put in a number of **Problems** accompanied by **Solutions**. Each such **Problem** examines a particular (usually, technical and, more often than not, relatively simple) theoretical issue.

---

*To be more exact, we discuss sometimes QED and also some hypothetical (not describing the real World) field theories, but we do it only to clarify and to illustrate what happens in QCD. Purely theoretical issues having no direct implications for QCD are not discussed.

The book does not contain many references. I have decided *not* to give references to classical original papers whose results are now well known and are discussed also in other books on QCD. Vanity-ridden, I sometimes do not refer to my own papers. I tried to quote, however, the *names* of the people who obtained the results I write about. Also, I give references to some less known papers, or to papers which are known well but are more recent and did not get into the textbooks yet. In other words, the references are quite chaotic and always (with one exception) have a nature of personal recommendation to *read* this or that particular book or a paper.

This book has grown up from the lecture course on the foundations of QCD which I have given for graduate students in high energy theory in ITEP in Moscow during the Spring of 1998, and which I later put in writing. Of course, when actually writing the book, I have added much more material than I was able to squeeze into the real time lectures and I also added some "off-shell" lectures which were never actually given. Thus, the book which you hold now in your hands differs from the original draft in an essential way. Still, I have decided to preserve the form of the virtual lecture course.

It is a pleasure to thank C. Bender, M. Chernodub, J. Hoppe, M. Lavelle, P. Mansfield, M. Peskin, K. Selivanov, M. Shifman, A. Slavnov, A. Vainshtein, and L. Yaffe for discussions and useful remarks concerning the book. Special thanks are due to M. Staudacher, who took pains to read carefully the whole manuscript, for many valuable remarks and comments and to my wife E. Savvina for her very essential help with editing the manuscript. I appreciate also the artistic contribution of Olga Feigelman who has drawn the headpieces for the parts.

Contents

PART 1: FOUNDATIONS

PART 3: NONPERTURBATIVE QCD

Notation and Conventions

Minkowskian notation mostly follows that of Bjorken and Drell [9]. We use the standard metric tensor

$$\eta^{\mu\nu} = \eta_{\mu\nu} = \text{diag}(1, -1, -1, -1) . \tag{N.1}$$

A certain amount of care is required when writing 4-vectors in components. For "normal" vectors like the coordinate x^μ, momentum p^μ, or current density j^μ, which are "naturally" contravariant, we write

$$x^\mu = (t, \boldsymbol{x}) , \quad p^\mu = (E, \boldsymbol{p}) , \quad j^\mu = (j_0, \boldsymbol{j}) , \tag{N.2}$$

while for the derivative operator $\partial_\mu \equiv \partial/\partial x^\mu$ and for the vector potential A_μ, which are "naturally" covariant, the convention

$$\partial_\mu = (\partial_0, \boldsymbol{\partial}) , \quad A_\mu = (A_0, \boldsymbol{A}) \tag{N.3}$$

is chosen. Note the difference in the sign of $\boldsymbol{A}$ compared to the standard convention [10]. The standard electric and magnetic field densities are

$$\boldsymbol{E} = \partial_0 \boldsymbol{A} - \boldsymbol{\partial} A_0 , \quad \boldsymbol{B} = -\boldsymbol{\partial} \times \boldsymbol{A} . \tag{N.4}$$

The scalar product of two "normal" 4-vectors X^μ, Y^μ is

$$X \cdot Y \equiv X_\mu Y^\mu = X_0 Y_0 - \boldsymbol{X} \cdot \boldsymbol{Y} , \quad \text{but}$$
$$\partial_\mu x^\mu = \partial_0 x_0 + \boldsymbol{\partial} \cdot \boldsymbol{x} = 4 \quad \text{and} \quad A_\mu dx^\mu = A_0 dt + \boldsymbol{A} \cdot d\boldsymbol{x} ,$$
$$\text{while} \quad \partial_\mu A^\mu = \partial_0 A_0 - \boldsymbol{\partial} \cdot \boldsymbol{A} . \tag{N.5}$$

The gauge boson and the fermion propagators are defined as

$$D_{\mu\nu}(x) = \langle A_\mu(x)A_\nu(0)\rangle_0 = \int \frac{d^4p}{(2\pi)^4} e^{-ip\cdot x} D_{\mu\nu}(p) \,,$$

$$G(x) = \langle\psi(x)\bar\psi(0)\rangle_0 = \int \frac{d^4p}{(2\pi)^4} e^{-ip\cdot x} G(p) \,. \tag{N.6}$$

The γ matrices satisfy the standard anticommutation relations

$$\gamma^\mu\gamma^\nu + \gamma^\nu\gamma^\mu = 2\eta^{\mu\nu} \,. \tag{N.7}$$

Whenever relevant, their explicit form is chosen in the spinor representation:

$$\gamma^0 = \begin{pmatrix} 0 & \mathbb{1} \\ \mathbb{1} & 0 \end{pmatrix}, \quad \gamma^i = \begin{pmatrix} 0 & -\sigma_i \\ \sigma_i & 0 \end{pmatrix},$$

$$\gamma^5 \stackrel{\text{def}}{=} i\gamma^0\gamma^1\gamma^2\gamma^3 = \begin{pmatrix} \mathbb{1} & 0 \\ 0 & -\mathbb{1} \end{pmatrix}, \tag{N.8}$$

with $\mathbb{1}$ being the unit 2×2 matrix, so that

$$\text{Tr}\{\gamma^5\gamma^\mu\gamma^\nu\gamma^\alpha\gamma^\beta\} = -4i\epsilon^{\mu\nu\alpha\beta} \tag{N.9}$$

with the convention

$$\epsilon^{0123} = -\epsilon_{0123} = 1 \,. \tag{N.10}$$

We will also often use the notation

$$\sigma_{\mu\nu} = \frac{1}{2}(\gamma_\mu\gamma_\nu - \gamma_\nu\gamma_\mu) \,. \tag{N.11}$$

The left-handed and right-handed spinors

$$\psi_{L,R} = \frac{1}{2}(1\mp\gamma^5)\psi \,, \quad \bar\psi_{L,R} = \frac{1}{2}\bar\psi(1\pm\gamma^5) \,, \tag{N.12}$$

$\bar\psi = \psi^\dagger\gamma^0$, represent, correspondingly, the lower and the upper components of the bispinor ψ. Obviously, $\gamma^5\psi_{L,R} = \mp\psi_{L,R}$. One says that $\psi_{L,R}$ have, correspondingly, negative and positive *chirality*. Also, the left-handed and right-handed solutions of the free massless Dirac equation have correspondingly, the negative and the positive *helicity* $h = \Sigma\cdot\mathbf{p}/E$ $[(1/2)\Sigma = (1/2)\text{diag}(\sigma,\sigma)$ is the operator of spin]．

The notation $\not{p} \equiv p_\mu\gamma^\mu$ is used, while the hats are reserved for operators and, in some cases, for color matrices.

The Euclidean γ matrices are chosen to be anti-Hermitian

$$\gamma_i^E = (\gamma^i)^M, \qquad \gamma_0^E = i\gamma_0^M \tag{N.13}$$

with the property

$$\gamma_\mu^E \gamma_\nu^E + \gamma_\nu^E \gamma_\mu^E = -2\delta_{\mu\nu} . \tag{N.14}$$

We define also $\gamma_5^E \equiv \gamma_5^M$ and

$$\epsilon_{0123} = -1 \tag{N.15}$$

(or $\epsilon_{1234} = 1$ if labelling the Euclidean time direction by the index 4).

When discussing the Schwinger model, we will need also the form of the γ matrices in two dimensions. The (Minkowskian) conventions are

$$\gamma^0 = \sigma_1 = \begin{pmatrix} 0 & 1 \\ 1 & 0 \end{pmatrix}, \ \gamma^1 = -i\sigma_2 = \begin{pmatrix} 0 & -1 \\ 1 & 0 \end{pmatrix},$$

$$\gamma^5 = \sigma_3 = \begin{pmatrix} 1 & 0 \\ 0 & -1 \end{pmatrix}, \tag{N.16}$$

so that the upper and lower components of a 2D spinor have, correspondingly, positive and negative chirality as in 4 dimensions. The property

$$\mathrm{Tr}\{\gamma^\mu \gamma^\nu \gamma^5\} = 2\epsilon^{\mu\nu} \qquad \text{with} \qquad \epsilon^{01} = 1 \tag{N.17}$$

holds. Then Euclidean convention is $\epsilon_{01} = -\epsilon_{12} = -1$.

When discussing general and nonperturbative aspects of the theory, the coupling constant g is included in the definition of the gauge field so that the covariant derivative is $\mathcal{D}_\mu = \partial_\mu - iA_\mu$ and the non-Abelian field strength tensor is $F_{\mu\nu} = \partial_\mu A_\nu - \partial_\nu A_\mu - i[A_\mu, A_\nu]$. When perturbation theory and Feynman graphs are discussed, the convention is $\mathcal{D}_\mu = \partial_\mu - igA_\mu$, $F_{\mu\nu} = \partial_\mu A_\nu - \partial_\nu A_\mu - ig[A_\mu, A_\nu]$. For QED (spinor and scalar, four- and two-dimensional), the convention is also $\mathcal{D}_\mu = \partial_\mu - ieA_\mu$, bearing in mind that $e = -\sqrt{4\pi\alpha}$ for the physical electron.

The Poisson bracket is defined with the convention $\{p, q\}_{P.B.} = 1$.

Introduction: Some History

Quantum Chromodynamics is the fundamental theory of the strong interactions. In everyday life, the strong interactions have the important task of binding protons and neutrons inside the nucleus. Back in 1932, immediately after the neutron had been discovered, it was realized that the forces keeping together the nucleons in nuclei have nothing to do with the electromagnetic forces, which are much weaker. The official birthdate of QCD is 1974 — it took therefore more than 40 years to understand what is the fundamental theory describing the dynamics of the nuclear forces and to explain the immense bulk of the accelerator data accumulated by that time.

The gradual progress in our understanding of the dynamics of nucleons and other strongly interacting particles, called *hadrons*,[†] cannot be visually represented by a reasonably straight line. It may be better described by a kind of fractal which, starting from Ignorance and passing through innumerable Fallacies, ends up eventually at the predestined final point of Truth.

Maybe we would still be far from the truth were it not for the constant pressure exerted by experimental data. QCD was created while constantly profiting from genuinely physical feedback. Unfortunately, this was probably the last example of such an efficient cooperation between theorists and experimentalists. Those fundamental laws of Nature which are still not known refer to a scale of energies currently not accessible to experiment. As for the most intriguing question of what is going on at the Planck scale $\sim 10^{19}\text{GeV}$ — it presumably will *never* be answered experimentally. We

---

[†]This term coined by L. Okun is derived from the Greek *hadros*, which means "bulky" as opposed to *leptos* meaning "small".

will have to guess the fundamental Theory of Everything without much experimental feedback. The situation is difficult but, of course, not hopeless. Everybody knows an example when such a purely theoretical quest was successful: the (classical) theory of gravitation (or general relativity). It was created by Einstein in 1916 by *fiat*, confirmed by experiment soon thereafter, and, since then, there has been little essential progress in further understanding gravity.

Going back to the "fractal" history of QCD, this fractal smoothens out in our perception as time passes by and the details which are no longer relevant are forgotten. It is still wiggly enough, but its length seems to be finite and the main line of development can (and will) be described in the few pages that follow.

The first theory of strong interactions was constructed by Yukawa in 1935. By analogy with QED, where the interaction between charged particles is mediated by electromagnetic fields whose quanta are photons, Yukawa assumed that the interaction of the nucleons is mediated by some new field whose quanta (the *mesons*) represent a new kind of particle. This new theory differed from QED in three points:

- The electromagnetic field is vectorial in nature. This leads to the repulsion of same-sign charges, and only charges of opposite signs attract. However, the nucleus involves only one kind of particle (in two isotopic forms) and the interaction is attractive. That can only be realized if the quanta of the mediating field have even spin. The simplest possibility is that the spin is zero.
- Photons are massless and the Coulomb interaction is long-range. But the nuclear forces have the short range of order $1\,\mathrm{fm}$. That means that the "meson" should be massive, with mass μ being of order $(1\,\mathrm{fm})^{-1} \sim 200\,\mathrm{MeV}$.
- We are dealing with *strong* interactions, and this means that the meson–nucleon coupling constant (the analog of electric charge) is large.

The theory of Yukawa was brilliantly confirmed in 1947 when the meson with the required properties was discovered in cosmic rays.[‡] That was, of course, the π meson. Three charged states of the π meson (π^+, π^0, and π^-)

[‡]Before that the muon was discovered and the attempts to associate it with the Yukawa mesons lead to considerable confusion. This is one of the "wiggles" we disregard here.

were found which meant that it is an isovector rather than an isoscalar (the apparatus of the isotopic symmetry was already well developped by that time). Also, it turned out to be pseudoscalar rather than scalar. Bearing all this in mind, the Lagrangian of the meson–nucleon theory takes the form

$$\mathcal{L}_{\text{meson}} = i\bar{N}\partial\!\!\!/N - m\bar{N}N + \frac{1}{2}(\partial_\mu\phi)^2 - \frac{\mu^2}{2}\phi^2 - ig\bar{N}\gamma^5\tau_i N\phi_i \ , \quad (0.1)$$

where $\tau_{1,2,3}$ are the Pauli matrices.

Soon after that, it was understood that, even though the theory (0.1) certainly correctly describes some features of nuclear interactions, it cannot be the final truth. First of all, after the discovery of the π mesons many more mesons and baryons were discovered, first in cosmic rays and then in accelerator experiments. The simple Lagrangian (0.1) had no place for them.

Secondly, the theory (0.1) had severe problems of a purely theoretical nature. The main problem was that the coupling constant g was rather big ($g^2/4\pi \sim 14$ which is ~ 1000 times larger than $e^2/4\pi \equiv \alpha$). As a result, practically no calculations were possible. Indeed, at that time (we are in the fifties now) the only known way to work with quantum field theory was to use the perturbative expansions. This works marvellously well for QED, where the expansion parameter is small, but fails completely in the case of the meson theory. Furthermore, at the beginning of the fifties, the so called zero charge phenomenon was discovered by Landau, Abrikosov, and Khalatnikov. It was found that the effective charge grows with the characteristic momentum. This means that the "continuum limit", where the ultraviolet cutoff is sent to infinity, cannot be reached: if the physical charge is kept fixed, we run into the *Landau pole* and the bare coupling constant becomes infinite at some finite Λ_{ultr}. On the other hand, if we keep g^2_{bare} finite, the physical charge goes to zero.

Thus, it became clear that a theory like (0.1), as well as QED, is sick and has no meaning beyond perturbation theory. In QED the latter works and, if we are not afraid of the inevitable trouble which we would meet if we tried to explore the unattainably high energies $E \sim M_{\text{pl}}$, we can obtain nontrivial theoretical predictions and compare them with experiment. In the meson theory, we cannot do that and, since the theory just makes no sense when the coupling is big, nonperturbative methods do not work either. This means that the Lagrangian (0.1) should be put into the waste basket.

Actually, the prevailing attitude of physicists for almost 20 years, from the mid-fifties to the mid-seventies, was that not only *that* Lagrangian, but also all other Lagrangians, the entire Lagrangian method of field theory should be discarded. It was assumed that the correct theory should be founded on some other principles. The idea of the *bootstrap* (that there are no fundamental particles and no fundamental fields whatsoever, and everything depends on itself in some self-consistent way) was popular. The theory of the S-matrix was being intensely developped. We now understand that the purely kinematical requirements of unitarity, causality, space–time and internal symmetries are not restrictive enough to determine the form of the S-matrix in a four-dimensional theory, but that *was* the hope. One of the positive developments in this "medieval" period was Regge theory. It is perceived now as good phenomenological effective theory describing hadron scattering at high energies and small tranverse momenta. But, in old days, this was considered *the* theory, the most refined and ingenious product of kinematical methods.

Thus, physicists, leaving aside the ambition to reveal a fundamental field theory, occupied themselves with a more pragmatic task — to handle the rapidly growing zoo of new "elementary" particles and resonances, if not in a refined theoretical fashion, then at least in some phenomenological way. It was basically the same job as had been done by Mendeleev for the zoo of chemical elements a century earlier. The "Mendeleev Table" of hadrons had been constructed by the early sixties by the efforts of many people, of which M. Gell-Mann probably deserves most credit. It was found that all known hadrons can be grouped into some octets and decuplets representing multiplets of $SU(3)$. This $SU(3)$ symmetry [now called *flavor $SU(3)$*] was not exact, but slightly broken in a controllable way. The model resulted in a lot of non-trivial predictions. The most famous one was the prediction of the existence of the Ω^--hyperon, including an estimate of its mass; this prediction was soon confirmed by experiment. This discovery played the same role as the discovery of missing chemical elements which filled out the empty cells in Mendeleev's Table. One could recall as well the classic story of the discovery of the planet Neptune.

We observe in experiment the octets and the decuplets, but where are the triplets and anti-triplets, the particles belonging to the fundamental representation of $SU(3)$ and its conjugate? These mysterious unobserved particles were called *quarks*.[§] The constituent quark model for hadrons was created. From a purely algebraic result concerning tensor products of several fundamental representations: $\mathbf{3} \otimes \bar{\mathbf{3}} = \mathbf{8} + \mathbf{1}$ and $\mathbf{3} \otimes \mathbf{3} \otimes \mathbf{3} = \mathbf{10} + \mathbf{8} + \mathbf{8} + \mathbf{1}$, it was inferred that mesons (grouped in octets) represent bound states of a quark and an antiquark, while the baryons (grouped in octets and decuplets) consist of three quarks. Assuming that the quarks are fermions and ascribing them the proper quantum numbers: the fractional electric charges $e_u = 2/3$, $e_d = e_s = -1/3$, the strangeness -1 and the isospin 0 for the strange quark and the strangeness 0 and the isospin 1/2 for the u and d quarks, and assuming that the hadrons are composed of quarks in roughly the same manner as the nuclei are composed of nucleons, the quantum numbers of all observed hadrons were correctly reproduced.

That was, of course, an important success, but the dynamics of the forces binding together the quarks in hadrons remained completely unknown. Even if one did not try to answer this question in precise terms, there were two points of trouble. First, it was absolutely unclear why the quarks are always bound and not present in Nature in free unbound form. (A *lot* of experimental efforts were devoted to the search of fractionally charged particles in accelerators, in cosmic rays, in meteorites and in the ice of Greenland: to no avail.) The second difficulty was more technical. Take e.g. the Δ^{++}-isobar which, according to the constituent quark model, is made of three u quarks in S wave with the same spin orientation. But the Pauli principle forbids it (by the same token, it forbids the existence of all other baryons including nucleons).

The remedy for the second trouble was found rather soon. It was assumed that the quark of each flavor (u, d, and s) is found in Nature in three *color*[¶] forms. Thus, on top of $SU(3)$ flavor, we have $SU(3)$ color, which is exact. Pauli's principle is no longer violated because we can write the

---

[§]At first, there were three of them: "up", "down", and "strange" quarks (or just u, d, and s quarks). Now we know three more: "charmed", "bottom" (or "beautiful"), and "top" quarks. The experimental values for the quark masses are

$$m_u \approx 4 \text{ MeV} , \quad m_d \approx 7 \text{ MeV} , \quad m_s \approx 150 \text{ MeV} ,$$
$$m_c \approx 1.35 \text{ GeV} , \quad m_b \approx 4.8 \text{ GeV} , \quad m_t \approx 170 \text{ GeV} . \tag{0.2}$$

[¶]This is not the first time the word "color" appeared in this book. It is actually present right in its title in a disguised form. The Greek word for "color" is *chroma* and *chromo*dynamics means the dynamics of color.

wave function for the baryons in the required antisymmetric form $\epsilon^{ijk}q^i q^j q^k$ ($i, j, k = 1, 2, 3$ are the color indices).

It was postulated that by some miraculous reason only the color–singlet states are allowed to exist in Nature in unbound form. The quarks are color–triplets and are therefore doomed to stay confined within the hadrons. The word *confinement* was coined, but no satisfactory explanation for this phenomenon was suggested. In the middle of the sixties (and even much later) many theorists found this picture so crazy that they did not believe in the existence of quarks. Their position was pretty much the same as the position of Copernicus who said that the astronomical data could be explained more easily if adopting a mathematical model that the Earth and other planets go around the Sun. He never said it *is* the case, indeed (and that is why Copernicus in contrast to Galilei was never bothered by the Church).

This viewpoint actually prevailed until 1969 when experiments on deep inelastic electron scattering were performed at SLAC. It was shown that the total cross section of the deep inelastic scattering of virtual photons on nucleons (recall that the deep inelastic scattering is the inclusive process $e + N \rightarrow e +$ anything or, which is the same, $\gamma^* N \rightarrow$ anything, where γ^* is the virtual photon emitted by an electron) behaved in the same way as the cross section of the scattering of γ^* on a charged pointlike particle. That meant that there *are* some pointlike constituents within the proton. They were at first called *partons*, but it soon became clear that everything works wonderfully well if one assumes that these partons have the quantum numbers of quarks. Thus, quarks surpassed the status of an amusing mathematical model and displayed themselves as dynamical entities which interact and, hence, are really there (*interago ergo sum*).

But what about field theory? Was it indeed true that theorists, being frustrated by the Landau pole problem, abandoned all field theory studies? Not, of course. The latter were no longer streamline, but the nature of science is such that one can always find some people standing aside the crowd, doing unpopular things. Sometimes, this happens to be the right thing to do.

The field theory on whose basis QCD was eventually built, Yang–Mills theory, had actually been invented a long time ago, in 1954. Yang and Mills generalized the principle of local gauge invariance to the non-Abelian gauge group $SU(2)$ and wrote the celebrated Lagrangian $\mathcal{L}_{YM} = -(1/4)F^a_{\mu\nu}F^{\mu\nu\,a}$.

The spectrum of this theory involves massless self-interacting vector particles belonging to the triplet of $SU(2)$. Yang and Mills added the mass term $-m^2(A_\mu^a)^2/2$ to the Lagrangian and tried to associate the resulting massive vector particles with the ρ mesons. This attempt did not work. As a result, Yang–Mills theory was not considered to be a realistic physical theory during the following 10–15 years.

But the model as such was interesting and many theorists continued to play with it. It turned out that the quantization of such a theory is far from being easy. Proceeding in a naïve way, one gets nonsensical results, including violation of unitarity. At the beginning of the sixties, Feynman tried to resolve this problem (he was interested in Yang–Mills theory as a model for quantum gravity: the nonlinear interactions of gauge bosons in the former carry some features of the nonlinear interactions of gravitons in the latter). Feynman did not succeed by himself, but he had some important insights, which helped Faddeev and Popov to develop their *ghost* method in 1967. Including the ghosts made perturbation theory self-consistent and it became possible to calculate scattering amplitudes and other quantities at, in principle, any order in the coupling constant.

While some people tried to construct a consistent theoretical scheme for the quantization of Yang–Mills theory, some other researchers contemplated on its possible phenomenological applications. Nobody was too much in a hurry.

When the idea to describe ρ mesons as Yang–Mills quanta was abandonned, people did not further think about the possibility of applying this theory to the physics of strong interactions, but rather turned to the weak interactions. Fermi theory, involving four-fermion vertices with dimensionful coupling, was sick: nonrenormalizable and nonunitary. The idea was to consider it as the effective low-energy Lagrangian of some other theory involving only dimensionless couplings of fermions with massive charged vector bosons. Charged fundamental vector particles are most naturally described as Yang–Mills particles (ρ mesons are not well described in this way, but neither are they fundamental). Of course, in pure Yang–Mills theory the quanta are massless and one should have guessed how to give them mass in some benign way in order not to spoil renormalizability and unitarity of the massless theory (which, as a matter of fact, were not yet proven at that time!)

An almost correct model of the weak interactions was constructed along these lines by Glashow back in 1961 and the fully correct one, which unites weak and electromagnetic interactions, by Weinberg and Salam in 1967. The $W^\pm$ and Z^0 bosons were endowed with masses by the non-Abelian Higgs mechanism. The renormalizability and unitarity of this model were proven later by 't Hooft and Veltman.

The idea that Yang–Mills theory describes not only the physics of weak interactions, but also the strong interaction dynamics was put forward in 1973 in three independent papers by Pati and Salam; Fritzsch, Gell-Mann, and Leutwyler; and Weinberg. They suggested that the quarks interact with each other by the exchange of *gluons*, massless vector particles belonging to the octet representation of $SU(3)$. The gluons also interact between themselves, as is dictated by the Yang–Mills Lagrangian with the $SU(3)$ color gauge group. A real breakthrough occurred when the charge renormalization in this theory was calculated by Gross, Wilczek, and Politzer, and the phenomenon of asymptotic freedom was discovered: the effective charge does not grow with energy as is the case for Landau-pole-plagued theories like QED, but rather decreases.

History is rather dramatic at this point. The first person who correctly calculated the renormalization of the coupling constant in pure Yang–Mills theory was Khriplovich as early as 1969. He calculated the 1-loop gauge boson polarization operator $\Pi_{\mu\nu}$ in Coulomb gauge, which does not involve ghosts. In Coulomb gauge, $\Pi_{\mu\nu}$ is the *only* source of the charge renormalization; the contribution of the vertices is irrelevant. Khriplovich did not understand, however, the meaning of his result, and did not think of applying Yang–Mills theories to physics of strong interaction. There was an even earlier paper in 1965 by Terent'ev and Vanyashin, who found indications that charge may decrease with energy, but they did the calculation in a nonrenormalizable theory (without ghosts, which were not yet known at that time) and ascribed the strange behavior of the charge to this. Finally, 't Hooft also calculated it about a year before Gross, Wilczek, and Politzer, but did not, at first, understand the overwhelming significance of his result and did not publish it.[||]

At any rate, by the end of 1974 the dust settled, and the picture became clear. The asymptotic freedom of the theory:

(1) Made it self-consistent: there *is* a continuum limit where the bare charge falls off as the ultraviolet cutoff increases in such a way that

---

[||]The interested reader can read about all that in more details in a historical review article by M. Shifman written for the volume [11].

the effective charge defined at a given energy scale is kept fixed.

(2) Explained SLAC deep inelastic data. When the energy is large, the interaction between quarks is relatively weak, we can treat them as free and that is what was seen in experiment. Moreover, a quantitative theory of "scaling violations", i.e. the theory of deviations from the free quark–parton model due to interaction between quarks was developped by Politzer in 1974. These deviations turned out to be rather weak. Now experiments are precise enough to see them, and they perfectly well agree with the theory.

(3) Explained why strong interactions are strong. If the coupling falls off at small distances, it grows at large distances, becoming roughly of order 1 at the distances of order 1 fm, which are relevant to the nuclear physics.

(4) Did not explain the phenomenon of confinement (to derive it from the first principles of QCD is still a major unresolved task), but made it probable. If the interaction is strong at large distances, who knows how it behaves there? It is quite conceivable that the spectrum of asymptotic physical states does not resemble the set of the fundamental fields and, in particular, does not include colored states. Some modern ideas on the possible mechanisms of confinement will be discussed at the end of this book.

Since that time, many other experimental tests of QCD were carried out, and there is not a slightest doubt anymore that QCD is correct. Probably, the most spectacular confirmation comes from jet physics. Consider the process of e^+e^- annihilation with creation of a bunch of hadrons in the final state. If the energy of colliding electron and positron is large, the process goes, according to the theory, in two stages. At the first stage, the fundamental particles, the quarks and gluons are produced. This fundamental process can be $e^+e^- \rightarrow q\bar{q}$, $e^+e^- \rightarrow q\bar{q}g$, $e^+e^- \rightarrow q\bar{q}q\bar{q}$, etc. The cross section of these processes can be calculated in the framework of QCD perturbation theory.

But since quarks and gluons do not exist in free form, we observe only hadrons in the final state. Thus, at the second stage each energetic quark or gluon creates a *jet* of hadrons going roughly in the same direction as the initial colored particles (see Fig. 0.1).

Such jet processes are indeed seen. The most common is the process when only two jets are created. The underlying fundamental process is

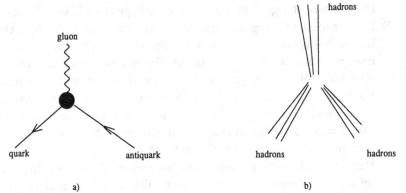

Fig. 0.1 Production of three jets in e^+e^- annihilation. *a)* quark and gluon stage, *b)* hadron stage.

$e^+e^- \to q\bar{q}$. The three-jet process (corresponding to the QCD process $e^+e^- \to q\bar{q}g$) is less probable because the cross section involves an extra power of the strong coupling constant α_s (which is small at high energies). An event where four jets are created (corresponding to the QCD processes $e^+e^- \to q\bar{q}gg$ and $e^+e^- \to q\bar{q}q\bar{q}$) is still less probable, etc. Experimentalists can measure jet cross sections, *viz.* the probabilities of producing, say, 3 jets with given total energies (E_{jet} is the sum of the energies of all particles forming the jet) and forming given angles. There is quantitative agreement between such data and the theoretical result for the differential cross section of the process $e^+e^- \to \bar{q}qg$.

There are many other tests of QCD: a series of tests for heavy quarkonium systems, for the decays of W and Z bosons, for the masses, widths, static and dynamical characteristics of low lying hadron resonances (they were calculated in QCD numerically by doing the functional integrals on the lattice and semianalytically using the ingenious *ITEP sum rules* technique) as well as many other things. The voices of the last "heretics" who have still expressed doubts whether QCD is the true theory of strong interactions, faded away by the end of the seventies.

PART 1

FOUNDATIONS

Lecture 1

Yang–Mills Field

1.1 Path Ordered Exponentials. Invariant Actions.

We start from the well known Lagrangian of electrodynamics involving a charged scalar field $\phi(x)$:

$$\mathcal{L} = -\frac{1}{4e^2} F_{\mu\nu} F^{\mu\nu} + \mathcal{D}_\mu \phi \mathcal{D}^\mu \phi^* \, , \tag{1.1}$$

where $F_{\mu\nu} = \partial_\mu A_\nu - \partial_\nu A_\mu$ and $\mathcal{D}_\mu = \partial_\mu - iA_\mu$ is the covariant derivative (a more conventional definition for A_μ differs from ours by the factor e). The Lagrangian (1.1) is invariant under *gauge transformations*

$$\begin{aligned} A_\mu &\rightarrow A_\mu + \partial_\mu \chi(x) \, , \\ \phi &\rightarrow e^{i\chi(x)} \phi \, . \end{aligned} \tag{1.2}$$

These transformations form a one-parameter Abelian group which is $U(1)$. Yang–Mills theory is a generalization of QED to the case of non-Abelian gauge groups. We will now suggest a geometric interpretation for the action (1.2) which is of little use for QED but is very useful for deriving the Yang–Mills action in a natural way.

Suppose we define an element of the $U(1)$ group $U(x) = e^{i\alpha(x)}$ at each point of our Minkowski space–time. One can realize it in many different ways. One such way can be obtained from another by multiplying each $U(x)$ by an element of the group $e^{i\chi(x)}$, i.e. by a gauge transformation. Mathematicians would say that a *principal fiber bundle* over R^4 with the group $U(1)$ as a fiber is defined. The function $U(x)$ is called the *section* of this fiber bundle.* Instead of considering an element of the group itself, one could consider equally well a representation of the group on which the elements of the group act and which, in the trivial case of $U(1)$, is simply a charged field $\phi(x)$. The geometric meaning of the gauge field A_μ is clarified if the operation of *parallel transport* is defined. For each point x we define

$$U_{\Rightarrow x+dx}(x) = \Phi(x+dx, x)U(x) \equiv e^{iA_\mu(x)dx^\mu}U(x) , \qquad (1.3)$$

where dx here is assumed to be small. The notation is not standard, and it probably makes sense to express it also in words: $U_{\Rightarrow x+dx}(x)$ is the function U taken at the point x and then transported parallelly with the factor Φ to an infinitesimally close point $x+dx$. If we perform a parallel transport at a finite distance along some path P parametrized as $\xi^\mu(s)$, we obtain

$$U_{\underset{\Rightarrow y}{P}}(x) = \Phi(y, x; P)U(x) \equiv e^{\int_x^y iA_\mu(\xi)d\xi^\mu}U(x) , \qquad (1.4)$$

where the integral is performed along P. $U_{\underset{\Rightarrow y}{P}}(x)$ depends on the path and does not coincide, of course, neither with $U(x)$ nor with $U(y)$. What is however remarkable is that it transforms under the gauge transformations in the same way as $U(y)$. Indeed, substituting $A_\mu \to A_\mu(x) + \partial_\mu\chi(x)$, we see that the exponential $\Phi(y, x; P)$ (P marks the path) is changed as

$$\Phi(y, x; P) \to e^{i\chi(y)}\Phi(y, x; P)e^{-i\chi(x)} . \qquad (1.5)$$

If substuting also $U(x) \to e^{i\chi(x)}U(x)$, we see that $U_{\Rightarrow y}(x) \to e^{i\chi(y)}U_{\Rightarrow y}(x)$. Note that for any *closed* contour C, the quantity $\Phi_C = \exp\{i\oint_C A_\mu(\xi)d\xi^\mu\}$ is gauge invariant.

To understand $F_{\mu\nu}$ in these terms, consider the operator Φ for parallel transport along the perimeter of a small square $\square$ with the side a in, say, the (12) plane. Consider first parallel transport along one of the sides:

$$\Phi(x^\mu + a\delta_1^\mu, x^\mu) = e^{ia\int_0^1 dsA_1(x_\mu + as\delta_{\mu1})} = e^{iaA_1(x^\mu + a\delta_1^\mu/2)} + o(a^2) .$$

---

*One must say here that the theory of gauge fields was developed independently in pure mathematics at roughly the same time as physicists developed it for their purposes. Later it was realized that one and the same thing was considered in different terms.

We have[†]

$$\Phi_\square = e^{-iaA_2[x^\mu + a\delta_2^\mu/2]} e^{-iaA_1[x^\mu + a(\delta_1^\mu/2 + \delta_2^\mu)]} e^{iaA_2[x^\mu + a(\delta_1^\mu + \delta_2^\mu/2)]}$$

$$\times e^{iaA_1[x^\mu + a\delta_1^\mu/2]}$$

$$= e^{ia^2(\partial_1 A_2 - \partial_2 A_1)} + o(a^2) = e^{ia^2 F_{12}} + o(a^2) . \tag{1.6}$$

A parallel transport over a finite closed contour is

$$\Phi_C = e^{i \iint F_{\mu\nu} d\sigma^{\mu\nu}} ,$$

where the integral goes over any surface bounded by the contour C. This, of course, could be derived immediately from the definition of Φ_C and the Stokes theorem.

A striking analogy with general relativity is now seen. The parallel transport on our fiber bundle $U(x)$ is fully analogous to the parallel transport of vectors on a curved Riemann surface. The gauge potential A_μ plays the role of the Christoffel symbols. The tensor $F_{\mu\nu}$ appeared as the operator or parallel transport over a small closed contour and is the analog of the Riemann curvature. Finally, one can define a *covariant derivative* in the same way as in general relativity: we define

$$n^\mu \mathcal{D}_\mu \phi \equiv \lim_{\epsilon \to 0} \frac{\phi(x + \epsilon n) - \phi_{\Rightarrow x + \epsilon n}(x)}{\epsilon} = n^\mu(\partial_\mu - iA_\mu)\phi(x) . \tag{1.7}$$

Thus, to compare $\phi(x)$ in closed spatial points, we should first transport $\phi(x)$ with the operator $\Phi(x + \epsilon n, x)$ to the adjacent point and only after that form a difference. That is why an alias for the gauge potential A_μ used by matematicians is the *connection*, it kind of "connects" adjacent points in our fiber bundle. Under the gauge transformation, the covariant derivative transforms as

$$\mathcal{D}_\mu \phi(x) \to e^{i\chi(x)} \mathcal{D}_\mu \phi(x) ,$$

in contrast to the usual derivative, which does not transform in a nice way.

Let us now generalize this construction to the non-Abelian case. Suppose we have a principal fiber bundle based on an arbitrary Lie group G, and its section (or just group-valued function) $U(x)$. The gauge transformation is the transition from one section to another according to

$$U(x) \to \Omega(x)U(x) ,$$

[†]The Minkowskian conventions (N.1)–(N.5) are used.

with $\Omega \in G$. Define an infinitesimal parallel transporter

$$\Phi(x + dx, x) = e^{iA_\mu^a(x)T^a dx^\mu} \, , \tag{1.8}$$

where $A_\mu^a(x)$ is a non-Abelian version of the connection (i.e. gauge potential) belonging to the corresponding Lie algebra, and T^a are the generators of the group in some representation. We will only be interested in unitary groups and choose T^a in the fundamental representation $T^a \equiv t^a$, $\text{Tr}(t^a t^b) = \frac{1}{2}\delta_{ab}$.

Let us now *require* that $U_{\Rightarrow x+dx}(x) = \Phi(x + dx, x)U(x)$ is transformed under the gauge transformations in the same way as $U(x + dx)$:

$$U_{\Rightarrow x+dx}(x) \rightarrow \Omega(x + dx)U_{\Rightarrow x+dx}(x) \, ,$$

which means that

$$\Phi(x + dx, x) \rightarrow \Omega(x + dx)\Phi(x + dx, x)\Omega^\dagger(x) \, .$$

Substituting here $\Phi(x + dx, x)$ from Eq. (1.8) and comparing the terms $\propto dx^\mu$, we *derive* the transformation law for the gauge potential:

$$\hat{A}_\mu \rightarrow \Omega(x)\hat{A}_\mu\Omega^\dagger(x) - i\left[\partial_\mu\Omega(x)\right]\Omega^\dagger(x) \, , \tag{1.9}$$

where $\hat{A}_\mu = A_\mu^a t^a$. For infinitesimal gauge transformations, $\Omega(x) = e^{i\hat{\chi}(x)} \approx 1 + i\hat{\chi}(x)$,

$$\hat{A}_\mu \rightarrow \hat{A}_\mu + \partial_\mu\hat{\chi}(x) + i[\hat{\chi}(x), \hat{A}_\mu] \, . \tag{1.10}$$

For the Abelian $U(1)$ group, we reproduce, of course, the law of transformation (1.2) of electrodynamics.

The parallel transporter (1.8) can also act on the fields in an arbitrary representation. In the full analogy with what we had earlier, we can define the covariant derivative of the field as

$$\mathcal{D}_\mu\phi(x) = (\partial_\mu - iA_\mu^a T^a)\phi(x) \, .$$

Under the gauge transformation, the covariant derivative $\mathcal{D}_\mu\phi(x)$ is multiplied from the left by the factor $e^{i\chi^a(x)T^a}$ in the same way as $\phi(x)$.

One can define also a (path-dependent) parallel transporter along a finite path P. It is given by a so-called P−ordered exponential which is defined as the product of an infinite number of infinitesimal transporters (1.8) along the path:

$$\Phi(y, x; P) = P \exp\left\{ i \int_x^y \hat{A}_\mu(\xi) d\xi^\mu \right\}$$

$$\overset{def}{=} \lim_{n \to \infty} \Phi(y, \xi_n) \Phi(\xi_n, \xi_{n-1}) \cdots \Phi(\xi_1, x) . \tag{1.11}$$

Note that the product is formed starting from the *end* point y. It is easy to see that the law of gauge transformation for $\Phi(y, x; P)$ is the same as for $\Phi(x + dx, x)$.

In the Abelian case, the parallel transporter along a closed path is gauge invariant. This is no longer true in the non-Abelian cases where it depends in general on an initial $\equiv$ final point x that we fix on the contour: $\Phi_C \to \Omega(x)\Phi_C\Omega^\dagger(x)$. However, the trace $W(C) = \mathrm{Tr}\{\Phi_C\}$ is gauge invariant. This is called the *Wilson loop*.

To define a non-Abelian field density tensor $\hat{F}_{\mu\nu}$, consider as before the operator of parallel transport along the perimeter of a small square (anticipating further lattice applications, we can call it *plaquette* right now). Substituting $A_\mu \to \hat{A}_\mu$ in (1.6) and taking into account the fact that $\hat{A}_\mu$ do not commute anymore (thereby, the order of the exponential factors is now important), we obtain

$$\Phi_{\square_{\mu\nu}} = e^{ia^2 \hat{F}_{\mu\nu}} + o(a^2) , \tag{1.12}$$

with

$$\hat{F}_{\mu\nu} = \partial_\mu \hat{A}_\nu - \partial_\nu \hat{A}_\mu - i[\hat{A}_\mu, \hat{A}_\nu] = i[\mathcal{D}_\mu, \mathcal{D}_\nu] . \tag{1.13}$$

In "colored vector" notation,

$$F_{\mu\nu}^a = \partial_\mu A_\nu^a - \partial_\nu A_\mu^a + f^{abc} A_\mu^b A_\nu^c , \tag{1.14}$$

where f^{abc} are the structure constants of the group defined in Eq. (A.23). In the following, we will frequently go over from the matrix notation to colored vector notation and back without further comments. We will not mark matrices with hats anymore.

Under the gauge transformation,

$$F_{\mu\nu} \to \Omega(x) F_{\mu\nu} \Omega^\dagger(x) . \tag{1.15}$$

Thus, $F_{\mu\nu}$ has the geometric meaning of the curvature of our fiber bundle in color space.[‡]

Take now the trace of $\Phi_{\Box\mu\nu}$. To get a nontrivial result, one should expand the exponentials in the analog of Eq. (1.6) up to the order a^4. We obtain (check it!)

$$W_{\Box\mu\nu} = \frac{1}{N}\text{Tr}\left\{\Phi_{\Box\mu\nu}\right\} = 1 - \frac{a^4}{2N}\text{Tr}\{F_{\mu\nu}F_{\mu\nu}\} + O(a^6) . \qquad (1.16)$$

The factor $1/N$ (for the $SU(N)$ gauge group) is introduced for convenience; no summation over μ, ν is assumed. This quantity is gauge invariant and so is the quantity $\text{Tr}\{F_{\mu\nu}F_{\mu\nu}\}$ (this is seen also from the gauge transformation law (1.15) for $F_{\mu\nu}$).

The gauge and Lorentz invariant Lagrangian of pure Yang–Mills theory is defined as

$$\mathcal{L} = -\frac{1}{2g^2}\text{Tr}\{F_{\mu\nu}F^{\mu\nu}\} = -\frac{1}{4g^2}F^a_{\mu\nu}F^{a,\mu\nu} , \qquad (1.17)$$

where the summation over μ, ν is now performed with the standard Minkowskian metric (N.1). This is the direct analog of the Lagrangian of pure photodynamics.

The latter is a trivial theory involving no interaction. In the Abelian case, to make things interesting, we have to include charged matter fields. In the non-Abelian case, already pure gluodynamics involves nonlinear interactions (with the coupling constant g) and is highly non-trivial. It is not difficult to also write down invariant Lagrangians describing the interactions of colored matter with non-Abelian Yang–Mills fields. Suppose we have a scalar field in the fundamental representation of the group. The corresponding Lagrangian can be written as a gauge invariant combination $[\mathcal{D}_\mu\phi(x)]^\dagger\mathcal{D}^\mu\phi(x)$. For spinor fields, we can write

$$\mathcal{L}_\psi = i\bar\psi\mathcal{D}_\mu\gamma^\mu\psi + \text{optional mass term} . \qquad (1.18)$$

QCD is a theory involving the Yang–Mills fields based on the gauge group $SU(3)$, and six different Dirac spinor fields in the fundamental representation of the group. The quanta of the gauge fields are called *gluons*), and

[‡]We hasten to comment that general relativity, the analogy with which we explore, is a *more* complicated theory than just a gauge theory with local Lorentz group. For sure, the local Lorentz symmetry is there, being associated with local rotations of the vierbein. But the Einstein action is invariant under a much larger symmetry: the symmetry of general coordinate transformations.

the quanta of the fermion fields quarks. The Lagrangian reads

$$\mathcal{L}_{QCD} = -\frac{1}{2g^2}\text{Tr}\{F_{\mu\nu}F^{\mu\nu}\} + \sum_{f=1}^{6}\bar{\psi}_f(i\not{D} - m_f)\psi_f . \qquad (1.19)$$

However, for the rest of the present lecture as well as in the following one, we will forget about the existence of the quarks and will concentrate on pure Yang–Mills dynamics.

The dynamical equations of motion are obtained by varying the Lagrangian (1.17) with respect to A_μ and have the form

$$[\mathcal{D}_\mu, F^{\mu\nu}] = \partial_\mu F^{\mu\nu} - i[A_\mu, F^{\mu\nu}] = 0 , \qquad (1.20)$$

or

$$\partial_\mu F^{a,\mu\nu} + f^{abc}A_\mu^b F^{c,\mu\nu} = 0 . \qquad (1.21)$$

Bearing in mind that, according to (1.13), $F_{\mu\nu}$ is a commutator of covariant derivatives and using the Jacobi identity, we derive the Bianchi identity for our color curvature

$$[\mathcal{D}_\sigma, F_{\mu\nu}] + [\mathcal{D}_\mu, F_{\nu\sigma}] + [\mathcal{D}_\nu, F_{\sigma\mu}] = 0 , \qquad (1.22)$$

the "second pair of Maxwell equations".

Let us construct now the classical Hamiltonian of the Yang–Mills field. The canonical Hamiltonian is $\sum_i p_i\dot{q}_i - \mathcal{L}$, where q_i are generalized coordinates and p_i are their canonical momenta. Just as in electrodynamics, we meet here a difficulty that not all variables A_μ are dynamical. A_0 enters the Lagrangian (1.17) without time derivatives and has no associated canonical momentum. On the other hand, the spatial components A_i^a are dynamical and their canonical momenta P_i^a are just chromoelectric fields $P_i^a = \frac{1}{g^2}E_i^a = \frac{1}{g^2}F_{0i}^a$. Only a part of Lagrange's equations of motion, namely the equations

$$\frac{d}{dt}\frac{\delta\mathcal{L}}{\delta\dot{A}_i} - \frac{\delta\mathcal{L}}{\delta A_i} = 0 \qquad (1.23)$$

describe the time evolution of the system. The equations of motion obtained by varying the Lagrangian with respect to A_0,

$$\begin{aligned} G &= [\mathcal{D}_i, E_i] = 0 & \text{or} \\ G^a &= \partial_i E_i^a + f^{abc}A_i^b E_i^c = 0 \end{aligned} \qquad (1.24)$$

(the Gauss law) have the meaning of *constraints* to be imposed on the initial Cauchy data for the equation system (1.23). Actually, the fact that a part of the variables is not dynamical and that we thus deal with a constrained system is very closely related to the fact that the Lagrangian (1.17) enjoys the gauge invariance (1.9). We will explore this relation in details when discussing the quantum theory. Here we just write the system of Hamilton's equations of motion corresponding to the Lagrangian (1.17). In full analogy with electrodynamics, the Hamiltonian is

$$\mathcal{H} = \frac{1}{2g^2} \int d^3x [E_i^a E_i^a + B_i^a B_i^a] = \frac{1}{2} \int d^3x \left[g^2 P_i^a P_i^a + \frac{1}{g^2} B_i^a B_i^a \right], \quad (1.25)$$

where $B_i^a = -\frac{1}{2}\epsilon_{ijk} F_{jk}^a$ is the chromomagnetic field. Dynamical Hamilton's equations, describing the phase space dynamics of the system, are

$$\frac{\delta \mathcal{H}}{\delta P_i^a} = E_i^a = \dot{A}_i^a \,,$$

$$-\frac{\delta \mathcal{H}}{\delta A_i^a} = \frac{1}{g^2}[\partial_j F_{ji}^a + f^{abc} A_j^b F_{ji}^c] = \dot{P}_i^a = \frac{1}{g^2}\dot{E}_i^a \,, \quad (1.26)$$

and the constraints (1.24) have already been written. The time evolution of the system described by the Hamiltonian (1.25) is compatible with the constraints (1.24). This is guaranteed by the fact that the Poisson bracket $\{G, \mathcal{H}\}_{\text{P.B.}}$ is zero, as can be checked explicitly.

Problem. Calculate $\{G^a(x), G^b(y)\}_{P.B.}$.
Solution. We have

$$\{G^a(x), G^b(y)\}_{P.B.} = \{\partial_i E_i^a(x) + f^{acd} A_i^c(x) E_i^d(x), \partial_j E_j^b(y)$$

$$+ f^{bef} A_j^e(y) E_j^f(y)\}_{P.B.}$$

$$= -f^{abc} \left[\frac{\partial}{\partial x_i}\delta(x-y) E_i^c(y) + \frac{\partial}{\partial y_i}\delta(x-y) E_i^c(x) \right]$$

$$+ f^{acd} f^{bef} \left[\delta^{ed} A_i^c(x) E_i^f(y) - \delta^{cf} A_i^e(y) E_i^d(x) \right] \delta(x-y) \,. \quad (1.27)$$

The first term is transformed using the identity

$$\delta'(x-y)[f(y) - f(x)] = -\delta'(x-y)f'(x)(x-y) = \delta(x-y)f'(x)$$

and gives $-f^{abc}\partial_i E_i^c(x)\delta(x-y)$. The second term is transformed by use of the Jacobi identity for structure constants (A.24) and gives, respectively,

$-f^{abc}f^{cde}A_i^d(x)E_i^e(x)\delta(x-y).$

The relation obtained

$$\{G^a(x), G^b(y)\}_{P.B.} = -f^{abc}G^c(x)\delta(x-y) \qquad (1.28)$$

is rather important and will be used later to quantize the theory.

1.2   Classical Solutions.

We will concentrate here on the solutions in real Minkowski time. Euclidean solutions will be a subject of the next lecture.

The simplest ansatz is $A_\mu(x) = cf_\mu(x)$, where c is a colored matrix, one and the same for all components μ and for all points x. Then the commutators in $F_{\mu\nu}$ and in the equations of motion disappear and the classical dynamics of the system is the same as for the photodynamics, i.e. trivial. Any solution can be expressed as a superposition of the plane waves $A_\mu(x) = ce_\mu e^{-ikx}$. Their quantization gives gluons which will be discussed at length later.

However, the non-linear equations (1.20) admit a lot of other nontrivial solutions. Let us first discuss the solutions that our system does *not* have, namely, soliton solutions. Solitons, i.e. localized solutions of the equations of motion with finite energy, appear in many nonlinear systems. For instance, one of the first examples of nonlinear partial differential equation studied by physicists — the KdV equation $u_t + \alpha u_x + \beta u_{xxx} + \gamma u u_x = 0$ (it describes waves on the surface of shallow water and many other physical phenomena) admits soliton solutions. The well-known 't Hooft–Polyakov monopole is a soliton in a relativistic field theory. In general, a soliton may depend on $x - vt$, describing a quasiparticle moving with the speed v. In a Lorentz-invariant theory, a soliton can always be "stopped" by a proper Lorentz transformation, so that the solution becomes static, depending only on x.

Suppose the equations (1.20) admitted a soliton solution with a finite characteristic scale ρ and finite energy E. Note, however, that the equations do not involve a dimensionful constant and are thereby invariant under scale transformations $x \to \lambda x$. Thus, if a field configuration with scale ρ is a solution, so is a configuration with the same shape and scale $\lambda\rho$. By dimensional reasons, the energy of such a configuration should differ from the energy of the original configuration by the factor λ: $E \to E/\lambda$. If $\lambda > 1$, energy decreases. Thus, we have come to a logical contradiction. A static solution should correspond to a minimum of the energy functional. But, assuming that such a solution exists, we have just constructed configurations with lower energy, which invalidates the initial assumption.

Note that e.g. monopoles may exist because the corresponding equations *do* involve a dimensionful parameter (the Higgs vacuum expectation value). Localized static solutions may be present in pure Yang–Mills theory as well if we consider the field living on a finite spatial manifold (say, on S^3): the size of the manifold provides the scale and the above arguments do not apply. But no localized static solutions exist in flat, infinite 3-dimensional space. Physically, if an initial configuration with a finite scale is created, it will necessarily smear out as time evolves.

Nontrivial solutions of the equations (1.20) have no scale and their energy is infinite. Let us choose a rather restrictive ansatz with $A_0 = 0$ and A_i depending only on t, but not on x. Let us also restrict ourselves to the $SU(2)$ case. The Lagrangian acquires the form

$$\mathcal{L} = \frac{L}{V} = \frac{1}{2g^2}(\dot{A}_i^a)^2 - \frac{1}{4g^2}[(A_i^a A_i^a)^2 - (A_i^a A_j^a)^2] , \qquad (1.29)$$

where V is the spatial volume. This is just a *mechanical* system with a finite number of degrees of freedom. The equations of motion are

$$\ddot{A}_i^a + [A_i^a(A_j^b A_j^b) - A_j^a(A_j^b A_i^b)] = 0 . \qquad (1.30)$$

Let us look for solutions of the form $A_1^1(t) = A_2^2(t) = A(t)$, with all other components vanishing. Then only one degree of freedom is left and the equations (1.30) are reduced to $\ddot{A} + A^3 = 0$. This equation describes an unharmonical oscillator with the potential $U = A^4/4$. The solution can be easily found if one writes down the energy integral of motion,

$$\frac{1}{g^2}\left(\dot{A}^2 + \frac{1}{2}A^4\right) = \frac{E}{V} \equiv \epsilon . \qquad (1.31)$$

Equation (1.31) may be easily integrated:

$$\int \frac{dA}{\sqrt{2\epsilon g^2 - A^4}} = \frac{t - t_0}{\sqrt{2}} .$$

The integral can be expressed through elliptic functions, and we finally obtain

$$A(t) = C_0 \operatorname{cn}\left[C_0(t - t_0), \frac{1}{\sqrt{2}}\right], \qquad C_0 = \left(2\epsilon g^2\right)^{1/4} \qquad (1.32)$$

in the notation of Ryzhik and Gradshtein. The period of oscillation is inversely proportional to the amplitude: $T \propto 1/C_0$. This, of course, could be seen without solving the equation: the only scale in the Yang–Mills equations is provided by a characteristic amplitude of the field whose dimension is $[A] = m \sim 1/T$.

The solution (1.32) is a non-linear standing wave. By a Lorentz boost, one can obtain as well solutions describing nonlinear propagating waves. These solutions do not seem to have a particular physical significance, but maybe their meaning has not yet been unravelled.

An even more peculiar picture emerges if we choose a somewhat less restrictive ansatz $A_1^1(t) = A(t)$, $A_2^2(t) = B(t)$, leaving *two* dynamical degrees of freedom. For the fields of this class, the Lagrangian

$$g^2\mathcal{L} = \frac{1}{2}[\dot{A}^2 + \dot{B}^2] - \frac{A^2B^2}{2} \qquad (1.33)$$

describes the motion in the (AB)-plane with a potential $U(A, B) \propto A^2B^2$. The profiles of constant potential are hyperbolae, and the problem is very much similar to the problem of the so-called Sinai billiard with concave boards. This motion is *chaotic*, i.e., for general initial conditions, the integral trajectory does not have a decent regular form, but is ergodic and densely covers the whole surface of constant energy in the phase space of the system. After few oscillations (with characteristic time $\sim 1/A_{\text{char}}$), the initial conditions are completely forgotten (Lyapunov instability). This sea of chaos has, however, islands of some ordered behavior. For some special initial conditions, the phase space trajectory is closed and regular. One example is the initial conditions with $A = B, \dot{A} = \dot{B}$ in which case the solution is given by Eq. (1.32). There are many other such solutions. Some examples found by Matinian and Savvidy are shown in Fig. 1.1, which is taken from the book [12]. §

The main physical conclusion is that the Yang–Mills system is not exactly solvable (otherwise, the trajectories would have a regular rather than

---

§I acknowledge the courtesy of the authors who provided me with the corresponding postscript file.

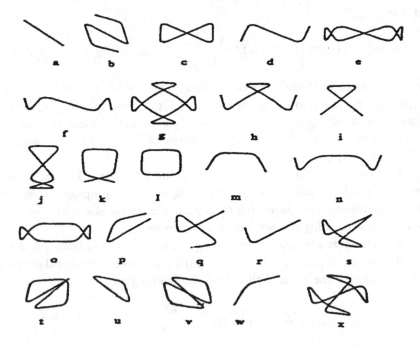

Fig. 1.1   Some closed trajectories in the (A, B) plane for the system (1.33).

chaotic form). This is in contrast to, say, the KdV equation and to some
early hopes.

Lecture 2

Instantons

In this lecture we will study classical Yang–Mills theory in Euclidean space–time. We will obtain some non-trivial solutions with finite action. As we will see later, these solutions play an important role in the quantum theory.

Thus, we change $t \to -i\tau$, where τ is the Euclidean time. The Euclidean Yang–Mills action is

$$S_E \stackrel{\text{def}}{=} -iS_M(t = -i\tau) = \frac{1}{4g^2} \int d^4x F_{\mu\nu}^a F_{\mu\nu}^a , \qquad (2.1)$$

where $d^4x = d^3x d\tau$ and $F_{\mu\nu}^a$ is an antisymmetric Euclidean tensor, so that S_E is positive definite.

2.1 Topological Charge.

Let us consider field configurations with a finite Euclidean action. The field strength $F_{\mu\nu}$ should fall off fast enough for the integral in Eq. (2.1) to converge. This does not necessarily mean that the vector potential A_μ vanishes at infinity as well. It suffices if it tends to a *pure gauge* potential

$$A_\mu = -i(\partial_\mu \Omega)\Omega^\dagger , \qquad (2.2)$$

for which the tensor $F_{\mu\nu}$ vanishes. In general, Ω and A_μ depend on the direction n_μ ($n_\mu^2 = 1$) in which Euclidean infinity is approached. The matrix $\Omega(n_\mu)$ realizes a mapping of S^3 onto the gauge group. The space of all such mappings represents an infinite set of isolated components such that the mappings belonging to different components cannot be continuously deformed into one another. Mathematicians call these components homotopy classes. For the mappings $S^3 \rightarrow G$, these classes are characterized by an integer number q, which we will call *topological charge*. A mathematical notation for this fact is $\pi_3(G) = \mathbb{Z}$, which is true for all simple Lie groups.

Consider the group $SU(2)$. Topologically, $SU(2) \equiv S^3$ and we are dealing with the mappings $S^3 \rightarrow S^3$. An example of a topologically nontrivial mapping with $q = 1$ is

$$\Omega(n_\mu) \equiv \Omega_0(n_\mu) = n_0 + in_i\sigma_i \ , \tag{2.3}$$

where σ_i are the Pauli matrices. Eq. (2.3) simply represents a parametrization of the distant Euclidean sphere by unitary matrices [see Eq. (A.12)]. Each $SU(2)$ matrix corresponds to a distinct point on S^3 and hence the *degree of mapping* is equal to one. An example of the mapping with $q = -1$ is $\Omega_0^\dagger(n_\mu)$. This mapping is also bijective, but the sphere is kind of "covered in the opposite direction": the corresponding Jacobian has the opposite sign. An example of a mapping with degree $q = 7$ is $\Omega_0^7(n_\mu)$ [seven different points n_μ on S^3 correspond to one and the same $SU(2)$ matrix], etc.

It is a very important fact that the topological charge of any mapping can be expressed as an integral over the distant 3-sphere. Before considering the four-dimensional theory of interest, let us discuss a very simple model: Abelian gauge theory in 2 Euclidean dimensions (the Schwinger model). Consider configurations with finite action which tend to the pure gauge $A_\mu = -i(\partial_\mu\Omega)\Omega^\dagger$, $\Omega(x) = e^{i\chi(x)}$ at Euclidean infinity. The field may depend on the direction ϕ along which infinity is approached. Now, $\Omega(\phi) = e^{ia(\phi)}$ realizes a mapping $S^1 \rightarrow U(1) \equiv S^1$. Obviously, the topological charge is the number of windings q: $\pi_1[U(1)] = Z$. For mappings of charge q, $a(\phi + 2\pi) = a(\phi) + 2\pi q$. It is easy to see now that q can be expressed in integral form,

$$q = \frac{1}{2\pi}\int_0^{2\pi} A_\phi d\phi = \frac{1}{2\pi}\oint_C \epsilon_{\mu\nu}A_\nu d\sigma_\mu \ , \tag{2.4}$$

where in the last formula $d\sigma_\mu$ is the length element of a distant contour C (not necessarily a "round" circle) multiplied by a unit normal vector pointing outside. Let us recall now that we are interested in the gauge field configurations defined on the whole two-dimensional plane. Using Gauss's theorem, we immediately obtain

$$q = \frac{1}{2\pi} \iint d^2x \, \partial_\mu(\epsilon_{\mu\nu}A_\nu) = \frac{1}{4\pi} \iint d^2x \, \epsilon_{\mu\nu}F_{\mu\nu} . \tag{2.5}$$

The fact that the integral on the right side of Eq. (2.5) is quantized to integer values depends crucially on the fact that we required $\Omega(\phi)$ to be a uniquely defined function, i.e. we assumed our gauge group to be compact. We are forced to do so if our theory includes *charged* fields besides the gauge field A_μ. In pure photodynamics, the only effect of a gauge transformation is the shift of A_μ by $\partial_\mu\chi$, and the compact nature of the group is not seen. But the charged fields are multiplied by the exponential factor $e^{i\chi} \in U(1)$. The quantization condition follows from the fact that $e^{i\chi(x)}$ is uniquely defined. When going to the standard normalization for A_μ, the integrals in Eq. (2.4, 2.5) are multiplied by the electric charge of the gauge field. The quantization condition (2.5) has the same origin as the widely known condition of the quantization of the magnetic flux $\Phi = \frac{e}{2\pi} \iint dx dy H_z(x,y)$ in superconductors. Mathematicians call the integral in Eq. (2.5) the *first Chern class*. The current $j_\mu = \frac{1}{2\pi}\epsilon_{\mu\nu}A_\nu$ is called the *Chern–Simons* current.

This mathematical construction can be generalized to fiber bundles on arbitrary spaces of even dimension. We are interested in non-Abelian fiber bundles in the four-dimensional world. Let us first write the result. For nonsingular gauge fields tending to a pure gauge (2.2) at infinity, the topological charge q of a mapping $\Omega(n_\mu)$ is expressed through the integral (called *second Chern class* or *Pontryagin index*)

$$q = \frac{1}{16\pi^2} \int d^4x \, \mathrm{Tr}\{F_{\mu\nu}\tilde{F}_{\mu\nu}\} , \tag{2.6}$$

where $\tilde{F}_{\mu\nu} = \frac{1}{2}\epsilon_{\mu\nu\alpha\beta}F_{\alpha\beta}$ [with the convention (N.15)]. To understand why this is correct, note that the integrand in (2.6) is a total derivative:

$$\frac{1}{16\pi^2}\mathrm{Tr}\{F_{\mu\nu}\tilde{F}_{\mu\nu}\} = \partial_\mu K_\mu = \partial_\mu \left[\frac{1}{8\pi^2}\epsilon_{\mu\nu\alpha\beta}\mathrm{Tr}\{A_\nu(\partial_\alpha A_\beta - \frac{2i}{3}A_\alpha A_\beta)\}\right] \tag{2.7}$$

(check it!). By Gauss' theorem,

$$q = \int_{S^3} d\sigma_\mu K_\mu = -\frac{1}{24\pi^2} \int d\sigma_\mu \, \epsilon_{\mu\nu\alpha\beta}\mathrm{Tr}\left\{(\partial_\nu\Omega)\Omega^\dagger(\partial_\alpha\Omega)\Omega^\dagger(\partial_\beta\Omega)\Omega^\dagger\right\} \tag{2.8}$$

[the asymptotic property $F_{\alpha\beta} \to 0$ and the form (2.2) for A_μ were used]. Substituting the particular mapping (2.3) with $q = 1$, one can explicitly check that the equality (2.8) is fulfilled, indeed. It is easy to understand why: the integrand in Eq. (2.8) is invariant with respect to the right or left multiplication of $\Omega(x)$ by an element of $SU(2)$ and is therefore nothing but the Haar measure [see Eqs. (A.6), (A.7)] on the group parametrized as in Eqs. (A.12), (2.3). The coefficient is chosen such that the volume of the group is normalized to 1. The integral (2.8) is not changed if our parametrization is deformed to a new one, $\Omega(x) \to \omega_0(x)\Omega(x)$, in such a way that a continuous family of gauge transformations $\omega(\alpha, x)$ with $\omega(0, x) = 1$, $\omega(1, x) = \omega_0(x)$ exists [i.e. when $\omega_0(x)$ is topologically trivial].* In general, $q[\Omega_1 \Omega_2] = q[\Omega_1] + q[\Omega_2]$.

The instanton is a configuration of topological charge $q = 1$ with the minimal action, i.e. a solution of the equations of motion with $q = 1$. The construction of such a solution is done in several steps. First, note that the action integral can be written as

$$
\begin{aligned}
S &= \frac{1}{2g^2} \int d^4x \, \mathrm{Tr}\{F_{\mu\nu} F_{\mu\nu}\} \\
&= \frac{1}{4g^2} \int d^4x \, \mathrm{Tr}\{(F_{\mu\nu} - \tilde{F}_{\mu\nu})(F_{\mu\nu} - \tilde{F}_{\mu\nu})\} + \frac{1}{2g^2} \int d^4x \, \mathrm{Tr}\{F_{\mu\nu} \tilde{F}_{\mu\nu}\} \\
&= \frac{1}{4g^2} \int d^4x \, \mathrm{Tr}\{(F_{\mu\nu} - \tilde{F}_{\mu\nu})(F_{\mu\nu} - \tilde{F}_{\mu\nu})\} + \frac{8\pi^2 q}{g^2} \geq \frac{8\pi^2 q}{g^2} . \quad (2.9)
\end{aligned}
$$

Thus, the action of *any* field configuration with a positive q has the absolute lower bound $8\pi^2 q/g^2$. It suggests that the minimal value of the action is realized on *self-dual* configurations

$$
F_{\mu\nu} = \tilde{F}_{\mu\nu} \qquad\qquad\qquad\qquad (2.10)
$$

and is just $8\pi^2 q/g^2$.

---

*One of the ways to see this is to take the variation of the integrand (2.7) and integrate it. We have

$$
\delta q \sim \int d\sigma_\mu \epsilon_{\mu\nu\alpha\beta} \mathrm{Tr} \, \{\delta A_\nu (\partial_\alpha A_\beta - iA_\alpha A_\beta)\} \sim \int d\sigma_\mu \mathrm{Tr} \, \{\delta A_\nu \tilde{F}_{\mu\nu}\} = 0
$$

due to the fact that our field is pure gauge at infinity.

For configurations with negative q, we can write

$$
\begin{aligned}
S &= \frac{1}{4g^2} \int d^4x \, \mathrm{Tr}\{(F_{\mu\nu} + \tilde{F}_{\mu\nu})(F_{\mu\nu} + \tilde{F}_{\mu\nu})\} - \frac{1}{2g^2} \int d^4x \, \mathrm{Tr}\{F_{\mu\nu}\tilde{F}_{\mu\nu}\} \\
&= \frac{1}{4g^2} \int d^4x \, \mathrm{Tr}\{(F_{\mu\nu} + \tilde{F}_{\mu\nu})(F_{\mu\nu} + \tilde{F}_{\mu\nu})\} - \frac{8\pi^2 q}{g^2} \geq \frac{8\pi^2 |q|}{g^2} \, , \quad (2.11)
\end{aligned}
$$

and the minimum $8\pi^2 |q|/g^2$ is realized on anti-self-dual configurations $F_{\mu\nu} = -\tilde{F}_{\mu\nu}$.

2.2   Explicit Solutions.

To prove that this is indeed the case, we should solve the self-duality equations (2.10)[†] explicitly. Let us do this for $q = 1$.

Note first that any self-dual antisymmetric tensor $S_{\mu\nu}^{\mathrm{s.d.}}$ can be put into the form

$$(S_{\mu\nu})^{\mathrm{s.d.}} = \eta_{\mu\nu}^a c^a \, ,$$

where $\eta_{\mu\nu}^a$ are the so-called 't Hooft symbols

$$\eta_{00}^a = 0 \, , \quad \eta_{ij}^a = \epsilon_{aij} \, , \quad \eta_{0i}^a = -\delta_{ai} \, , \quad \eta_{i0}^a = \delta_{ai} \, . \quad (2.12)$$

The objects (2.12) are not so mysterious. They can be understood if one notes that an antisymmetric self-dual tensor represents an irreducible representation $\begin{pmatrix} 1 \\ 0 \end{pmatrix}$ of the Lorentz group $\equiv SU(2) \times SU(2)$ and $\eta_{\mu\nu}^a$ are nothing but the Clebsch–Gordon coefficients for constructing $(S_{\mu\nu})^{\mathrm{s.d.}}$ out of two 4-vectors:

$$\begin{pmatrix} 1/2 \\ 1/2 \end{pmatrix} \otimes \begin{pmatrix} 1/2 \\ 1/2 \end{pmatrix} \rightarrow \begin{pmatrix} 1 \\ 0 \end{pmatrix} \oplus \cdots \, .$$

Likewise, an anti-self-dual antisymmetric tensor can be written as

$$(A_{\mu\nu})^{\mathrm{a.s.d.}} = \bar{\eta}_{\mu\nu}^a c^a \, ,$$

where $\bar{\eta}_{\mu\nu}^a$ differ from $\eta_{\mu\nu}^a$ by the sign of the components $\bar{\eta}_{0i}^a = -\bar{\eta}_{i0}^a = \delta_{ai}$. There are many beautiful relations involving the symbols $\eta_{\mu\nu}^a$. Here we will

---

[†]They are first-order and are thus much simpler than the general equations of motion (1.20).

need two of them:

$$\epsilon^{abc}\eta_{\mu\rho}^{b}\eta_{\nu\sigma}^{c} = \delta_{\mu\nu}\eta_{\rho\sigma}^{a} - \delta_{\mu\sigma}\eta_{\rho\nu}^{a} + \delta_{\rho\sigma}\eta_{\mu\nu}^{a} - \delta_{\rho\nu}\eta_{\mu\sigma}^{a} \, ,$$
$$\eta_{\mu\nu}^{a}\eta_{\mu\nu}^{a} = 12 \, . \tag{2.13}$$

The simplest ansatz for a self-dual field density tensor $F_{\mu\nu}^{a}(x)$ is

$$F_{\mu\nu}^{a}(x) = \eta_{\mu\nu}^{a}B(x) \, ,$$

where $B(x)$ is a scalar function. However, not any such tensor can play the role of a field density tensor expressible via potentials according to Equation (1.14). Let us adopt the ansatz

$$A_{\mu}^{a}(x) = \eta_{\mu\nu}^{a}\partial_{\nu}\ln\phi(x) \tag{2.14}$$

and try to find out under what conditions the corresponding field strength tensor would be self-dual. Using the definition (1.14) and the first identity in (2.13), we obtain

$$F_{\mu\nu}^{a} = \eta_{\nu\rho}^{a}[\partial_{\mu}\partial_{\rho}\ln\phi + (\partial_{\mu}\ln\phi)(\partial_{\rho}\ln\phi)] - (\mu \leftrightarrow \nu) + \eta_{\mu\nu}^{a}(\partial_{\rho}\ln\phi)^{2} \, . \tag{2.15}$$

In general, only the last term has a nice self-dual form. A sufficient condition for the whole expression to be self-dual is

$$[\ldots]_{\mu\rho} = \frac{\partial_{\mu}\partial_{\rho}\phi}{\phi} \propto \delta_{\mu\rho}$$

for the square bracket in Eq. (2.15).

The simplest way to fulfill this condition is to assume $\phi(x) = x^{2}/\rho^{2}$. If we, however, substitute such a $\phi(x)$ in Eq. (2.14) and calculate the corresponding $F_{\mu\nu}^{a}$, we find that $F_{\mu\nu}^{a} = 0$. This is not the solution we were looking for, but the corresponding field configuration,

$$A_{\mu}^{a} = 2\eta_{\mu\nu}^{a}x_{\nu}/x^{2} \, , \tag{2.16}$$

is also meaningful. It is a topologically nontrivial pure gauge configuration (2.2) with $\Omega = \Omega_{0}$ defined in Eq. (2.3), $n_{\mu} \equiv x_{\mu}/\sqrt{x^{2}}$.

The instanton solution is given by the ansatz (2.14) with

$$\phi(x) = 1 + x^{2}/\rho^{2} \, . \tag{2.17}$$

It has the form

$$A_\mu^a = \frac{2\eta_{\mu\nu}^a x_\nu}{x^2 + \rho^2} .$$ (2.18)

The field strength is

$$F_{\mu\nu}^a = -\frac{4\rho^2 \eta_{\mu\nu}^a}{(x^2 + \rho^2)^2} .$$ (2.19)

The explicit calculation of the action integral [using the second relation in Eq. (2.13)] gives the value $8\pi^2/g^2$ for the action as it should be.

The solution (2.18, 2.19) is localized in Euclidean space and thus represents an Euclidean soliton. At first sight, this is a counterexample to the no-go theorem proved in the previous lecture. But it is not the case. We have proven the theorem excluding the existence of *static* localized solutions in Yang–Mills theory. Such solutions would be characterized by an energy which is inversely proportional to the characteristic scale. On the other hand, Euclidean solutions are characterized by their *action*, which is dimensionless and *does* not change when varying the scale. The action of an instanton is always $8\pi^2/g^2$ independently of its characteristic size ρ, which is a free parameter of the solution. Another free parameter is the position of the instanton center: the solution (2.18) is centered at $x = 0$, but the problem is translationally invariant and, shifting $x \to x - x_0$, one can equally well obtain a solution centered at $x = x_0$.

The solution (2.18) is written in a particular gauge. Many other forms of the solution can be obtained by applying local gauge transformations to (2.18). Actually, one can prove that *all* the solutions of the Euclidean equations of motion in the topological sector $q = 1$ can be represented in the form (2.18), up to rescalings, translations, and gauge transformations.

All our reasoning can be transferred without essential changes to the case $q = -1$. One should only substitute $\bar{\eta}_{\mu\nu}^a$ for $\eta_{\mu\nu}^a$ in the ansatz (2.14). The dual 't Hooft symbols satisfy the same relations (2.13). Thus, the choice (2.17) provides for an anti-self-dual field strength tensor. We obtain the same solutions (2.18, 2.19) up to the change $\eta \to \bar{\eta}$. At large $\sqrt{x^2}$, the anti-instanton solution tends to a pure gauge configuration (2.2) with $\Omega = \Omega_0^+ = n_0 - i n_i \sigma_i$, $n_\mu = x_\mu / \sqrt{x^2}$.

The equations (2.10) can also be solved for higher q. This was done by Atiyah, Drinfeld, Hitchin, and Manin. The general solution looks complicated, but a restricted class of solutions with higher topological charge q, found by 't Hooft, has a very simple form (see **Problem 2**).

A final comment is that instanton solutions exist also for higher unitary (and not only unitary) groups. They all can be obtained, however, by gauge group rotations from an $su(2)$ solution with a trivial natural embedding $su(2) \subset su(N)$.[‡]

Problem 1. Apply a gauge transformation with $\Omega = \Omega_0^\dagger$ to the configuration (2.18). Such a transformation unwinds the singularity associated with a nontrivial mapping at infinity, but creates it at $x = 0$. Show that the gauge transformed field has the form

$$A_\mu^a = \frac{2\rho^2 \bar{\eta}_{\mu\nu}^a x_\nu}{x^2(x^2 + \rho^2)} \tag{2.20}$$

with the dual 't Hooft symbol.

Solution. Let us rewrite the instanton solution (2.18) in the form

$$
\begin{aligned}
A_\mu &= A_\mu^a t^a = \frac{x^2}{x^2 + \rho^2}[-i(\partial_\mu \Omega_0)\Omega_0^\dagger] \\
&= \frac{i\rho^2}{x^2 + \rho^2}(\partial_\mu \Omega_0)\Omega_0^\dagger - i(\partial_\mu \Omega_0)\Omega_0^\dagger .
\end{aligned}
\tag{2.21}
$$

We have

$$
\begin{aligned}
A_\mu^{\Omega_0^\dagger} &= \Omega_0^\dagger A_\mu \Omega_0 - i(\partial_\mu \Omega_0^\dagger)\Omega_0 = \frac{i\rho^2}{x^2 + \rho^2}\Omega_0^\dagger \partial_\mu \Omega_0 + 0 \\
&= -\frac{i\rho^2}{x^2 + \rho^2}(\partial_\mu \Omega_0^\dagger)\Omega_0 ,
\end{aligned}
\tag{2.22}
$$

which coincides with (2.20).

[‡]We have used here lower case letters to denote the Lie algebras (where the gauge fields A_μ belong) as opposed to the Lie groups, to make mathematicians happy. By a "natural" embedding, we understand e.g. the embedding realized by the generators t^1, t^2, t^3 living in the 2×2 block in the left upper corner of $su(N)$ [cf. Eq. (A.16)]. There are many other embeddings. For example, one of the embeddings $su(2) \subset su(3)$ is realized by the matrices t^2, t^5, t^7. An $su(3)$ field configuration (2.18) with $a = 2, 5, 7$ is a perfect Euclidean solution, but with $q = 4$ rather than $q = 1$. Also, other nontrivial embeddings lead to solutions with some higher q.

Problem 2. Show that the configuration

$$A_\mu^a = -\bar{\eta}_{\mu\nu}^a \partial_\nu \ln\left[1 + \sum_{i=1}^{q} \frac{\rho_i^2}{(x - x_{i0})^2}\right] \qquad (2.23)$$

is a solution of the equations of motion (though not the most general one) with the (positive) topological charge q. (*'t Hooft*).

Solution. Let $A_\mu^a = -\bar{\eta}_{\mu\nu}^a \partial_\nu \ln\phi(x)$. Then

$$F_{\mu\nu}^a = \bar{\eta}_{\nu\rho}^a[-\partial_\mu\partial_\rho \ln\phi + (\partial_\mu \ln\phi)(\partial_\rho \ln\phi)] - (\mu \leftrightarrow \nu) + \bar{\eta}_{\mu\nu}^a(\partial_\rho \ln\phi)^2 .$$

$$(2.24)$$

Using the relations (2.13) and the relation

$$\epsilon_{\mu\nu\alpha\beta}\bar{\eta}_{\beta\gamma}^a = \delta_{\alpha\gamma}\bar{\eta}_{\mu\nu}^a + \delta_{\mu\gamma}\bar{\eta}_{\nu\alpha}^a + \delta_{\nu\gamma}\bar{\eta}_{\alpha\mu}^a , \qquad (2.25)$$

we find

$$\tilde{F}_{\mu\nu}^a = \bar{\eta}_{\nu\rho}^a[-\partial_\mu\partial_\rho \ln\phi + (\partial_\mu \ln\phi)(\partial_\rho \ln\phi)] - (\mu \leftrightarrow \nu) - \bar{\eta}_{\mu\nu}^a \Box \ln\phi .$$

$$(2.26)$$

Thereby, the field is self-dual if the condition

$$\Box \ln\phi + (\partial_\rho \ln\phi)^2 = 0 \qquad (2.27)$$

is fulfilled. Eq. (2.27) implies $\Box\phi = 0$, and we see that the argument of the logarithm in Eq. (2.23) satisfies this equation everywhere except the points $x = x_{i0}$ where the gauge potential (but not the field strength!) is singular.

Let us calculate the Pontryagin index (2.6) for the field configuration (2.23). As before, we are integrating a total derivative so that the only contributions come from the boundary. The important distinction, compared to the regular gauge solution, is that the boundary does not now include a large sphere S^3 at Euclidean infinity: the gauge potential (2.23) falls off at infinity pretty fast, $\Omega^{\text{sing. gauge}}(\infty) = 1$, and the integral (2.8) is just zero in this case. Nonvanishing boundary contributions come, however, from small spheres surrounding the singularities at $x = x_{i0}$. At the vicinity of each such singularity, the configuration (2.23) behaves as a pure gauge configuration (2.2) with $\Omega = \Omega_0^\dagger$, $n_\mu = (x - x_{i0})_\mu/\sqrt{(x - x_{i0})^2}$. The contribution of each such small sphere to the Pontryagin index is 1 [the integral like in Eq. (2.8) with $\Omega = \Omega_0^\dagger$ would give -1, but one should bear in mind that the normal vector $d\sigma_\mu$ is now directed not outwards, but inwards]. We have q singular points, and the total Pontryagin index is equal to q.

One can think of 't Hooft's solution (2.23) as a combination of q instantons with the same orientation in group space (this is especially transparent when the instantons are far from each other). A general ADHM solution with topological charge q is also a combination of q instantons, but with arbitrary group orientations.

Lecture 3
Path Integral in Quantum Mechanics

3.1 Conventional Approach.

Quantum Chromodynamics is a quantum field theory. Quantum field theory is a quantum mechanical system with an infinite continuum number of degrees of freedom. The latter circumstance makes the theory complicated, but the basic ideology of QFT is the same as in conventional quantum mechanics: we have a Hilbert space of quantum states, observables, and an Hamiltonian operator which determines the time evolution of the wave function.* Solving the stationary Schrödinger equation, one could *in principle* find the spectrum of our QFT system (provided it exists and is well defined; as we will see later, not all classical field theories have consistent quantum counterparts.), etc.

This language is, however, not convenient for applications. Wave functions are actually not functions but functionals depending on an infinite number of variables. Also, the main object of interest in QFT applications is not the spectrum (though it is also important), but a set of scattering amplitudes, i.e. the S-matrix. The latter can be conveniently calculated in the framework of path integral techniques. In this lecture, we will develop this first for quantum mechanical systems, having a finite number of degrees of freedom. We will generalize to quantum field theories and in particular to QCD in subsequent lectures.

---

*Unfortunately, this approach is not possible in quantum gravity, because there time cannot be treated as a flat *independent* variable, but is intertwined with the other (dynamic) variables. This *is* the reason why quantum gravity does not yet exist today in the form of a consistent theory. This is the major challenge for the theorists today.

Classical mechanics can be described in the Hamiltonian or in the Lagrangian language. The former involves Hamilton's equations of motion, describing the dynamics of the system in phase space and implying conservation of energy. The latter involves Lagrange's equations, describing the dynamics in configuration space. Its underlying basic axiom is the minimal action principle. These two ways of description are equivalent to each other, and both can be generalized to the quantum case. The conventional description of the quantum mechanics by means of the Schrödinger equation represents the quantum version of the Hamiltonian description. The path integral formulation of quantum mechanics, developed originally by Feynman, is the quantum version of the Lagrange description. Like the minimal action principle, the Feynman path integral can be postulated as the basic axiom of the theory. Here we will derive it starting from the Schrödinger formulation.

Let us consider a quantum system involving just one dynamical degree of freedom q with the Hamiltonian

$$\hat{H} = \frac{\hat{p}^2}{2} + V(q) , \tag{3.1}$$

where $\hat{p} = -id/dq$ is the momentum operator (from now on, hats will indicate operators). The wave function of the system satisfies the Schrödinger equation,

$$i\frac{d\psi}{dt} = \hat{H}\psi . \tag{3.2}$$

If we fix the initial conditions $\psi(q, t_i)$, the formal solution of the equation (3.2) reads $\psi(q, t) = \hat{U}(t - t_i)\psi(q, t_i)$, where

$$\hat{U}(t - t_i) = \exp\{-i(t - t_i)\hat{H}\} \tag{3.3}$$

is the evolution operator. Let us consider the *kernel* of the evolution operator, i.e. the matrix element

$$\mathcal{K}(q_f, q_i; t_f - t_i) = \langle q_f|\hat{U}(t_f - t_i)|q_i\rangle \equiv \langle q_f, t_f|q_i, t_i\rangle . \tag{3.4}$$

It describes the probability amplitude that the system will find itself at point q_f at $t = t_f$ provided it was located at $q = q_i$ at $t = t_i$. The quantity $\mathcal{K}$ plays a fundamental role in the whole approach. The spectral decomposition of the kernel of the evolution operator reads:

$$\mathcal{K}(q_f, q_i; t_f - t_i) = \sum_k \psi_k(q_f)\psi_k^*(q_i)e^{-i\mathcal{E}_k(t_f - t_i)} , \qquad (3.5)$$

where $\psi_k(q)$ are the eigenstates of $\hat{H}$, with eigenvalues $\mathcal{E}_k$. Let $t_f - t_i = \Delta t$ be small. We have

$$\langle q_f|\hat{U}(\Delta t)|q_i\rangle = \langle q_f|1 - i\hat{H}\Delta t|q_i\rangle = \int dp\langle q_f|p\rangle\langle p|1 - i\hat{H}\Delta t|q_i\rangle .$$

Here $|p\rangle$ is a state with definite momentum p and wave function $\psi_p(q) = \frac{1}{\sqrt{2\pi}}e^{ipq}$. By definition,

$$\langle q|p\rangle = \frac{1}{\sqrt{2\pi}}e^{ipq}, \quad \langle p|q\rangle = \frac{1}{\sqrt{2\pi}}e^{-ipq} . \qquad (3.6)$$

If the Hamiltonian has the simple form (3.1), we can also write

$$\langle p|\hat{H}|q\rangle = \frac{1}{\sqrt{2\pi}}e^{-ipq}H(p, q) ,$$

where $H(p, q)$ is just the right side of Eq. (3.1), where p is now a c-number.[†] We have

$$\mathcal{K}(q_f, q_i; \Delta t) = \frac{1}{2\pi}\int dp \exp\{ip(q_f - q_i) - iH(p, q_i)\Delta t\} . \qquad (3.7)$$

To find the evolution operator over a finite time interval, divide the latter into a very large number n of very small intervals Δt. The evolution operator is then expressed as a convolution of an infinite number of infinitesimal

---

[†]If the Hamiltonian has a more intricate form involving terms of the type $\propto p^2q^2$ mixing momenta and coordinates, the situation is more difficult. To perform quantization, one should resolve the ordering ambiguity problem. As the Hamiltonian of QCD does not mix generalized coordinates and generalized momenta, we do not have this problem here and need not worry about it. Just note the following: (*i*) a path integral can be defined for *any* Hamiltonian no matter what particular way of the operator ordering is chosen (*ii*) very frequently, the latter can be fixed uniquely due to symmetry constraints.

evolution operators

$$\mathcal{K}(q_f, q_i; t_f - t_i) =$$

$$\lim_{n \to \infty} \int dq_{n-1} \cdots \int dq_1 \; \langle q_f | \hat{U}(\Delta t) | q_{n-1} \rangle \cdots \langle q_1 | \hat{U}(\Delta t) | q_i \rangle$$

$$= \lim_{n \to \infty} \int \exp\{ i p_n (q_f - q_{n-1}) + \ldots + i p_1 (q_1 - q_i) \;-$$

$$i \Delta t [H(p_n, q_{n-1}) + \ldots + H(p_1, q_i)] \} \frac{dp_n}{2\pi} \frac{dp_{n-1} dq_{n-1}}{2\pi} \cdots \frac{dp_1 dq_1}{2\pi} . \quad (3.8)$$

This can be formally written as

$$\mathcal{K}(q_f, q_i; t_f - t_i) = \int \exp\left\{ i \int_{t_i}^{t_f} [p\dot{q} - H(p,q)] dt \right\} \prod_t \frac{dp(t) dq(t)}{2\pi} , \quad (3.9)$$

with the boundary conditions

$$q(t_i) = q_i; \quad q(t_f) = q_f , \quad (3.10)$$

while the function $p(t)$ is quite arbitrary and is integrated over. For the Hamiltonian (3.1), the integral over all dp_n in Eq. (3.8) is Gaussian and can be done easily:

$$\mathcal{K}(q_f, q_i; t_f - t_i) =$$

$$\mathcal{N} \lim_{n \to \infty} \int \exp\left\{ i \left[\frac{(q_f - q_{n-1})^2}{2\Delta t} + \cdots + \frac{(q_1 - q_i)^2}{2\Delta t} \right. \right.$$

$$\left. - \; \Delta t V(q_{n-1}) - \cdots \Delta t V(q_i)] \right\} \prod_{j=1}^{n} dq_j$$

$$\sim \int \prod_t dq(t) \exp\left\{ i \int_{t_i}^{t_f} \left[\frac{\dot{q}^2}{2} - V(q) \right] dt \right\} . \quad (3.11)$$

The infinite normalization factor $\mathcal{N}$ plays no role for our purposes and will be dropped. Eq. (3.11) is the original Feynman form of the path integral, where the integral is done over trajectories in configuration space rather than in phase space as it is the case in Eq. (3.9). The boundary conditions (3.10) are assumed. The integral in the exponent is just the action on a trajectory $q(t)$ connecting the points q_i, t_i and q_f, t_f.

The (semi-)classical limit of quantum theory corresponds to large action (in units of $\hbar$). In this case, the main contribution to the integral (3.11) stems from the stationary points (minima) of the action, where many trajectories are added up with nearly identical phase factors.

All the results we have just derived are easily generalized to systems with several or many dynamical degrees of freedom. A slightly less trivial generalization, which we will need for some applications below, concerns the systems where the Hamiltonian represents not a scalar, but a matrix function on phase space. An example is the Pauli Hamiltonian $H = [\sigma_i(p_i - eA_i)]^2/(2m)$ for a spinning particle in an external magnetic field. The eigenfunctions $[\psi_k(q)]_p$ for such a system carry a spinor index p and the evolution operator also represents a matrix,

$$\mathcal{K}_{lp}(q_f, q_i; \Delta t) = \sum_k [\psi_k(q_f)]_l [\psi_k^*(q_i)]_p e^{-i\mathcal{E}_k \Delta t} . \tag{3.12}$$

Repeating all the steps of the derivation above, we can express (3.12) by a path integral which has the same form as Eq. (3.9), except that H and $\mathcal{K}$ are now assumed to be matrices and we have a *T-ordered* exponential,

$$T \exp\left\{ i \int_{t_i}^{t_f} [p\dot{q} - H(p, q)]dt \right\} ,$$

instead of an ordinary one.

A word of caution is in order here. Eqs. (3.8, 3.11) do not yet provide us with an operational definition of the quantum path integral, because the integrals over $\prod_j dq_j$ on the left side are not well defined: the integrand rapidly oscillates at large $|q_j|$ and does not fall off. To make the integrals convergent, we define the evolution operator through

$$\mathcal{K}(q_f, q_i; t_f - t_i) = \lim_{\epsilon \to +0} \mathcal{K}(q_f, q_i; t_f - t_i - i\epsilon) ,$$

i.e. as a convolution of a large number of infinitesimal operators over time $\widetilde{\Delta t} = (t_f - t_i - i\epsilon)/n$. For finite ϵ, the integrals converge and the only problem is to prove that the limit $\epsilon \to 0$ exists. As far as I understand, a mathematical proof of this fact is still absent (and that is why the very notion of Minkowski path integrals does not exist, for mathematicians, to date!). However, the existence of the limit follows from physical considerations. After all, we know that the finite time evolution operator (3.5) is well defined and, if the principles of quantum mechanics are correct, it *should* be represented as a convolution of many infinitesimal ones.

3.2   Euclidean Path Integral.

In the previous section we took the time variable to be slightly imaginary in order to make the integrals convergent. But we can equally well define and study the path integral assuming time to be *purely imaginary*. The Euclidean path integral thus defined is very much important in many quantum field theory applications.

Consider the evolution operator over an imaginary time interval $-i\beta$ ($t_i = 0$, $t_f = -i\beta$). According to Eq. (3.5), it is

$$\mathcal{K}(q_f, q_i; -i\beta) = \sum_k \psi_k(q_f)\psi_k^*(q_i)e^{-\beta\mathcal{E}_k} . \qquad (3.13)$$

Note that when β is large, the main contribution to the sum comes from the ground (i.e. vacuum) state. After the transformation $t \to -i\tau$ in Eq. (3.11), we obtain

$$\mathcal{K}(q_f, q_i; -i\beta) \sim \int \prod_\tau dq(\tau) \exp\left\{-\int_0^\beta \left[\frac{\dot{q}^2}{2} + V(q)\right] d\tau\right\} , \qquad (3.14)$$

with $q(0) = q_i$, $q(\beta) = q_f$. Introduce the quantity

$$Z(\beta) = \int dq\, \mathcal{K}(q, q; -i\beta) = \sum_k e^{-\beta\mathcal{E}_k} . \qquad (3.15)$$

This is nothing but the *partition function* of the system with temperature $T = 1/\beta$. Thereby, we have obtained a remarkable result: the partition function of a thermal system can be written as an Euclidean path integral over a finite imaginary time related to temperature.

In quantum mechanics and, especially, in quantum field theory, the ground state expectation values (vacuum averages) of various operators present a considerable interest. They can be expressed in the form

$$\langle 0|\hat{O}|0\rangle = \lim_{\beta\to\infty} \frac{\sum_k e^{-\beta\mathcal{E}_k}\langle k|O|k\rangle}{\sum_k e^{-\beta\mathcal{E}_k}} . \qquad (3.16)$$

This can be written as a ratio of two path integrals

$$\langle O\rangle_0 = \lim_{\beta\to\infty} Z^{-1}(\beta) \int dq\, \mathcal{K}(q, q; -i\beta)O(q)$$

$$= \lim_{\beta\to\infty} Z^{-1}(\beta) \int \prod_\tau dq(\tau)O[q(0)]e^{-\int_0^\beta d\tau \mathcal{L}_E[q(\tau)]} \qquad (3.17)$$

with periodic boundary condition $q(\beta) = q(0)$. Here the integral is performed over all closed trajectories. The boundary values are not fixed as before, but instead are integrated over, with the only restriction that $q(0)$ and $q(\beta)$ coincide. We have limited ourselves to the case where the operator O depends only on q. A path integral representation can also be found for the vacuum average of an arbitrary operator $O(\hat{p}, q)$, but it would be a phase space path integral like in Eq. (3.9).

It is often convenient to introduce a *generating functional*

$$Z[\eta(\tau)] = \int \prod_\tau dq(\tau) e^{-\int_0^\beta d\tau \{\mathcal{L}_E[q(\tau)] - \eta(\tau)q(\tau)\}} . \qquad (3.18)$$

Then

$$\langle O \rangle = Z^{-1}[0] O \left[\frac{\delta}{\delta\eta(0)} \right] Z[\eta(\tau)] \Big|_{\eta=0} . \qquad (3.19)$$

Generating functionals can be equally well defined for Minkowski path integrals.

The developed formalism works for all quantum mechanical problems. As far as quantum field theory is concerned, it is not yet complete for two reasons. The first reason is that later on we are going to construct perturbation theory, where we will be interested in transition amplitudes connecting *in* and *out* asymptotic states describing free particles. The transition amplitude $\mathcal{K}_A = \langle A_f(x), t_f | A_i(x), t_i \rangle$ connecting states with definite values of the fields $A(x)$, which is the immediate analog of (3.4), does not have a direct relevance to QFT applications. One-particle states are created from the vacuum by creation operators. They are described by non-trivial wave function $\Psi[A(x)]$, which the amplitude $\mathcal{K}_A$ should be convoluted with. It turns out that a convenient formalism can be developed which allows one to calculate path integrals for the amplitudes $\langle out|in \rangle$ directly, without the necessity to handle $\mathcal{K}_A$ on intermediate stages.

The second and most important reason is that quantum field theories (and, in particular, QCD) involve fermions, while standard quantum mechanics involves only bosonic degrees of freedom. We have got to learn how to write path integrals with fermionic dynamical variables. The adequate technique for this is the so-called *holomorphic representation*. Before turning to fermions, let us see how this technique works for purely bosonic systems.

3.3 Holomorphic Representation.

This is a very concise presentation. See e.g. the books [4; 13] for more details.

The Hamiltonian of a system of free fields (describing the asymptotic states) is reduced to the Hamiltonian of an infinite number of oscillators. The relevant formalism can be introduced by simply studying the problem of a quantum oscillator with only one dynamical degree of freedom. Introduce the *classical* variables

$$a^* = \frac{1}{\sqrt{2\omega}}(\omega q - ip) \ , \quad a = \frac{1}{\sqrt{2\omega}}(\omega q + ip) \ , \tag{3.20}$$

so that the classical Hamiltonian acquires the form

$$H^{\text{cl}}(p,q) = \frac{p^2}{2} + \frac{\omega^2 q^2}{2} = \omega a^* a \ . \tag{3.21}$$

The classical equations of motion are just $i\dot{a}^* = -\partial H^{\text{cl}}/\partial a = -\omega a^*$ (and its complex conjugate) with the trivial solution $a^*(t) = a_0^* e^{i\omega t}$.

In quantum mechanics, a^* and a are upgraded to the creation and annihilation operators $\hat{a}^\dagger$, $\hat{a}$, with standard commutation relation

$$[\hat{a}, \hat{a}^\dagger] = 1 \ . \tag{3.22}$$

The quantum Hamiltonian is $\hat{H}^{\text{qu}} = \omega \hat{a}^\dagger \hat{a}$ (with the zero-point energy $\omega/2$ subtracted). The oscillator eigenstates with energies $\mathcal{E}_k = \omega k$ are

$$|k\rangle = \frac{1}{\sqrt{k!}}(\hat{a}^\dagger)^k |0\rangle \ .$$

Usually, we write wave functions either in the coordinate or the momentum representation. Let us write them in the *holomorphic* representation, namely as a linear superposition of the wave functions

$$\psi_k(a^*) = \frac{1}{\sqrt{k!}}(a^*)^k \tag{3.23}$$

of the oscillator eigenstates. A generic wave function would be an entire (or holomorphic) function $\psi(a^*)$ of the variable a^*. The operator $\hat{a}^\dagger$ acting on $\psi(a^*)$ just multiplies it by a^* and

$$\hat{a}\psi \overset{\text{def}}{=} d\psi(a^*)/da^* \ .$$

The operators $\hat{a}^\dagger$ and $\hat{a}$ thus defined satisfy the commutation relation (3.22). To be able to calculate the matrix elements of various operators, we have to define the scalar product in the Hilbert space with basis (3.23) in such a way that the states (3.23) are orthonormal $\langle n|m\rangle = \delta_{nm}$. One can check that the definition

$$(f,g) \stackrel{\text{def}}{=} \int [f(a^*)]^* g(a^*) e^{-a^* a} \frac{da^* da}{2\pi i} \tag{3.24}$$

satisfies this criterium (do this by introducing polar variables $a = re^{i\phi}$, and using $da^* da = 2ir\,dr\,d\phi$). The action of various operators is defined as

$$(\hat{A}f)(a^*) = \int A(a^*, b) f(b^*) e^{-b^* b} \frac{db^* db}{2\pi i} . \tag{3.25}$$

The function $A(a^*, b)$ is called the *kernel* of the operator $\hat{A}$. The kernel of the product of two operators is the convolution of their kernels:

$$(\hat{A}\hat{B})(a^*, c) = \int A(a^*, b) B(b^*, c) e^{-b^* b} \frac{db^* db}{2\pi i} . \tag{3.26}$$

Using the identity

$$\int (bb^*)^n e^{-b^* b} \frac{db^* db}{2\pi i} = n! ,$$

it is not difficult to check that *(i)* the kernel of the unit operator is $e^{a^* b}$, *(ii)* the kernel of the Hamiltonian is $e^{a^* b} w a^* b$, and, in general, the kernel $A(a^*, b)$ of any operator equals $e^{a^* b}$ times its *normal symbol*, i.e. the function obtained from the normal-ordered quantum operator by substituting $\hat{a}^\dagger \to a^*$, $\hat{a} \to b$. The holomorphic kernel of the evolution operator for the oscillator is

$$U(a^*, b; t) = \exp\{a^* b e^{-i\omega t}\} . \tag{3.27}$$

Indeed, using the definitions (3.23, 3.25), it is easy to check that

$$\left[\hat{U}(t)\psi_k\right](a^*) = \psi_k(a^*) e^{-i\mathcal{E}_k t} ,$$

as one expects.

The holomorphic representation is particularly transparent and simple for the harmonic oscillator (a simple problem in any approach), but it can be used for *any* dynamical system. For a general system, the finite time evolution operator is complicated. But the kernel of the infinitesimal evolution operator is simple:

$$U(a^*, b; t) = \exp\{a^*b - i\mathcal{H}(a^*, b)\Delta t\} , \tag{3.28}$$

where $\mathcal{H}(a^*, b)$ is the normal symbol of the Hamiltonian. As before, the kernel of the finite time evolution operator is the convolution of many infinitesimal kernels:

$$U(a^*, b; t_f - t_i) =$$

$$\lim_{n \to \infty} \int e^{-c_n^* c_n} \frac{dc_n^* dc_n}{2\pi i} \cdots \int e^{-c_1^* c_1} \frac{dc_1^* dc_1}{2\pi i} U(a^*, c_n; \Delta t) \cdots U(c_1^*, b; \Delta t)$$

$$= \lim_{n \to \infty} \int \exp \{ [a^* c_n - c_n^* c_n + \ldots - c_1^* c_1 + c_1^* b]$$

$$- i[\mathcal{H}(a^*, c_n) + \ldots + \mathcal{H}(c_1^*, b)]\Delta t\} \prod_{s=1}^{n} \frac{dc_s^* dc_s}{2\pi i} . \tag{3.29}$$

Adding and subtracting b^*b in the exponent, this can be formally written as

$$U(a^*, b; t_f - t_i) =$$

$$\int \exp\{c^*(t_i)c(t_i)\} \exp \left\{ \int_{t_i}^{t_f} dt[\dot{c}^*c - i\mathcal{H}(c^*, c)] \right\} \prod_t \frac{dc^*(t)dc(t)}{2\pi i}, \tag{3.30}$$

with the boundary conditions $c^*(t_f) = a^*$, $c(t_i) = b$. Note that $c(t_f)$ and $c^*(t_i)$ are *not* fixed. They are completely arbitrary variables one should integrate over.

The spectral representation for the evolution operator (3.30) is quite analogous to (3.5):

$$U(a^*, b; t_f - t_i) = \sum_k \psi_k(a^*)[\psi_k(b^*)]^* e^{-i\mathcal{E}_k(t_f - t_i)} . \tag{3.31}$$

3.4   Grassmann Dynamic Variables.

Consider the quantum mechanical problem of a *fermionic oscillator*. The spectrum of the system includes just two states: the ground (or vacuum) state $|0\rangle$ and the excited state $|1\rangle = \hat{a}^\dagger |0\rangle$, where $\hat{a}^\dagger$ is the fermion creation operator. The fermionic creation and annihilation operators $\hat{a}^\dagger$ and $\hat{a}$ satisfy the algebra

$$\hat{a}^\dagger \hat{a} + \hat{a}\hat{a}^\dagger = 1, \quad \hat{a}^{\dagger 2} = \hat{a}^2 = 0 . \tag{3.32}$$

The Hamiltonian of the system is $\hat{H} = \omega \hat{a}^\dagger \hat{a}$ (with positive ω). Then the energy of the ground state is zero and the energy of the excited state is ω.

A conventional realization of the algebra (3.32) is $\hat{a} \equiv \sigma_-$, $\hat{a}^\dagger \equiv \sigma_+$. Then

$$\hat{H} = \begin{pmatrix} \omega & 0 \\ 0 & 0 \end{pmatrix}, \quad |0\rangle = \begin{pmatrix} 0 \\ 1 \end{pmatrix}, \quad |1\rangle = \begin{pmatrix} 1 \\ 0 \end{pmatrix} .$$

We will use another realization based on the holomorphic representation of the wave functions. It is constructed in full analogy with the holomorphic representation for the bosonic oscillator, with the important difference that the holomorphic variables a^* and a are not ordinary numbers as in the bosonic case, but *Grassmann* anticommuting numbers: $aa^* + a^*a = a^2 = a^{*2} = 0$.

The use of Grassmann variables is the only known consistent way to define a quantum theory of interacting fermionic fields beyond the perturbative expansion. This method was developed by Berezin and described in details in his book [14]. The main definitions are the following:

- Let $\{a_i\}$ be a set of n basic anticommuting variables: $a_i a_j + a_j a_i = 0$. The elements of the Grassmann algebra can be written as the functions $f(a_i) = c_0 + c_i a_i + c_{ij} a_i a_j + \ldots$, where the coefficents $c_0, c_i, \ldots$ are ordinary (real or complex) numbers. The series terminates at the n-th term: the $(n{+}1)$-th term of this series would involve a product of two identical anticommuting variables (say, a_1^2), which is zero. Note that even though a Grassmann number like $a_1 a_2$ commutes with any other number, it still cannot be treated as a normal number but instead represents "an even element of the Grassmann algebra". There are, of course, also odd anticommuting elements.
- One can add functions $(f(a_i) + g(a_i) = c_0 + d_0 + (c_i + d_i)a_i + \ldots)$ as well as multiply them. For example, $(1 + a_1 + a_2)(1 + a_1 - a_2) = 1 + 2a_1 - 2a_1 a_2$ (the anticommutation property of a_i was used).
- One can also differentiate the functions $f(a_i)$ with respect to Grassmann variables: $d/da_i \, 1 \stackrel{\text{def}}{=} 0$, $d/da_i \, a_j \stackrel{\text{def}}{=} \delta_{ij}$, and the derivative of a product of two functions satisfies the Leibnitz rule, except that

the operator d/da_i should be thought of as a Grassmann variable, which sometimes leads to the sign change when d/da_i is pulled in to annihilate a_i in the product. For example, $d/da_1(a_2a_3a_1) = a_2a_3$, but $d/da_1(a_2a_1a_3) = -a_2a_3$.

- Finally, one can integrate over Grassmann variables. In contrast to the usual bosonic case, the integral cannot be obtained here as a limit of integral sums, cannot be calculated numerically through "finite limits" (this makes no sense for Grassmann numbers) with the Simpson method, etc. What one can do, however, is to integrate over a Grassmann variable in the "whole range" ("from $-\infty$ to ∞", if you wish, though this is, again, meaningless). The definition is $\int da_j f(a_i) \overset{\text{def}}{=} d/da_j f(a_i)$.

- If the Grassmann algebra involves an even number of generators $\{a_i\}$, one can introduce an operation of involution, which we will associate with the complex conjugation. In our fermionic oscillator example, we have a pair of variables a^*, a, which are exchanged under complex conjugation. We will further assume that ordinary numbers c_0, etc. are replaced by their complex conjugates under such transformations:

$$f(a) = c_0 + c_i a_i + d_i a_i^* + \ldots \longrightarrow f^*(a) = \bar{c}_0 + \bar{c}_i a_i^* + \bar{d}_i a_i + \ldots$$

(even though complex conjugation of ordinary numbers and involution of Grassmann numbers could be defined independently). It is convenient to assume $[f(a_i)g(a_i)]^* = g^*(a_i)f^*(a_i)$ like for Hermitian conjugation. Then the product a^*a is real.

Let us return to the fermionic oscillator. The classical Hamiltonian is $\omega a^* a$. Like earlier, the classical equations of motion are $i\dot{a}^* = -\omega a^*$ with the solution $a^*(t) \propto e^{i\omega t}$ (needless to say, in this case we cannot plot the solution on a graph or numerically solve the equations of motion).

The quantum operators $\hat{a}^\dagger$ and $\hat{a}$ are represented by $\hat{a}^\dagger \equiv a^*$ and $\hat{a} \equiv d/da^*$. The wave functions of the system are holomorphic functions $\psi(a^*) = c_0 + c_1 a^*$. The functions "1" and "a^*" are the eigenstates $|0\rangle$, $|1\rangle$ of the quantum Hamiltonian. One can check that these states are orthonormal if the scalar product on our Hilbert space is defined in the following way:

$$(f,g) \overset{\text{def}}{=} \int [f(a^*)]^* g(a^*) e^{-a^* a} da^* da . \tag{3.33}$$

The analogy with the bosonic case is pretty close. We can again define the holomorphic kernels of the operators such that

$$(\hat{A}f)(a^*) = \int A(a^*, b)f(b^*)e^{-b^*b}db^*db , \qquad (3.34)$$

and calculate the kernel of the finite time evolution operator. For a single pair of fermionic variables, the Hamiltonian is bound to have the oscillator form $\omega a^*a + E_0$ (all higher products are zero, so that unharmonic interactions are not allowed). As in the bosonic case, the finite time evolution operator has the form (3.27). If the system involves several pairs of fermionic variables a_r^*, a_r, unharmonic terms in the Hamiltonian are allowed. Making the same transformations as earlier, the finite time evolution operator can be expressed as a convolution of a large number of infinitesimal evolution operators $U(\{a_r^*\}, \{b_r\}; \Delta t) = \exp\{\sum_r a_r^*b_r - i\mathcal{H}(\{a_r^*\}, \{b_r\})\Delta t\}$, giving rise to the path integral

$$U(\{a_r^*\}, \{b_r\}; t_f - t_i) =$$

$$\int \exp\left\{\sum_r c_r^*(t_i)c_r(t_i)\right\} \exp\left\{\int_{t_i}^{t_f} dt \left[\sum_r \dot{c}_r^* c_r - i\mathcal{H}(c_r^*, c_r)\right]\right\}$$

$$\prod_{rt} dc_r^*(t)dc_r(t) \qquad (3.35)$$

with the boundary conditions $c_r(t_i) = b_r$, $c_r^*(t_f) = a_r^*$. As in the bosonic case, $c_r^*(t_i)$ and $c_r(t_f)$ are not fixed. Also, at intermediate times $t_i < t < t_f$, $c_r(t)$ and $c_r^*(t)$ are *independent* Grassmann variables to be integrated over.

Let us now consider the Euclidean evolution operator. Its spectral representation is

$$U(a^*, b; -i\beta) = \sum_k \psi_k(a^*)[\psi_k(b^*)]^* e^{-\beta\mathcal{E}_k} = 1 + a^*be^{-\beta\omega} \qquad (3.36)$$

[confer Eq. (3.31); for simplicity, we restricted ourselves to just one fermionic degree of freedom]. Let us try to construct the partition function of the system $Z = 1 + e^{-\beta\omega}$ in the same way as we did it for the bosonic case. A first guess is

$$Z \stackrel{?}{=} \int U(a^*, a; -i\beta)e^{-a^*a}da^*da . \qquad (3.37)$$

However, the explicit calculation of the integral gives $1 - e^{-\beta\omega}$, which is not what we want. The minus sign in the second term appeared due to the anticommutative nature of the Grassmann numbers. For a general system with several fermionic degrees of freedom, the integral generalizing (3.37) gives

$$\sum_n (-1)^F e^{-\beta \mathcal{E}_n} \; , \tag{3.38}$$

where F if the fermion number of the eigenstate $|n\rangle$, i.e. the number of Grassmann factors in the holomorphic wave function of the eigenstate $|n\rangle$.[‡] The true answer for the standard partition function of a fermionic system is found to be

$$Z = \int U(\{a_r^*\}, \{-a_r\}; -i\beta) e^{-\sum_r a_r^* a_r} \prod_r da_r^* da_r \; . \tag{3.39}$$

It can be written as a path integral

$$Z = \int \exp\left\{ \sum_r c_r^*(0) c_r(0) \right\} \prod_{r\tau} dc_r^*(\tau) dc_r(\tau)$$

$$\exp\left\{ \int_0^\beta d\tau \left[\sum_r \dot{c}_r^* c_r - \mathcal{H}(c_r^*, c_r) \right] \right\} \; , \tag{3.40}$$

where the trajectories $c_r(\tau)$, $c_r^*(\tau)$ are antiperiodic in a sense that $c_r^*(\beta) = -[c_r(0)]^*$.

Let us briefly comment on the practical way to calculate fermionic path integrals. In theories like QCD the action is quadratic in the fermionic fields and the fermionic path integral is always Gaussian. The integral can be done using the very important identity (prove it!)

$$\int \prod_r da_r^* da_r \exp\{-A_{ij} a_i^* a_j\} = \det \|A\| \; . \tag{3.41}$$

Compare this to

$$\int \prod_r \frac{da_r^* da_r}{2\pi i} \exp\{-A_{ij} a_i^* a_j\} = \frac{1}{\det \|A\|} \tag{3.42}$$

for the bosonic case.

[‡]The sum (3.38) has a name: *Witten index*. It plays an important role for supersymmetric theories, but we do not need it.

Lecture 4

Quantization of Gauge Theories

4.1 Dirac Quantization Procedure.

The technique we have developed so far allows one to perform the quantization of an ordinary (non-gauge) field theory like $\lambda\phi^4$ or Yukawa theory. What we need, however, is to learn how to quantize gauge theory. This problem has a number of important subtleties, and some additional methods need to be elaborated.

We will briefly touch on some general aspects of the theory of gauge (or, in another language, *constrained*) systems and will test general methods in the simple example of a quantum mechanical gauge system where the physics and the results are most transparent. We finally apply these methods to the Yang–Mills system of interest.

We start with the classical theory. Suppose, we have a Hamiltonian $\mathcal{H}(p_i, q_i)$ depending on the phase space variables (p_i, q_i), $i = 1, \ldots, n$, and subject to a number of constraints $\phi^a(p_i, q_i) = 0$, $a = 1, \ldots, k < n$. Thus, we are searching for the classical trajectories satisfying the following system of equations:

$$\dot{p}_i + \frac{\delta\mathcal{H}}{\delta q_i} = 0 \,,$$

$$\dot{q}_i - \frac{\delta\mathcal{H}}{\delta p_i} = 0 \,,$$

$$\phi^a(p_i, q_i) = 0 \,. \tag{4.1}$$

Here the first two equations are the dynamical equations of motion and the last one expresses the constraints. The equations (4.1) are compatible, provided that, for any trajectory with initial conditions satisfying $\phi^a = 0$, this holds also at later times. This implies $\dot{\phi}^a = -\{\phi^a, \mathcal{H}\}_{\text{P.B.}} = 0$, which is true if

$$\{\phi^a, \mathcal{H}\}_{\text{P.B.}} = \sum_b C_{ab}\phi^b \ . \tag{4.2}$$

We want to quantize the theory in both the canonical and in the path integral formalism. For the latter, the action functional needs to be defined. A sufficient condition for the latter to exist is that the ϕ^a satisfy also the property

$$\{\phi^a, \phi^b\}_{\text{P.B.}} = \sum_c C_{abc}\phi^c \ . \tag{4.3}$$

Indeed, the constrained action is defined as

$$S = \int \left[\sum_i p_i \dot{q}_i - \mathcal{H}(p_i, q_i) - \sum_a \lambda^a \phi^a(p_i, q_i) \right] dt \ . \tag{4.4}$$

The action (4.4) depends on extra variables λ^a (the Lagrange multipliers) but not on their time derivatives. Variations of the functional S with respect to $\lambda^a(t)$ reproduce the constraints. Variations of S with respect to $p_i(t)$ and $q_i(t)$ give the dynamical equations

$$\dot{p}_i + \frac{\delta\mathcal{H}}{\delta q_i} + \lambda^a \frac{\delta\phi^a}{\delta q_i} = 0 \ ,$$
$$\dot{q}_i - \frac{\delta\mathcal{H}}{\delta p_i} - \lambda^a \frac{\delta\phi^a}{\delta p_i} = 0 \ . \tag{4.5}$$

This system is compatible with the constraints provided

$$\dot{\phi}^a = (\delta\phi^a/\delta p_i)\dot{p}_i + (\delta\phi^a/\delta q_i)\dot{q}_i = -\{\phi^a, \mathcal{H}\}_{\text{P.B.}} - \sum_b \lambda^b \{\phi^a, \phi^b\}_{\text{P.B.}} = 0 \ .$$

If we want this to be true for any λ^a, the conditions (4.2, 4.3) follow. Expressing the momenta $p_i(t)$ via $q_i(t)$ and $\dot{q}_i(t)$ by use of the second equation in Eq. (4.5) and substituting it into the action (4.4), we arrive at the standard Lagrange formulation.

A system with constraints satisfying the criteria (4.2, 4.3) is called the Hamiltonian system with constraints of the first kind. As was shown before, the Yang–Mills system with the Hamiltonian (1.25) and the constraints

(1.24) enjoys the properties (1.28) and $\{G^a, \mathcal{H}\}_{\text{P.B.}} = 0$ and hence belongs to this class.*

Note that the Lagrange multipliers $\lambda^a(t)$ entering the equations (4.5) are completely arbitrary: for any $\lambda^a(t)$ the solutions for $p_i(t)$ and $q_i(t)$ can be found. The freedom of choice for the $\lambda^a(t)$ is precisely a gauge invariance! A distinguished choice is $\lambda^a = 0$ where the equations of motion (4.5) coincide with original Hamilton's equations (4.1).

It is easier to understand the general theory in a concrete example, which we choose to be very simple (may be, the simplest possible one). Consider a two-dimensional oscillator with the Hamiltonian

$$\mathcal{H} = \frac{1}{2}(p_x^2 + p_y^2) + \frac{\omega^2}{2}(x^2 + y^2) \,. \tag{4.6}$$

Impose the constraint

$$p_\phi = x p_y - y p_x \equiv \epsilon_{ij} x_i p_j = 0 \tag{4.7}$$

$[x_i \equiv (x, y), \ p_i \equiv (p_x, p_y)]$. Obviously, $\{p_\phi, \mathcal{H}\}_{\text{P.B.}} = 0$ and we have a constrained system of the first kind. The canonical Lagrangian is

$$\mathcal{L} = p_i \dot{x}_i - \mathcal{H} - \lambda p_\phi \,. \tag{4.8}$$

The equations of motion (4.5) in phase space are

$$\dot{p}_i + \omega^2 x_i + \lambda \epsilon_{ij} p_j = 0 \,,$$
$$\dot{x}_i - p_i + \lambda \epsilon_{ij} x_j = 0 \,. \tag{4.9}$$

Solving for the momenta from the lower pair of equations in Eqs. (4.9), and substituting them in Eq. (4.8), we obtain the canonical Lagrangian

$$\mathcal{L} = \frac{1}{2}(\dot{x}_i + \lambda \epsilon_{ij} x_j)^2 - \frac{\omega^2}{2} x_i^2 \,. \tag{4.10}$$

We see that different $\lambda(t)$ physically correspond to different choices for the reference frame, which may rotate with a time-dependent angular velocity $\lambda(t)$. The freedom of choice of $\lambda(t)$ is the gauge invariance. Explicitly, the Lagrangian (4.10) is invariant under the transformations

$$\begin{cases} x_i' = O_{ij}(\chi) x_j \\ \lambda' = \lambda + \dot{\chi} \end{cases} , \tag{4.11}$$

---

*This also follows from the fact that the Yang–Mills system has a Lagrangian formulation which was our starting point. It *implies* that the conditions (4.2, 4.3) are satisfied.

with

$$O_{ij} = \begin{pmatrix} \cos\chi, & -\sin\chi \\ \sin\chi, & \cos\chi \end{pmatrix} .$$

The best way to handle our system is, of course, to realize that the constraint $p_\phi = 0$ physically means that the system is not allowed to move in the angular direction. Then we simply can go to polar coordinates, set $\phi = 0$ and obtain the equations of motion $\dot{p}_r + \omega^2 r = 0$, $\dot{r} - p_r = 0$ or $\ddot{r} + \omega^2 r = 0$ describing radial motion of the system. Imposing the additional constraint $\phi = 0$ on top of the initial constraint $p_\phi = 0$ is called *fixing the gauge*. The constraint $\phi = 0$ picks out one representative from each *gauge orbit* defined as a set of points in the configuration space $\{q_i\}$ related to each other by a time-independent gauge transformation. In our case, the gauge orbits are just the circles of constant r.

In the general case, fixing the gauge amounts to imposing, on top of the constraints $\phi^a = 0$, also additional constraints $\chi^a = 0$ [the functions $\chi^a(p_i, q_i)$ are not quite arbitrary and should satisfy some natural requirements, which we will not discuss here]. After explicitly resolving, whenever it can be done practically, the equation system $\phi^a = 0$, $\chi^a = 0$ in terms of a restricted set of *gauge invariant* phase space variables p_i^* and q_i^*, with $i = 1, \ldots, n - k$ (p_r and r in our case), the Hamiltonian $\mathcal{H}(p^*, q^*)$ describes an ordinary unconstrained system, which may be handled classically and, if needed, quantized in the usual way. Note that *two* phase space degrees of freedom are eliminated for each constraint $\phi^a = 0$ originally imposed. In our case, both p_ϕ and ϕ are eliminated.

The analogy with gauge field theories is now seen. Consider an Abelian theory without matter fields (pure photodynamics). The dictionary of correspondences is

$$\lambda \equiv A_0/2 , \quad p_\phi \equiv \partial_i E_i , \quad \phi \equiv \partial_i A_i .$$

Imposing the constraint $\partial_i A_i = 0$ amounts to going over to Coulomb gauge. The Hamiltonian $\mathcal{H}_{\text{photons}} = 1/2 \int d\boldsymbol{x}(E_i^2 + B_i^2)$ depends then only on the two transverse field polarizations $\boldsymbol{A}_\perp$ and their canonical momenta $\boldsymbol{E}_\perp$.

However, it is not always practically possible to explicitly solve the constraints. It is difficult and practically inconvenient for QED, which includes apart from photons also charged particles. A complication is that the constraints $\partial_i E_i = \rho_{\text{matter}}$, $\partial_i A_i = 0$ dictate in this case the appearance of a nonvanishing $A_0 = -\Delta^{-1}\rho$, and the Hamiltonian with resolved constraints involves a nonlocal term $\propto \rho\Delta^{-1}\rho$. For non-Abelian Yang–Mills theory, it is hopeless. Coulomb gauge does not really work here due to the presence of the so-called *Gribov copies* to be discussed later. We are forced to learn how to construct and handle quantum gauge theory without explicitly resolving constraints at the classical level.

There are two ways to quantize: *(i)* via the Schrödinger equation and *(ii)* via the path integral. Let us first describe the first way, i.e. the *Dirac quantization procedure*. We are going to solve the equation system

$$\begin{aligned}
\mathcal{H}(\hat{p}_i, q_i)\Psi_k(q_i) &= \mathcal{E}_k\Psi_k(q_i) \,, \\
\phi^a(\hat{p}_i, q_i)\Psi_k(q_i) &= 0 \,,
\end{aligned} \qquad (4.12)$$

where $\hat{p}_i = -i\partial/\partial q_i$, and $\mathcal{H}, \phi^a$ are the same functions as before (we disregard here possible ordering ambiguities as this problem does not arise for the Yang–Mills Hamiltonian and constraints). The conditions (4.2, 4.3) are promoted to

$$[\hat{\phi}^a, \hat{\mathcal{H}}] = \sum_b C_{ab}\hat{\phi}^b, \quad [\hat{\phi}^a, \hat{\phi}^b] = \sum_c C_{abc}\hat{\phi}^c \,. \qquad (4.13)$$

The conditions (4.13) render the system (4.12) self-consistent, i.e. the Hilbert space of definite energy eigenstates and with definite (zero!) eigenvalues for all operators $\hat{\phi}^a$ is not empty.[†] The procedure (4.12) *defines* our quantum gauge system.

In the toy model (4.6, 4.7), the situation is very simple and transparent. The spectrum of the unconstrained system is $\mathcal{E}_{n_r,m} = \omega(1 + |m| + 2n_r)$, where m is the eigenvalue of the momentum operator $\hat{p}_\phi$ and n_r is the radial quantum number. The Hilbert space (4.12) involves only the states with zero angular momentum. The spectrum is

$$\mathcal{E}_k = \omega(1 + 2n_r) \,. \qquad (4.14)$$

It coincides, up to an irrelevant overall energy shift, with the spectrum of the Hamiltonian of a one-dimensional oscillator describing radial motion on the half-line $0 \leq r < \infty$. Note that the wave functions of our system do

---

[†]These conditions are sufficient but not really necessary; there are some other self-consistent gauge systems with weaker restrictions: the constrained systems of the second kind, etc. We are not going to discuss these complications here as for QCD the conditions (4.13) are duly fulfilled.

not depend on the gauge variable ϕ. In other words, the wave functions are invariant under the gauge transformation (4.11) (with constant χ).

The Hilbert space of the quantum Yang–Mills field system is defined through

$$\mathcal{H}[\hat{E}_i^a(x), A_i^a(x)]\Psi_k[A_i^a(x)] = \mathcal{E}_k\Psi_k[A_i^a(x)] \,,$$
$$G^c[\hat{E}_i^a(x), A_i^a(x)]\Psi_k[A_i^a(x)] = 0 \,, \qquad (4.15)$$

where $\hat{E}_i^a = -ig^2\delta/\delta A_i^a(x)$. The system (4.15) has non-trivial solutions due to the commutation relations

$$[\hat{G}^a, \hat{\mathcal{H}}] = 0 \,, \quad [\hat{G}^a(x), \hat{G}^b(y)] = if^{abc}\hat{G}^c(x)\delta(x - y) \,. \qquad (4.16)$$

Just like in our toy quantum mechanical model, the constraint $\hat{G}^a\Psi = 0$ requires for the wave functions to be invariant under the gauge transformation (1.9) of the dynamical variables A_i^a, with time-independent $\Omega(x)$.

We will see a bit later that, in the Yang–Mills case, the Hilbert space defined in (4.15) is actually too large. One needs, on top of the constraint $G^a\Psi = 0$, to impose an *additional* constraint called *superselection rule*, which is closely associated with the instanton Euclidean field configurations. Let us, however, postpone the discussion of this issue until we learn another approach to the quantization of gauge theories, the path integral approach.

Let us concentrate on the path integral representation of the constrained partition function

$$Z_{\text{cons.}}(\beta) = \sum_k P_k e^{-\beta\mathcal{E}_k} \qquad (4.17)$$

(P_k is the projector: $P_k = 1$ for gauge-invariant states and $P_k = 0$ for noninvariant ones) and again first discuss our toy oscillator model. The partition function of the unconstrained two-dimensional oscillator can be expressed via its Euclidean evolution operator (3.13) according to Eq. (3.15). Now we want, however, to include in the sum only the gauge (rotationally) invariant states. This can be done if one defines

$$Z_{\text{cons.}}(\beta) = \int_0^{2\pi} \frac{d\chi_0}{2\pi} \int dx \, \mathcal{K}(x^{\chi_0}, x; -i\beta) \,, \qquad (4.18)$$

where $\mathcal{K}$ is still the unconstrained evolution operator, but the final and initial points do not coincide anymore. x^{χ_0} differs from x by rotation by the angle χ_0, which is to be integrated over. Indeed, all states in the sum

(3.13) with non-zero angular momenta give zero after integration over χ_0, and only gauge invariant states survive.

Substitute now into Eq. (4.18) the path integral representation of the evolution operator $\mathcal{K}$. We obtain

$$Z_{\text{cons.}}(\beta) = \int_0^{2\pi} \frac{d\chi_0}{2\pi} \int \prod_\tau d\boldsymbol{x}(\tau) \, e^{-\int_0^\beta d\tau \, \mathcal{L}^E_{\text{uncons.}}[\boldsymbol{x}(\tau)]} , \qquad (4.19)$$

with $\mathcal{L}^E_{\text{uncons.}} = (\dot{\boldsymbol{x}})^2/2 + \omega^2 \boldsymbol{x}^2/2$. The inner integral is done over all the trajectories with the "twisted" boundary conditions $x_i(\beta) = O_{ij}(\chi_0)x_j(0)$. Let us now change variables in the integral (4.19):

$$x_i(\tau) = O_{ij}[\chi(\tau)]x'_j(\tau) , \qquad (4.20)$$

where the *only* condition imposed on the otherwise arbitrary function $\chi(\tau)$ is

$$\chi(0) = 0, \quad \chi(\beta) = \chi_0 . \qquad (4.21)$$

Equation (4.20) corresponds to changing to a reference frame which rotates in Euclidean time with variable angular velocity $\lambda(\tau) = \dot{\chi}(\tau)$. The condition (4.21) requires that the overall rotation angle is fixed. However, as we also have to integrate over χ_0, *nothing* is in fact fixed and $\chi(\tau)$ is quite arbitrary. The variables $x'_i(\tau)$ are now periodic. Let us substitute Eq. (4.20) into the integral (4.19) and upgrade the integration over χ_0 to a path integral over $\prod_\tau d\chi(\tau) \sim \prod_\tau d\lambda(\tau)$. As the integral does not depend on the choice of variables, this only gives an irrelevant (albeit infinite) overall factor. The final result is very natural:

$$Z_{\text{cons.}}(\beta) \sim \int \prod_\tau d\boldsymbol{x}'(\tau) d\lambda(\tau) \, e^{-\int_0^\beta d\tau \, \mathcal{L}^E[\boldsymbol{x}'(\tau), \lambda(\tau)]} , \qquad (4.22)$$

where $\mathcal{L}^E[\boldsymbol{x}'(\tau), \lambda(\tau)]$ is the Lagrangian (4.10) with the sign of the potential reversed. As was mentioned, $\boldsymbol{x}'(\tau)$ is periodic in τ. Periodicity of $\lambda(\tau)$ is not necessary, but, as imposing this condition is convenient and does not change the result, we will assume it. As the Lagrangian (4.10) is invariant under the gauge transformations (4.11), this is also true for the partition function (4.22).

The result (4.22) can also be obtained in a different, somewhat heuristic way. Let us write the Minkowski path integral in the Hamiltonian form, implementing the constraints by inserting the delta-function $\delta(p_\phi)$ into the integrand:

$$\int \prod_t \frac{dp(t)dx(t)}{2\pi} \delta(p_\phi) \exp\left\{i \int dt[\boldsymbol{p} \cdot \dot{\boldsymbol{x}} - \mathcal{H}(\boldsymbol{p}, \boldsymbol{x})]\right\} =$$
$$\int \prod_t \frac{dp(t)dx(t)}{2\pi} \frac{d\lambda(t)}{2\pi} \exp\left\{i \int dt[\boldsymbol{p} \cdot \dot{\boldsymbol{x}} - \mathcal{H}(\boldsymbol{p}, \boldsymbol{x}) - \lambda(t)p_\phi(t)]\right\}. \quad (4.23)$$

The integral over momenta is Gaussian and can be done, after which the momenta are replaced by their saddle point values, which are found as the solution of Hamilton's equations of motion [the second equation in Eq. (4.9)]. We arrive at the Minkowskian version of the integral (4.22). The Euclidean partition function (4.22) may be obtained by analytic continuation.

Everything we have done for the oscillator can be applied without essential change to any system with the constraints of the first kind. In particular, in full analogy with Eq. (4.22), the partition function for Yang–Mills theory can be expressed as

$$Z = \int \prod_{\boldsymbol{x},\tau} dA_\mu^a(\boldsymbol{x},\tau) \exp\left\{-\frac{1}{2g^2} \int_0^\beta d\tau \int d\boldsymbol{x} \; \mathrm{Tr}\{F_{\mu\nu}F_{\mu\nu}\}\right\} . \quad (4.24)$$

This expression is gauge invariant. In addition, the $O(4)$ invariance is manifest. The Minkowski path integral is obtained from Eq. (4.24) by inserting the factor i in the exponent:

$$Z_M = \int \prod_x dA_\mu^a(x) \; \exp\left\{-i\frac{1}{2g^2} \int d^4x \; \mathrm{Tr}\{F_{\mu\nu}F^{\mu\nu}\}\right\} . \quad (4.25)$$

Consider now the theory with quarks. It involves also the path integrals over Grassmann field variables. In Minkowski space, we have

$$\int \prod_x d\bar{\psi}_f(x)d\psi_f(x) \; \exp\left\{i \int d^4x \; \bar{\psi}_f(i\not{\partial} - m_f)\psi_f\right\} \quad (4.26)$$

for each quark flavor. In Euclidean space, the expression (4.26) turns into

$$\int \prod_x d\bar{\psi}_f(x)d\psi_f(x) \exp\left\{\int d^4x \; \bar{\psi}_f(i\not{\partial}^E - m_f)\psi_f\right\} , \quad (4.27)$$

where $\not{\partial}^E = \gamma_\mu^E \mathcal{D}_\mu$ is Hermitian with the convention (N.13).

This is the right place to discuss the rather confusing issue of *fermions in Euclidean space*. After the integration over fermionic variables in Eqs.(4.26) and (4.27) is done, we obtain, up to an irrelevant factor i^∞, $\det\| -i\slashed{\mathcal{P}} + m_f\|$ for the integral in Eq. (4.26) and $\det\| -i\slashed{\mathcal{P}}^E + m_f\|$ for the integral in Eq. (4.27). Thus, if we are interested in the partition function only, we can use the following rule of analytic continuation

$$\det\| -i\slashed{\mathcal{P}} + m_f\| \longrightarrow \det\| -i\slashed{\mathcal{P}}^E + m_f\| \qquad (4.28)$$

and forget about fermionic variables alltogether. The recipe (4.28) alone is not sufficient, however, if we want to calculate also certain Green's functions involving fermionic fields. To calculate Euclidean Green's functions, Euclidean fermionic fields need to be properly defined.

Our main message is that the symbols $\bar{\psi}_f$ in Eq. (4.26) and Eq. (4.27) have different meaning and, if you will, a different status. In Minkowski space, there exists an alternative operator approach to quantization. The Heisenberg field operators $\hat{\bar{\psi}}$ and $\hat{\psi}$ are related to each other by Dirac conjugation: $\hat{\bar{\psi}} = \hat{\psi}^\dagger \gamma^0$. Speaking of the path integral (4.26), ψ and $\bar{\psi}$ are independent Grassmann variables. It is convenient, however, to assume that they are also related to each other by Dirac conjugation. Then the Lagrangian in Eq. (4.26) is a real Lorentz scalar. Indeed, $\delta^{\text{Lorentz}}\psi = \omega_{\mu\nu}\sigma_{\mu\nu}\psi$ $[\sigma_{\mu\nu} = \frac{1}{2}(\gamma_\mu\gamma_\nu - \gamma_\nu\gamma_\mu)]$ and hence $\delta^{\text{Lorentz}}\bar{\psi} = \omega_{\mu\nu}\bar{\psi}\gamma^0\sigma^+_{\mu\nu}\gamma^0 = -\omega_{\mu\nu}\bar{\psi}\sigma_{\mu\nu}$. It follows then that $\bar{\psi}\psi$ is a (real) Lorentz scalar and $\bar{\psi}\gamma_\mu\psi$ is a (real) Lorentz vector.

In Euclidean space the situation is different. First, in the operator language, the very notion of Hermiticity ceases to be useful: the Euclidean Heisenberg operators

$$\hat{A}(\tau) = \exp\{\hat{H}\tau\}\hat{A}(0)\exp\{-\hat{H}\tau\} \qquad (4.29)$$

cannot be rendered Hermitian for all τ. Second, in the path intergal formalism, the transformation law of the fermionic variables under $SO(4)$ rotations is $\delta^{SO(4)}\psi = \omega_{\mu\nu}\sigma^E_{\mu\nu}\psi$ with $\left(\sigma^E_{\mu\nu}\right)^\dagger = -\sigma^E_{\mu\nu}$ and, for $\bar{\psi}\psi$ to be an $SO(4)$ scalar, $\bar{\psi}$ should transform as a hermitially conjugated spinor $\psi^\dagger$ (without the extra γ^0 factor). We could introduce the notation $\psi^\dagger$ instead of $\bar{\psi}$ in Eq. (4.27), but prefer not to do so and not to assume that $\bar{\psi}$ is hermitially conjugated to ψ.[‡] The proper way to think of the relationship

---

[‡]We will see in Lecture 13 that, to define a viable lattice approximation to the Euclidean fermionic path integral, keeping ψ and $\bar{\psi}$ independent is not only an option, but a

between $\bar{\psi}(x)$ and $\psi(x)$ is to write

$$\begin{cases} \psi(x) = \sum_k c_k u_k(x) \\ \bar{\psi}(x) = \sum_k \bar{c}_k u_k^\dagger(x) \end{cases} , \qquad (4.30)$$

where $\{u_k(x)\}$ is a complete basis in the corresponding Hilbert space. It is conveniently choosen as a set of eigenfunctions of the Dirac operator: $\mathcal{D}^E u_k(x) = \lambda_k u_k(x)$; c_k, $\bar{c}_k$ are independent Grassmann integration variables.

In contrast to the Minkowskian fermionic action, the Euclidean action in the exponent of Eq. (4.27) cannot be rendered real even if the variables $\bar{\psi}$ and ψ are assumed to be conjugate to each other; the term $i\bar{\psi}_f \mathcal{D}^E \psi_f$ changes sign under complex conjugation while the mass term does not. But this actually does not cause a trouble, because the full fermionic path integral

$$\det \| - i\mathcal{D}^E + m \| = \prod_k (-i\lambda_k + m) \qquad (4.31)$$

(the product is performed over all eigenmodes of the massless Dirac operator in a given gauge field background) is real! To see this, let us assume that the theory is somehow regularized both in the infrared (so that the spectrum of the Dirac operator is discrete) and in the ultraviolet [so that large eigenvalues λ_k do not contribute to the product (4.31)].[§] Note now that the spectrum of the massless Dirac operator enjoys the following symmetry: for any eigenfunction u_k of the operator $\mathcal{D}^E$ with eigenvalue λ_k, the function $u_k' = \gamma^5 u_k$ is also an eigenfunction with the eigenvalue $-\lambda_k$ ($\mathcal{D}^E$ and γ^5 anticommute). Therefore,

$$\prod_k (-i\lambda_k + m) = m^q \prod_k^{'} (\lambda_k^2 + m^2) , \qquad (4.32)$$

where q is the number of possible exact zero modes [we will show in Lecture 12 that q coincides with the absolute value of the topological charge (2.6) of the gauge field] and the product $\prod_k^{'}$ is performed over non-zero eigenvalues only. We see that the expression (4.32) is real and the Euclidean partition function is real, too.

necessity.

[§]There are many such regularization schemes and we will begin the discussion of this very important issue in the next section.

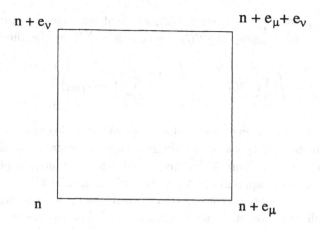

Fig. 4.1 An elementary plaquette.

4.2 Path Integral on the Lattice.

The path integrals (4.24–4.27) are symbols which require an operational definition. For ordinary quantum mechanical systems, we divided a finite time interval into $n + 1$ small intervals and defined the path integral as the limit of a finite dimensional integral (3.11), with n tending to ∞. For field theories, we need to discretize not only time, but also space. Furthermore, for gauge theories, we want to do this in a way that preserves gauge invariance at intermediate steps. The corresponding gauge invariant discretization procedure for Yang–Mills theory was worked out by Wilson. We will discuss it for the Euclidean version of the theory only. A discretized gauge invariant path integral can also be defined in Minkowski spacetime, but there is not much use of it: to date, only Euclidean path integrals are feasible for numerical calculations.

Let us introduce a four-dimensional hyper-cubic lattice. The nodes of the lattice are labelled by an integer 4-vector n. Let us define on each *link* of the lattice a unitary matrix $U_{n+e_\mu, n} \in SU(N)$, where $e_1 = (1, 0, 0, 0)$, etc. For each *plaquette* (or two-dimensional face) of the lattice labelled by its corner n_α with all the components less or equal than the corresponding components for other three corners, and by two directional vectors e_μ, e_ν (see Fig. 4.1), we define

$$W_{n,\mu\nu} = \frac{1}{N} \text{Re Tr} \left\{ U_{n,n+e_\nu} U_{n+e_\nu, n+e_\mu+e_\nu} U_{n+e_\mu+e_\nu, n+e_\mu} U_{n+e_\mu, n} \right\}, \quad (4.33)$$

with $U_{n,n+e_\nu} \equiv U^\dagger_{n+e_\nu,n}$. For $SU(2)$, we need not take the real part in Eq. (4.33) as the trace of any $SU(2)$ matrix is real. Consider the integral

$$Z = \int \prod_{\text{links}} \mathcal{D}U_{\text{link}} \exp\left\{-\frac{2N}{g^2}\sum_{\text{plaq}}(1-W_{\text{plaq}})\right\}, \qquad (4.34)$$

where $\mathcal{D}U$ is the Haar measure on the group. We are going to show that the exponent in Eq. (4.34) is a correct discrete approximation to the Euclidean action of continuum Yang–Mills theory and hence the integral (4.34) is a reasonable discrete approximation of the path integral (4.24).

We assume the actual spacing of our Euclidean lattice a be small compared to all characteristic scales of the theory.[¶] Let us associate $U_{n+e_\mu,n}$ with the parallel transporter (1.8) along the corresponding link: $U_{n+e_\mu,n} \equiv \exp\{iaA_\mu e_\mu\}$, where the A_μ are defined at the middle of the link. Then the quantity under the trace in Eq. (4.33) is the operator of the parallel transport along the plaquette and $W_{n,\mu\nu}$ is nothing but a Wilson loop operator, which for small boxes can be expressed as in Eq. (1.16). Thus, we see that, indeed,

$$S_{\text{lat}} \to \frac{2N}{g^2}a^4\sum_n \frac{1}{2}\sum_{\mu\nu}\frac{1}{2N}\text{Tr}\{F^2_{\mu\nu}\} + O(a^6) \to \int d^4x \,\frac{1}{2g^2}\text{Tr}\{F^2_{\mu\nu}\}$$

in the continuum limit (the identity $\sum_{\text{plaq}} = \sum_n\sum_{\mu>\nu} = \frac{1}{2}\sum_n\sum_{\mu\nu}$ was used). The integral (4.34) is gauge invariant as was desired. Indeed, the action (and also the Haar measure) are invariant under the transformations

$$U_{n+e_\mu,n} \to \Omega_{n+e_\mu}U_{n+e_\mu,n}\Omega^\dagger_n, \qquad (4.35)$$

where $\{\Omega_n\}$ is a set of unitary matrices defined in the nodes of the lattice. The definition (4.34) is not the only possible one. Other gauge-invariant lattice actions can be constructed. They all reproduce the structure $\text{Tr}\{F^2_{\mu\nu}\}$ to leading order in a, but differ by terms $\propto a^6$. A very probable (but not rigorously proven) statement is that all lattice path integrals with various such Yang–Mills actions give the *same* physical results in the continuum limit, which means that the continuum theory is well defined. Note that,

---

[¶] As the coupling constant g in the Yang–Mills action is dimensionless, we do not understand yet what the characteristic scale is. That will be discussed at length in Lecture 9.

in the theories like QED or $\lambda\phi^4$ theory, continuum limit exists, but corresponds to a *free* theory, with interactions being switched out. This is the known zero charge problem. As we will see and discuss at length later, this problem is *absent* in QCD.

The integral in (4.34) involves a discrete but still infinite number of variables. To make it finite, our lattice should have finite size in both spatial and Euclidean time directions. In practice, it is convenient to implement this by imposing periodic boundary conditions on the matrix link variables:

$$U_{n+L_\alpha e_\alpha+e_\mu,\,n+L_\alpha e_\alpha} = U_{n+e_\mu,\,n} , \qquad (4.36)$$

where the set of integers $L_\alpha = \{L_x, L_y, L_z, L_\tau\}$ characterizes the size of the lattice (the number of nodes in the corresponding direction); no summation over α is assumed. With the conditions (4.36), the theory is effectively defined on a discrete four-dimensional torus.

Toroidal boundary conditions are easier to handle than boundary conditions with rigid walls: finite size effects, which are present in all practical numerical calculations are less prominent for the torus. For these boundary effects not to be important, the physical length of the torus aL_μ should be larger than the characteristic scale of the theory, while a should be kept much smaller. In practical calculations, "much" usually means at most 4–5 times. Calculating the path integral (4.34) numerically on an asymmetric lattice with $L_x = L_y = L_z \gg L_\tau$ in the regime where boundary effects due to finite Euclidean time extension are important, while effects due to a finite spatial size are not, one finds the partition function of the system at finite temperature $T = 1/(aL_\tau)$. The properties of the vacuum wave functional are explored in calculations on large symmetric lattices.

Up to now, we discussed only pure Yang–Mills theory without fermions. QCD involves quarks and, to define the path integral in QCD, we need to handle fermionic fields on the lattice. Let us define to this end Grassmann variables $\bar\psi_n$, ψ_n in the nodes of the lattice for each quark flavor (color and Lorentz indices are not displayed). As a first and natural guess, let us write the extra term in the action as follows,

$$S^{\text{ferm.lat.}} = -\frac{ia^3}{2} \sum_{n,\mu} [\bar\psi_n U_{n,n+e_\mu} \gamma_\mu^E \psi_{n+e_\mu} - \bar\psi_n U_{n,n-e_\mu} \gamma_\mu^E \psi_{n-e_\mu}]$$
$$+ ma^4 \sum_n \bar\psi_n \psi_n . \qquad (4.37)$$

We see that the action (4.37) reproduces the action

$$S_{\text{ferm}}^E = -\int d^4x[i\bar{\psi}\gamma_\mu^E(\partial_\mu - iA_\mu)\psi - m\bar{\psi}\psi] \qquad (4.38)$$

in the continuum limit. Indeed, for free fermions

$$-\frac{ia^3}{2}\sum_{n,\mu}\bar{\psi}_n\gamma_\mu^E[\psi_{n+e_\mu} - \psi_{n-e_\mu}] \to -i\int d^4x\bar{\psi}\gamma_\mu^E\partial_\mu\psi .$$

Expanding $U_{n,n+e_\mu} \equiv 1 - iaA_\mu + O(a^2)$, we also restore the interaction term, and the last term in Eq. (4.37) turns into the continuum mass term. The action (4.37) is invariant under gauge transformations when the U are transformed according to Eq. (4.35), while $\psi_n \to \Omega_n\psi_n$.

Equation (4.37) is called the "naïve lattice fermion action", and I have to say that if the reader was convinced by the above reasoning that, in the continuum limit, it goes over to Eq. (4.38), he was naïve, too. Our implicit assumption was that the fermion fields ψ_n depend on the lattice node n in a smooth manner, so that the finite difference $\psi_{n+e_\mu} - \psi_{n-e_\mu}$ goes over to the continuum derivative. It turns out, however, that fermion field configurations which behave as $\psi_n \sim (-1)^{n_1}$ or $\psi_n \sim (-1)^{n_2+n_4}$ and change significantly at the microscopic lattice scale are equally important. After carefully performing continuum limit, these wildly oscillating modes give rise to 15 extra light fermion species with the same mass, the so-called *doublers*.

Simple-minded modifications of this naïve action, which leave only one light fermion for each flavor in the continuum limit, do not respect the chiral invariance of the theory — an issue which we did not yet discuss. We postpone a detailed discussion of these problems to Lecture 13. We will see that all these difficulties can be overcome, and a good consistent definition of the QCD path integral exists.

It is important that the action (4.37), as well as its sophistications, are bilinear in $\bar{\psi}_n$, ψ_n. The fermionic part of the path integral has the form

$$\int \prod_i d\bar{\psi}_i d\psi_i \exp\{-\mathfrak{D}_{ij}\bar{\psi}_i\psi_j\} , \qquad (4.39)$$

where $i \equiv (n, \alpha)$, with α marking both the color and Lorentz spinor index, and $\mathfrak{D}_{ij}$ is a matrix turning into the Euclidean Dirac operator $i\slashed{\mathcal{P}}^E - m$ in the continuum limit. We have learned how to perform integrals like (4.39) in the previous lecture. According to Eq. (3.41), the answer is just $\det \|\mathfrak{D}\|$. Thus, the full path integral for QCD has the form

$$Z = \int \prod_{\text{links}} \mathcal{D}U_{\text{link}} \prod_f \det \|\mathfrak{D}_f\| \exp\left\{ -\frac{2N}{g^2} \sum_{\text{plaq}} (1 - W_{\text{plaq}}) \right\}, \quad (4.40)$$

where $\prod_f$ runs over all quark flavours.

Numerical calculations of the integrals like (4.40) are technically very difficult: not only one has to do a multidimensional integral, but also the *integrand* becomes very complicated involving the determinant of a large matrix. But they are possible, and this problem is being tackled now.

Lecture 5

θ–Vacuum

It was already mentioned that, for Yang–Mills theory, the Dirac quantization procedure giving the equation system (4.15) is not yet the full story. An additional constraint called *superselection rule* is required. To understand it better, consider first a simple quantum mechanical model.

5.1 Quantum Pendulum.

The classical Hamiltonian of a pendulum is

$$H = \frac{p_\phi^2}{2} + \omega^2(1 - \cos\phi) \tag{5.1}$$

($\omega^2 = g/l$, $m = l = 1$). Motion is finite $0 \le \phi \le 2\pi$, with the points 0 and 2π identified.

We want now to quantize the system. The easiest way to do it is to substitute $-i\partial/\partial\phi$ for p_ϕ into the Hamiltonian (5.1) and to impose the boundary condition $\Psi(2\pi) = \Psi(0)$ on the wave functions. Then the spectrum of the system is discrete. When there is no gravity and $\omega = 0$, the solution is especially simple:

$$\Psi_k(\phi) = \frac{1}{\sqrt{2\pi}}e^{ik\phi}, \qquad \mathcal{E}_k = \frac{k^2}{2} \tag{5.2}$$

with integer k. However, one is not forced to impose the requirement of *strict* periodicity on the wave function. The boundary condition

$$\Psi(\phi + 2\pi) = e^{i\theta}\Psi(\phi) \tag{5.3}$$

with any $\theta \in [0, 2\pi)$ is admissible as well. The condition (5.3) selects a different set of states which is, again, discrete. When $\omega = 0$, we have

$$\Psi_k(\phi) = \frac{1}{\sqrt{2\pi}} e^{i(k+\theta/2\pi)\phi}, \qquad \mathcal{E}_k = \frac{(k+\theta/2\pi)^2}{2} . \qquad (5.4)$$

Anticipating field theoretic generalizations, let us express this transparent result in more general terms. In our case, the operator rotating the pendulum by one turn counterclockwise,

$$\hat{U}\Psi(\phi) = \Psi(\phi + 2\pi) , \qquad (5.5)$$

commutes with the Hamiltonian. This is true in the free case, and also when a periodic in ϕ potential term is present. We can then simultaneously diagonalize the Hamiltonian and the operator $\hat{U}$. Eigenvalues of $\hat{U}$ are certain complex numbers z lying on the unit circle: $z \equiv e^{i\theta}$ [this condition follows from unitarity: the operator (5.5) preserves the normalization of the wave function]. Bearing all this in mind, we can subdivide the *large* Hilbert space of the Hamiltonian (5.1), including all the states without any particular boundary condition on the wave function imposed*, into the sectors characterized by definite eigenvalues $e^{i\theta}$ of the operator $\hat{U}$. A small Hilbert space (or the θ-sector) for the free pendulum is described by Eq. (5.4). The spectrum (5.4) is drawn in Fig. 5.1 as a function of θ. Thick lines mark the ground state of the system (the vacuum). We see that the dependence $\mathcal{E}_{\text{vac}}(\theta)$ is discontinuous at $\theta = \pi(2k+1)$. When $\omega \neq 0$, the functions $\Psi_n^\theta(\phi)$ and $\mathcal{E}_n(\theta)$ are more complicated, but the picture is qualitatively the same.

The quantity θ and the picture in Fig. 5.1 have a very natural physical interpretation. We need to solve the problem of quantum motion in a periodic potential $V(\phi + 2\pi) = V(\phi)$. The problem has been solved at the dawn of quantum mechanics by Bloch who was interested in the quantum behaviour of electrons in crystals. Consulting standard textbooks, one finds that an alias for the quantity θ entering Eq. (5.3) is *quasi-momentum*. The analog of the spectrum (5.4) for a periodic potential in interest gives a zone structure of our "crystal". Each such zone corresponds to the interval $0 \leq \theta \leq 2\pi$ with fixed k. In the free case, the zones cover the whole range

---

*A basis of this large Hilbert space is formed by the plane waves $\Psi(\phi) = (1/\sqrt{2\pi})e^{ip_\phi\phi}$ with arbitrary p_ϕ.

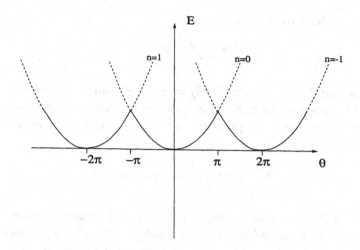

Fig. 5.1 Spectrum of the quantum pendulum as a function of θ. The solid line is the true ground state energy $\mathcal{E}_{\text{vac}}(\theta) = \min_n\{\mathcal{E}_n(\theta)\}$. It is periodic in θ with the period 2π.

of energy, but when $\omega \neq 0$ and, especially, when ω is large, the zones are separated by gaps.

Let us find now the "effective mass" of our "Bloch electron"

$$\chi = m_{\text{eff}}^{-1} = \frac{\partial^2 \mathcal{E}_{\text{vac}}(\theta)}{\partial \theta^2}\Big|_{\theta=0} \tag{5.6}$$

for a true pendulum in a gravity field ($\omega \neq 0$). Heuristically, the larger ω, the more difficult it is for the pendulum to overcome the potential barrier, and to perform a rotation by 2π. Thus, m_{eff}, measuring the inertia of the system with respect to such rotations, should be large (and χ should be small) for large ω. We will find χ analytically in the region $\omega \gg 1$, using path integral methods. This will teach us many important lessons, which we will later use for QCD.

As a first step, substitute the definition

$$\mathcal{E}_{\text{vac}}(\theta) = -\lim_{\beta \to \infty} \frac{1}{\beta} \ln Z(\theta)$$

into Eq. (5.6). Noting that $\partial Z(\theta)/\partial\theta|_{\theta=0} = 0$ due to the symmetry $Z(\theta) = Z(-\theta)$, we obtain

$$\chi = -\lim_{\beta \to \infty} \frac{1}{\beta Z(0)} \frac{\partial^2 Z(\theta)}{\partial \theta^2}\Big|_{\theta=0} . \tag{5.7}$$

Next, we have to express

$$Z(\theta) = \sum_k{}^{\theta} e^{-\beta \mathcal{E}_k(\theta)} \tag{5.8}$$

(here the sum $\sum^{\theta}$ is performed only over states from sectors with a given value of θ) as a path integral. This is similar to what we have done in the previous lecture, when we projected onto the subspace of gauge-invariant wave functions:[†]

$$Z(\theta) = \frac{1}{2\pi} \sum_{q=-\infty}^{\infty} e^{-iq\theta} \int_0^{2\pi} d\phi \, \mathcal{K}(\phi + 2\pi q, \, \phi; \, -i\beta) \,, \tag{5.9}$$

where $\mathcal{K}$ is the Euclidean evolution operator of the "unconstrained" system defined in the large Hilbert space, which includes all eigenstates of the Hamiltonian without any boundary condition on the wave function specified. The evolution operator $\mathcal{K}(\phi + 2\pi q, \, \phi; \, -i\beta)$ describes the amplitude for the system to rotate counterclockwise q turns ($|q|$ turns clockwise for negative q) over an Euclidean time interval β. The equivalence of Eq. (5.8) and Eq. (5.9) can be explicitly seen if one substitutes into Eq. (5.9) the spectral decomposition (3.13) for the Euclidean evolution operator, and uses the identity

$$\sum_{q=-\infty}^{\infty} e^{-iq(\theta-\theta')} = 2\pi\delta(\theta - \theta') \,. \tag{5.10}$$

Thirdly, the evolution operator can be represented as a path integral according to Eq. (3.14). Substituting it into Eq. (5.9), we obtain

$$Z(\theta) \sim \sum_{q=-\infty}^{\infty} e^{-iq\theta} \int \prod_\tau d\phi(\tau) \exp\left\{ -\int_0^\beta \left[\frac{\dot{\phi}^2}{2} + \omega^2(1 - \cos\phi) \right] d\tau \right\} \,, \tag{5.11}$$

where each term of the sum involves the path integral over all the trajectories with boundary conditions $\phi(\beta) = \phi(0) + 2\pi q$. As actually $\phi \equiv \phi + 2\pi$, one should rather speak of *topologically nontrivial* closed trajectories describing rotation of the pendulum by a specific number of turns.

The fourth and final step is to calculate the path integrals entering the sum (5.11). When $\omega \gg 1$, the action of all topologically non-trivial

[†]We will soon see that the similarity here is far from being accidental.

trajectories is large. Hence the path integrals Z_q with $q \neq 0$ are suppressed, compared to the topologically trivial integral with $q = 0$. If we kept in the sum only the latter, we would not detect any dependence of the partition function on θ. The susceptibility $\chi^{q=0}$ would vanish, and the effective mass would be infinite. To get a non-zero χ, we have to take into account small nonzero Z_q. For large ω, the integral can be estimated by the saddle point (semi-classical) approximation: $Z_q \propto \exp\{-S_q\}$, where S_q is the *minimal* possible action in the class of all topologically non-trivial Euclidean trajectories with q turns, i.e. the instanton action.

The reader has already realized, of course, that we are using now the terminology of Lecture 2, where topologically non-trivial Euclidean configurations and solutions in Yang–Mills field theory were discussed. The analogy is, indeed, very precise, as we will shortly see. But let us first finish our calculation.

We are interested in the limit $\beta \to \infty$. Before taking it, it is convenient to rewrite the integrals in Eq. (5.11) in the symmetric form, $\int_{-\beta/2}^{\beta/2} \to \int_{-\infty}^{\infty}$. A saddle point in the integral with, say, $q = 1$ is the solution of the Euclidean equations of motion with boundary conditions

$$\phi(-\infty) = 0 \, , \quad \phi(\infty) = 2\pi \, . \tag{5.12}$$

The equations admit an integral of motion (normally, it has the meaning of energy, cf. Eq. (1.31), but here we are solving the equations in Euclidean time and such an interpretation is absent) which should be zero due to the boundary conditions (5.12):

$$\frac{\dot{\phi}^2}{2} - \omega^2(1 - \cos\phi) = 0 \, .$$

This equation is easily solved:

$$\phi = 4\arctan e^{\omega(\tau - \tau_0)} \, . \tag{5.13}$$

This tunneling trajectory is the quantum mechanical analog of an instanton. τ_0 is the position of the instanton center in Euclidean time. The action of the solution (5.13) is [‡]

$$S_I = 4\omega^2 \int_{-\infty}^{\infty} \frac{d\tau}{\cosh^2(\omega\tau)} = 8\omega \, . \tag{5.14}$$

[‡]Note in passing that the solution we have just derived describes also a soliton in the two-dimensional Sine–Gordon model.

That means that the full contribution to the partition function from the topologically nontrivial trajectories with unit winding q is

$$Z_1 \propto (\beta\omega)\exp\{-S_I\} = (\beta\omega)\exp\{-8\omega\} , \qquad (5.15)$$

where the large factor β comes from the integral over the collective coordinate τ_0, on which the action does not depend. The contribution from trajectories with $q = -1$ is exactly the same. Consider now the contribution from the sector with $q = 2$. The functional integral is saturated by the classical solution with boundary conditions $\phi(-\infty) = 0$, $\phi(\infty) = 4\pi$. It is not difficult to realize that this is actually a combination of two solutions (5.14) separated by a large Euclidean time interval. The corresponding contribution to the action is $Z_2 \propto (\beta\omega)^2 \exp\{-16\omega\}$. We see that, as far as the exponential factors are concerned, $Z_2 = Z_{-2}$ is suppressed compared to Z_1. On the other hand, Z_2 involves the extra factor β, which we agreed to keep large. Let it first be large, but not exponentially large so that the "fugacity"

$$\kappa = \frac{Z_1}{Z_0} = C\beta\omega\exp\{-8\omega\} \qquad (5.16)$$

is much less than 1. In this case, all terms with higher q in the sum (5.11) may be neglected:

$$Z(\theta) \approx Z_0[1 + 2\kappa\cos\theta] , \qquad (5.17)$$

and we derive for the topological susceptibility (5.7)

$$\chi = \frac{2\kappa}{\beta} = 2C\omega\exp\{-8\omega\} . \qquad (5.18)$$

The effective mass is inversely proportional to the amplitude of quantum jump, which is the exponential of the action of the classical tunneling trajectory: a very natural and transparent result. The pre-exponential constant C in Eq. (5.18) can in principle be determined by calculating the functional integral in the Gaussian approximation. An analogous calculation for the instantons in QCD was done by 't Hooft.

Methodologically, it is very interesting to study also the opposite limit $\kappa \gg 1$. For large κ, all the terms in the series (5.11) are relevant. And not only that. Even in the topologically trivial sector with $\phi(2\pi) = \phi(0)$, a characteristic contribution to the path integral involves a certain number of instantons n_+ compensated by an equal amount of anti-instantons n_-:

our pendulum may make a number of counterclockwise turns and the same number of clockwise turns. Likewise, in the sector with nonvanishing net winding number q, the functional integral has the form of a double sum over n_+ and n_- with the restriction $n_+ - n_- = q$.

The leading contribution comes from the region where the instantons and anti-instantons are very well separated from each other. Then the functional integral presents roughly a product of the factors associated with individual (anti-)instantons, and only a combinatorial factor appears:

$$Z_q = \sum_{n_+ - n_- = q} \frac{\kappa^{n_+ + n_-}}{n_+! n_-!} = I_{|q|}(2\kappa) . \qquad (5.19)$$

(The Bessel function $I_\nu(x)$ grows exponentially at large x.) The full partition function is

$$Z(\theta) = \sum_{q=-\infty}^{\infty} Z_q e^{-iq\theta} = \sum_{q=-\infty}^{\infty} I_{|q|}(2\kappa) e^{-iq\theta} = e^{2\kappa \cos\theta} . \qquad (5.20)$$

Sunstituting it into (5.7), we again obtain (5.18), which universally holds both for small and large κ, with the only condition that the action $S_E = 8\omega$ is large and that β should be large compared to the characteristic oscillation period of the pendulum: $\beta \gg \omega^{-1}$.

This exercise was not idle. It teaches a lot concerning vacuum dynamics of QCD: the *instanton gas model* for the QCD vacuum state is introduced in a very similar manner. We will return to the discussion of this question in Lecture 15. Before abandoning this beautiful toy model, let us do one more exercise and rewrite the path integral (5.11) as

$$Z(\theta) \sim \int \prod_\tau d\phi(\tau) \exp\left\{ -\int_0^\beta \left[L_E + i\frac{\theta}{2\pi}\dot\phi \right] d\tau \right\} , \qquad (5.21)$$

where the integral is done over all closed, topologically trivial as well as nontrivial trajectories. The factor $\exp\{-iq\theta\}$ is traded here for an extra term in the Lagrangian called *θ-term*. Usually, adding a total time derivative to the Lagrangian changes nothing, but in quantum theory this remains true only if no topologically nontrivial trajectories are involved. If $q \neq 0$, also $\int_0^\beta \dot\phi \equiv 2\pi q \neq 0$. The functional integral is modified and a total derivative leads to quite observable effects like a modified spectrum, etc. The analytic continuation of the integral (5.21) to a real time gives a standard Minkowski path integral with $L_M \to L_M - (\theta/2\pi)\dot{\phi}$; the extra term in the Minkowski Lagrangian is real.

5.2   Large Gauge Transformations in Non-Abelian Theory.

Wave functionals representing the solutions of the equation system (4.15) are annihilated by the constraints $\hat{G}^a$, which means that they stay invariant under the action of the operator

$$\exp\left\{-i\int d^3x \; \chi^a(x)\hat{G}^a(x)\right\}$$

with small $\chi^a(x)$. This is tantamount to saying that the wave functionals are invariant under infinitesimal gauge transformations (1.10) of their arguments, $A_i \to A_i^\chi$, and hence also under finite *topologically trivial* gauge transformations (1.9), i.e. the transformations which can be reduced to a continuous series of infinitesimal gauge transformations (1.10). However, not all gauge transformations are topologically trivial. An example of a topologically nontrivial one is

$$\Omega_*(x) \;=\; \exp\left\{\frac{i\pi x^a \sigma^a}{\sqrt{x^2+\rho^2}}\right\} \tag{5.22}$$

with an arbitrary finite parameter ρ. We have $\Omega_*(0) = 1$. As $|x| \to \infty$, $\Omega_*(x) \to -1$ irrespectively of the direction $x/|x|$ along which infinity is approached. For any other $|x|$, the function $\Omega_*(x)$ gives an element of $SU(2)$, with every such element appearing only once. Therefore the function (5.22) represents a topologically nontrivial mapping of S^3 (R^3 compactified at infinity) onto the gauge group $SU(2)$. Indeed, the Chern–Simons number of the pure gauge configuration $A_i = -i\partial_i\Omega_*\Omega_*^\dagger$,

$$\begin{aligned}
N_{C.S.} &= \int K_0 d^3x \\
&= \frac{1}{24\pi^2}\int dx \; \epsilon_{ijk}\mathrm{Tr}\left\{(\partial_i\Omega_*)\Omega_*^\dagger(\partial_j\Omega_*)\Omega_*^\dagger(\partial_k\Omega_*)\Omega_*^\dagger\right\},
\end{aligned} \tag{5.23}$$

with K_μ being defined in Eq. (2.7), is equal to one. The gauge transformation (5.22) is topologically nontrivial, which means that one cannot find a family of gauge transformations $\Omega(\alpha, x)$ continuous in α such that

each $\Omega(\alpha, x)$ represents a smooth function on S^3 (the smoothness require-
ment means in particular that the limiting value of $\Omega(\alpha, x)$ at $|x| \to \infty$
exists and is uniquely defined; the necessity of this restriction will be seen
a little bit later), $\Omega(0, x) = 1$, and $\Omega(1, x) = \Omega_*(x)$. $\Omega_*(x)$ may be called a
"large" gauge transformation. The Gauss law constraint says nothing about
what happens with the wave functional after its arguments are transformed
by $\Omega_*(x)$.

Along with $\Omega_*(x)$, there are further distinct topologically nontrivial
transformations $\Omega^{(q)}(x)$ belonging to the same homotopy class as (i.e. continuously deformable to) $[\Omega_*(x)]^q$ with integer q. Two $\Omega^{(q)}(x)$ with different
q cannot be smoothly transformed into one another. The classes $\{\Omega^{(q)}(x)\}$
form what is called a homotopy group $\pi_3[G] = \mathbb{Z}$.

Note now that the following large gauge transformation operator,

$$\hat{U}\Psi[A_i(x)] = \Psi[A_i^{\Omega_*}(x)], \tag{5.24}$$

commutes with the Hamiltonian and with all other physical gauge-invariant
operators. This means that the operator (5.24) has the same meaning and
can be handled in the same way as the operator (5.5) of the rotation by 2π
for the pendulum. Thus, we diagonalize the Hamiltonian and the operator
of the large gauge transformation (5.24) simultaneously, and divide the large
Hilbert space spanned by all the eigenstates of the Hamiltonian satisfying
the Gauss law constraints into sectors with a definite eigenvalue of $\hat{U}$. The
latter is $e^{i\theta}$, as a large gauge transformation does not change the norm of
the wave function. We are allowed and, moreover, are forced to consider
only one such distinct sector because no physical operator has nonvanishing
matrix elements between the states with different θ. Once finding ourselves
in a state characterized by some θ, we can go over to a different state in the
same sector as a result of some perturbation, but the value of θ *cannot* be
changed. It is a fundamental constant of our World, fixed once and forever.
The dynamics of quantum Yang–Mills theory is simply not specified until
the value of θ is given.

The analog of Eq. (5.11) reads

$$Z(\theta) \sim \sum_{q=-\infty}^{\infty} e^{-iq\theta} \int \prod_{\tau, x} dA_\mu^a(\tau, x) \exp\left\{ -\frac{1}{4g^2} \int_0^\beta d\tau \int dx \left[(F_{\mu\nu}^a)^2 \right] \right\}, \tag{5.25}$$

with the boundary conditions

$$A_\mu(\boldsymbol{x}, \beta) = [A_\mu(\boldsymbol{x}, 0)]^{[\Omega_*^q(\boldsymbol{x})]} \tag{5.26}$$

for the $q^{\underline{th}}$ term in the sum. The expression (5.25) defines the thermal partition function. Tending $\beta \to \infty$, and substituting $\int_0^\beta \equiv \int_{-\beta/2}^{\beta/2} \to \int_{-\infty}^\infty$, we obtain the partition function concentrated on the vacuum state.

Let us look at the $q = 1$ term in Eq. (5.25). The important fact is that the Pontryagin index (2.6) of Euclidean field configurations in the corresponding path integral is equal to 1. To see that, consider the configuration $A_i(-\infty, \boldsymbol{x}) = 0$, $A_i(\infty, \boldsymbol{x}) = -i\partial_i \Omega_* \Omega_*^\dagger$. Present our Euclidean space as the cylinder $S^3 \times R$ where R is Euclidean time, and S^3 is 3-dimensional space with the point at infinity added. Recall that the Pontryagin density $\sim F\tilde{F}$ is a total derivative, and that the 4-volume integral of $F\tilde{F}$ can be written as a surface integral according to Eq. (2.8). As the base S^3 of our cylinder has no boundary, the surface of $S^3 \times R$ consists of two spheres S^3 at $\tau = -\infty$ and $\tau = \infty$. The relation (2.7) is reduced to

$$\int d^4x \, \frac{1}{32\pi^2} F_{\mu\nu}^a \tilde{F}_{\mu\nu}^a = N_{C.S.}(\tau = \infty) - N_{C.S.}(\tau = -\infty) \,. \tag{5.27}$$

Using compact S^3 instead of R^3 as a base is quite appropriate in our case, but may seem at first a bit confusing. Alternatively, one could consider a disk D^3 with large, finite radius, and an integral of the Pontryagin density over $D^3 \times R$. One can then show that the integral of the Chern–Simons current in Eq. (2.7) over the side surface of the cylinder vanishes if the radius of the disk is much larger than the characteristic scale ρ.

When the coupling constant g is small (for the time being, we do not yet understand under what conditions this is true, this will be explained in Lecture 9, devoted to asymptotic freedom), the action of all topologically non-trivial configurations is large. The integral can be done semi-classically and $Z_1 = Z_{-1} \sim \exp\{-8\pi^2/g^2\}$, where $8\pi^2/g^2$ is the action of the instanton solution studied in Lecture 2.

Thus, we understand now the physical meaning of the instanton. It is a semi-classical tunneling trajectory connecting the classical, trivial vacuum $A_i = 0$ and the configuration $-i\partial_i \Omega_* \Omega_*^\dagger$ obtained from the trivial vacuum when applying a large gauge transformation. When the coupling constant g is small, the tunneling amplitude is suppressed as $\exp\{-S_I\}$, where $S_I = 8\pi^2/g^2$ is the instanton action. Everything is the same as for quantum pendulum and the physical interpretation of θ — the quasi-momentum describing the drift of the system between the degenerate vacua characterized by different definite Chern–Simons numbers — is also the same.

We are now able to explain why the requirement that $\Omega(|x| \to \infty)$ be uniquely defined was imposed in the first place. One could, of course, consider gauge transformations which tend to different values when $x \to \infty$ along different directions. For example, so-called Gribov copies with $\tilde{\Omega}(x) \sim \sqrt{\Omega_*(x)}$ have this property (see Lecture 7 for more details). It turns out, however, that the action on a tunneling trajectory interpolating between the trivial vacuum $A_i = 0$ and the configuration $-i\partial_i \tilde{\Omega}(x)\tilde{\Omega}^\dagger(x)$ has an *infinite* action and therefore does not contribute to the path integral. In fact, it is just the presence of this infinite barrier which justifies the topological classification.

This tunneling trajectory (the so-called *meron* solution) may still present some interest, however (see Lecture 7 for further discussion). For reference purposes, let us write here the explicit form of the meron solution in a covariant gauge:

$$A_\mu^a = \eta_{\mu\nu}^a x_\nu / x^2 . \tag{5.28}$$

It differs from the pure gauge field configuration (2.16) by the absense of the overall factor 2 and, in contrast to (2.16), has a nonzero field strength

$$F_{\mu\nu}^a = \frac{x_\alpha(x_\mu\eta_{\alpha\nu}^a - x_\nu\eta_{\alpha\mu}^a) - x^2\eta_{\mu\nu}^a}{x^4} \tag{5.29}$$

and action density. The latter is proportional to $1/x^4$, and the action integral diverges logarithmically both at small and large $|x|$.

In the same way as we did it for the pendulum, one can rewrite the integral (5.25) in the form

$$Z(\theta) \sim$$

$$\int \prod_{\tau,x} dA_\mu^a(\tau,x) \exp\left\{ -\int_0^\beta d\tau \int dx \left[\frac{1}{4g^2} F_{\mu\nu}^a F_{\mu\nu}^a + i\frac{\theta}{32\pi^2} F_{\mu\nu}^a \tilde{F}_{\mu\nu}^a \right] \right\},$$

$$\tag{5.30}$$

where the integral is performed over all field configurations which are periodic in τ up to a large gauge transformation. The second term in the integrand is a total derivative and is relevant when topologically nontrivial configurations come into play. It is the famous θ-term of QCD. Analytical continuation of the path integral to Minkowski spacetime gives

$$\mathcal{L}_M \;\rightarrow\; \mathcal{L}_M + \frac{\theta}{32\pi^2} F^a_{\mu\nu} \tilde{F}^{a\ \mu\nu} \;, \tag{5.31}$$

and the θ-term is real.

What is the value of θ in the World we live? It is experiment rather than theory which can answer this question. Note that the θ-term breaks both P and T invariance. (Indeed, $F^a_{\mu\nu} \tilde{F}^{a\ \mu\nu}$ is the same as $\boldsymbol{E}^a \cdot \boldsymbol{B}^a$, where $\boldsymbol{E}^a$ is chromoelectric and $\boldsymbol{B}^a$ is chromomagnetic field. Now, $\boldsymbol{E}^a$ is a vector, which is odd under time reflection, while $\boldsymbol{B}^a$ is a pseudo-vector, even under time reflection.) We know, however, that neither parity nor T invariance are broken in strong interactions. This means that θ is either just zero or very small. The best experimental restriction comes from the measurements on the electric dipole moment d_n of the neutron. We know that $d_n \lesssim 10^{-25} e \cdot cm$ which means that $\theta \lesssim 10^{-9}$.[§]

The fact that θ is zero or very small has no explanation in the framework of QCD. Maybe the problem will be eventually solved when we understand the nature of the sought unified theory of all interactions, incorporating also QCD. The status of this problem is roughly the same as the status of the cosmological term problem: experiment tells us that the cosmological term is zero or very small, but we do not understand why.[¶]

[§] There is a subtlety here. All physical effects noninvariant with respect to P and T transformations are actually proportional to $\sin\theta$. Thus, the symmetry requirements alone restrict θ to be close to *either* 0 *or* π. We simply note here that the second possibility *is* excluded by what we know from experiments on the properties of light pseudoscalar mesons. This can be shown in the framework of the *effective chiral Lagrangian* describing light meson dynamics to be introduced in Lecture 12. We will discuss more of the physics of the imaginary world with $\theta \approx \pi$ in Lecture 16.

[¶] Probably, the analogy between θ and the cosmological term is even deeper. In the Ogievetsky–Sokachev formulation of supergravity, the cosmological term is also written as a total derivative [15]. But this question is far beyond the scope of this book.

PART 2

PERTURBATION THEORY

Lecture 6
Diagram Technique in Simple and Complicated Theories

As was repeatedly emphasized earlier, quantum field theory is nothing but quantum mechanics with an infinite number of degrees of freedom. We know three main methods of solving quantum mechanical problems: *(i)* explicit solution of the Schrödinger equation, *(ii)* semi-classical approximation, and *(iii)* perturbation theory. In principle, the first "brute force" approach is possible, but not very practical for a system of interacting fields. We cannot solve the Schrödinger equation analytically, while solving it numerically is extremely difficult (though possible in principle with lattice methods). We have made an excursion to semi-classical methods in the previous lecture.

The technique most widely used, which allows one to obtain a lot of nontrivial results for physically observable effects, is, of course, perturbation theory. It is especially fruitful for the theories like QED, where the coupling constant is small and the perturbative series converges rapidly.* Perturbation theory also works well for many problems in QCD.† For quantum field systems, an alias for perturbation theory is the *Feynman diagram technique*. In this and in the following lecture we will construct the diagram technique for QCD and we will learn how to calculate the simplest Feynman graphs in QCD.

---

*To be more precise, the coefficients of α^n in this series grow as $n!$ at large n, the series is asymptotic and starts to diverge when n becomes larger than $n^* \sim 1/\alpha$. This brings about an uncertainty $\sim \exp\{-C/\alpha\}$, which is absolutely irrelevant for all practical purposes.

†We will see later that it is so for processes with large characteristic energy transfer.

6.1   Feynman Rules from Path Integral

We assume that the reader is familiar with the standard operator method to derive the Feynman rules. Here we will give a brief sketch how this is done with path integrals. Consider $\lambda\phi^4$ theory, the simplest field theory with a nontrivial interaction. Its Lagrangian is

$$\mathcal{L} = \frac{1}{2}(\partial_\mu\phi)^2 - \frac{m^2}{2}\phi^2 - \frac{\lambda}{24}\phi^4 \ . \tag{6.1}$$

Our task is to find the elements of the S-matrix — the matrix elements $\langle out|in \rangle$. The first step is to make use of the *reduction formula*, which relates the scattering amplitudes to the residues at the poles of the vacuum expectation value of the time-ordered product of the Heisenberg field operators $\hat{\phi}_H(x) = e^{i\hat{H}t}\phi_H(x)e^{-i\hat{H}t}$. For example, for the scattering $p_1 p_2 \to p_3 p_4$ ($p_{j0} > 0$), we may write

$$\int d^4x_1\, d^4x_2\, d^4x_3\, d^4x_4\ e^{-ip_1\cdot x_1 - ip_2\cdot x_2 + ip_3\cdot x_3 + ip_4\cdot x_4}$$

$$\times\ \langle 0|T\{\hat{\phi}_H(x_1)\hat{\phi}_H(x_2)\hat{\phi}_H(x_3)\hat{\phi}_H(x_4)\}|0\rangle$$

$$= \left(\prod_{j=1,\dots,4} \frac{i\sqrt{Z_j}}{p_j^2 - m_R^2 + i0}\right) iM_{12\to34}\ (2\pi)^4\delta^{(4)}(p_1 + p_2 - p_3 - p_4)$$

$$+\ \text{less singular terms} \ , \tag{6.2}$$

where Z is the residue of the exact propagator at the pole:

$$\int d^4x\, e^{ipx} \langle 0|T\{\hat{\phi}_H(x)\hat{\phi}_H(0)\}|0\rangle \ \sim\ \frac{iZ}{p^2 - m_R^2 + i0}$$

when $p^2 \sim m_R^2$, and m_R is the renormalized physical mass which does not generally coincide with the bare mass m in the Lagrangian (6.1). The result (6.2) can be derived both in the path integral language (as it is done in Faddeev's and Slavnov's book [4]) and in the operator language (see e.g. [2], Chapt. 7.2). We skip it.

Secondly, we must express the vacuum expectation value on the left side of Eq. (6.2) as a path integral. We already know the formula (3.17) expressing the v.e.v. $\langle \hat{O} \rangle$ as an Euclidean path integral. In our case, we need to find the average of the T-product of the Heisenberg operators depending explicitly on the real Minkowski times $x_{j0} = t_j$. The answer is

$$\langle 0|T\{\hat{\phi}_H(x_1)\ldots\hat{\phi}_H(x_n)\}|0\rangle =$$

$$\lim_{\epsilon\to+0}\lim_{T\to\infty}\frac{\int \mathcal{D}\phi \, \phi(x_1)\ldots\phi(x_n)\exp\left\{i\int_{-T(1-i\epsilon)}^{T(1-i\epsilon)}dtd\boldsymbol{x}\mathcal{L}(x)\right\}}{\int \mathcal{D}\phi \, \exp\left\{i\int_{-T(1-i\epsilon)}^{T(1-i\epsilon)}dtd\boldsymbol{x}\mathcal{L}(x)\right\}}, \quad (6.3)$$

where the small factor $-i\epsilon$ ensures the dominance of the vacuum contribution and also the convergence of the path integral (cf. the discussion in Lecture 3). The boundary values of $\phi(-T(1 - i\epsilon), \boldsymbol{x})$ and $\phi(T(1 - i\epsilon), \boldsymbol{x})$ need not be specified because they do not affect the result.

To make things clearer, let us derive Eq. (6.3) first for the correlator of two coordinate Heisenberg operators $\langle 0|T\{\hat{q}_H(t_1)\hat{q}_H(t_2)\}|0\rangle$ in a quantum mechanical system. Note that the vacuum state $|0\rangle$ can be obtained from *any* state $|\psi\rangle$ with nonzero matrix element $\langle 0|\psi\rangle$ by the action of the operator $\exp\{-\beta\hat{H}\}$ with large enough β. The identity

$$|0\rangle = \langle 0|\psi\rangle^{-1}\lim_{\beta\to\infty} e^{\beta\mathcal{E}_{\mathrm{vac}}}e^{-\beta\hat{H}}|\psi\rangle \quad (6.4)$$

holds. Assume for definiteness that $t_1 \geq t_2$. Then the T-product coincides with the usual one. Choose $|\psi\rangle = |q\rangle$, where $|q\rangle$ is the state with a definite value of the coordinate q. Using (6.4), the definition $\hat{q}_H(t) = e^{i\hat{H}t}qe^{-i\hat{H}t}$, and inserting two times the spectral decomposition of unity, $1 = \int dq \, |q\rangle\langle q|$, we obtain up to an irrelevant numerical factor

$$\langle 0|T\{\hat{q}_H(t_1)\hat{q}_H(t_2)\}|0\rangle \propto \lim_{\beta\to\infty}\int dq_1 \, dq_2 \, \langle q_f|e^{-i\hat{H}(-i\beta-t_1)}|q_1\rangle \, q_1$$

$$\times \langle q_1|e^{-i\hat{H}(t_1-t_2)}|q_2\rangle q_2 \langle q_2|e^{-i\hat{H}(t_2-i\beta)}|q_i\rangle$$

$$= \lim_{\beta\to\infty}\int q_1 dq_1 \int q_2 dq_2 \, \mathcal{K}(q_f, q_1, -i\beta - t_1)\mathcal{K}(q_1, q_2, t_1 - t_2)$$

$$\times \mathcal{K}(q_2, q_i, t_2 - i\beta)$$

$$= \lim_{\beta\to\infty}\int \prod_t dq(t)q(t_1)q(t_2)\exp\left\{i\int_C dt\mathcal{L}[q(t)]\right\}, \quad (6.5)$$

where the contour C connects the points $i\beta \to t_2 \to t_1 \to -i\beta$ as shown in Fig. 6.1a, and the boundary conditions $q(i\beta) = q_i$, $q(-i\beta) = q_f$ are chosen. It is convenient to deform the contour by adding to the initial time $i\beta$ a large real negative constant $-T$ and to the final time $-i\beta$ — a large positive constant T as shown in Fig. 6.1b. The change $\beta \to \beta + iT$ does

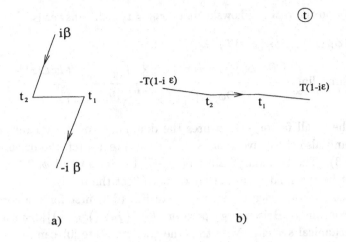

Fig. 6.1 The contour C in Eq. (6.5) and its deformation.

not affect the projection property (6.4), and the relation (6.5) is still valid. We choose this constant T to be much larger than β.

Taking the limit $T, \beta \to \infty$ so that the ratio $\epsilon = \beta/T$ tends to zero, we arrive at the quantum-mechanical analog of Eq. (6.3) for $n = 2$. The generalizations to the case $n > 2$ and to field theory are obvious.

On the third step we have to calculate actually the path integrals in Eq. (6.3). For theories with nontrivial interaction, that can be done only approximately. We will do it perturbatively by representing the result as a series in the coupling constant λ. To this end, we first express $\mathcal{L} = \mathcal{L}_0 + \mathcal{L}_{\text{int}}$, where $\mathcal{L}_0$ involves at most second powers of the fields and $\mathcal{L}_{\text{int}}$ contains all the rest. In our case, $\mathcal{L}_{\text{int}} = -\lambda \phi^4/24$. Next, we expand

$$\exp \left\{ -i\lambda \int \frac{\phi^4}{24} d^4x \right\} \tag{6.6}$$

as a series in powers of λ and express our vacuum average as a ratio of the integrals

$$\langle \phi_1 \ldots \phi_m \rangle_0 = \frac{\int \mathcal{D}\phi \; \phi(x_1) \ldots \phi(x_m) \exp \left\{ i \int d^4x \mathcal{L}_0(x) \right\}}{\int \mathcal{D}\phi \; \exp \left\{ i \int d^4x \mathcal{L}_0(x) \right\}} . \tag{6.7}$$

Here, $m = n + 4k$ if the kth term in the expansion of the exponential (6.6) is taken into account. The point is that the integrals (6.7) have the

Gaussian form and can be calculated analytically.[‡] It is convenient to do it via the generating functional which we already introduced when discussing quantum mechanics path integrals [see Eqs. (3.18, 3.19)]. In our case we may write

$$\langle \phi_1 \ldots \phi_m \rangle_0 = \frac{1}{Z[0]} \left(-i \frac{\delta}{\delta J(x_1)} \right) \cdots \left(-i \frac{\delta}{\delta J(x_m)} \right) Z[J] \bigg|_{J=0} , \quad (6.8)$$

where

$$Z[J] = \int \mathcal{D}\phi \exp \left\{ i \int d^4 x [\mathcal{L}_0(x) + J(x)\phi(x)] \right\} =$$
$$\lim_{\epsilon \to 0} \int \mathcal{D}\phi \exp \left\{ i \int d^4 x \left[\frac{1}{2}\phi(x)(-\partial^2 - m^2 + i\epsilon)\phi(x) + J(x)\phi(x) \right] \right\} .$$

$$(6.9)$$

In the last line we performed the integration by parts; the term $i\epsilon$ should be inserted to take into account the fact that, according to Eq. (6.3), the integral over dt was originally done over a path in Fig. 6.1b with a small negative slope $dt_C = dt_R(1 - i\epsilon)$ (in the last line, we integrate just over the real axis $dt \equiv dt_R$). The shift $i\epsilon$ makes the integral (6.9) convergent. The integral can be explicitly done by defining

$$\phi = \phi' + i \int d^4 y D_F(x - y) J(y) , \quad (6.10)$$

where $D_F(x - y)$ is the Feynman scalar Green's function,

$$(\partial^2 + m^2 - i0) D_F(x - y) = -i\delta(x - y) .$$

Its Fourier image has a familiar form

$$D_F(p) = \frac{i}{p^2 - m^2 + i0} . \quad (6.11)$$

The change of variables (6.10) kills the linear term, the integral over $\mathcal{D}\phi'$ gives an irrelevant constant, and we obtain

$$Z[J] = Z[0] \exp \left\{ -\frac{1}{2} \int d^4 x d^4 y J(x) D_F(x - y) J(y) \right\} . \quad (6.12)$$

[‡]The trick we use here is exactly equivalent to going into the interaction representation in the operator approach.

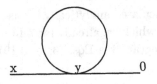

Fig. 6.2   1-loop correction to scalar propagator.

Varying it according to Eq. (6.8), we easily obtain

$$\langle \phi(x)\phi(0) \rangle_0 = D_F(x) \,,$$
$$\langle \phi(x_1)\phi(x_2)\phi(x_3)\phi(x_4) \rangle_0 = D_F(x_1 - x_2)D_F(x_3 - x_4)$$
$$+ D_F(x_1 - x_3)D_F(x_2 - x_4) + D_F(x_1 - x_4)D_F(x_2 - x_3), \quad (6.13)$$

and so on, i.e. the average of the product of $2p$ factors $\phi(x_i)$ presents a sum of $(2p - 1)!!$ terms, each terms being the product of p Feynman Green's functions $\langle \phi(x_i)\phi(x_j) \rangle_0$ with different arguments. The property (6.13) (the absence of higher nontrivial correlators) is a well-known property of Gaussian stochastic ensembles. In deriving the Feynman rules using path integral techniques, this property plays the same role as the Wick contraction rules used in operator formalism.

To be absolutely clear, we consider an example. Let us find the one-loop correction to the propagator $\langle \phi(x)\phi(0) \rangle$. To first order in λ, the propagator is given by the ratio

$$\frac{\int \mathcal{D}\phi \ \phi(x)\left(1 - \frac{i\lambda}{24}\int d^4y\phi^4(y)\right)\phi(0)\exp\left\{i\int d^4x\mathcal{L}_0(x)\right\}}{\int \mathcal{D}\phi \left(1 - \frac{i\lambda}{24}\int d^4y\phi^4(y)\right)\exp\left\{i\int d^4x\mathcal{L}_0(x)\right\}}. \quad (6.14)$$

Consider the numerator. The average of the product of 6 fields involves many terms corresponding to different pairing among the fields. First, let $\phi(x)$ be paired directly with $\phi(0)$. Then we have $D_F(x)$ as a common factor, and whatever it multiplies, the same factor appears also in the denominator and cancels exactly the corresponding λ-dependent terms in the numerator. This phenomenon is known as *cancellation of vacuum loops*. A nontrivial perturbative correction comes from the terms involving the products $\langle \phi(x)\phi(y) \rangle\langle \phi(y)\phi(y) \rangle\langle \phi(y)\phi(0) \rangle$. There are 12 such terms, and we finally obtain

$$\langle\phi(x)\phi(0)\rangle = D_F(x) - \frac{i\lambda}{2}\int d^4y D_F(x-y)D_F(0)D_F(y) + O(\lambda^2) .$$

(6.15)

The second term here corresponds, of course, to the simplest diagram in Fig. 6.2, with $\frac{1}{2}$ being the symmetry factor. The integral diverges quadratically at small distances as we can best see by transferring to momentum space:

$$\int e^{ipx}\langle\phi(x)\phi(0)\rangle d^4x =$$

$$D_F(p) - \frac{i\lambda}{2}[D_F(p)]^2 \int \frac{d^4k}{(2\pi)^4} D_F(k) + O(\lambda^2) .$$

(6.16)

$D_F(k_0, \mathbf{k})$ has poles at $k_0 = \pm\sqrt{\mathbf{k}^2 + m^2} \mp i0$, so that we can perform the *Wick rotation* $k_0 \to ik_0^E$, after which the integral has only the ultraviolet singularities at large Euclidean momenta. In this particular case, $\int d^4k_E/(k_E^2 + m^2) \sim \Lambda_{UV}^2$, where Λ_{UV} is the ultraviolet cutoff (see Lecture 8 for detailed discussion), and the divergence is quadratric as we asserted above.

One of the technical steps in the derivation above was the introduction of a finite temperature β^{-1}, which was taken to zero afterwards [see Eq. (6.4)]. But we can also keep β finite and develop the *thermal* diagram technique for the correlators $\langle T\{\hat{\phi}_H(x_1)\dots\hat{\phi}_H(x_n)\}\rangle_\beta$, etc. Consider the free *Euclidean* correlator $D_E(\tau, \mathbf{x}) = \langle\phi(\tau, \mathbf{x})\phi(0)\rangle_\beta^{\text{free}}$. It is periodic in τ with period β. Hence, it is developed into a Fourier series rather than Fourier integral, and the Fourier image of $D_E(\tau, \mathbf{x})$ has the same form as the Euclidean counterpart of Eq. (6.11), but is defined at a discrete set of points $p_0^{(n)} = 2\pi i n T$ with integer n. The loop integrals are transformed according to the rule

$$\int \frac{d^4k_E}{(2\pi)^4} f(k_E) \to T\sum_n \int \frac{dk}{(2\pi)^3} f(2\pi i n T, \mathbf{k}) .$$

(6.17)

For fermions, the recipe is modified a little bit. Fermionic fields are *antiperiodic* in Euclidean time [see Eq. (3.40)], and the Euclidean fermion propagator is defined at the points $p_0 = \pi i(2n + 1)T$.

The *Matsubara diagram technique* just described will be employed in Lecture 18 to determine thermodynamic and some other static characteris-

tics of the QCD heat bath. To study the *kinetics* of hot media, it is better
to use the real time thermal diagram technique developed by Bakshi, Ma-
hanthappa, and Keldysh. This technique is a little more tricky, and we will
not discuss it here.

6.2   Fixing the Gauge

To treat QCD perturbatively, we must ensure that characteristic field fluctuations are small. This is not so for the path integrals (4.24, 4.25) — a
gauge transformation which can change the values of the fields substantially
leaves the integrand invariant. In other words, the theory has *gauge zero
modes*. All such gauge copies, which are not necessarily close in function
space to each other and to the perturbative vacuum $A_i = 0$, contribute to
path integrals on an equal footing. To cope with this unwanted effect, we
should somehow *fix the gauge*, that is we should impose a constraint that
picks up only one representative from the whole gauge orbit, a set of field
configurations related to each other by a gauge transformation.

In the toy oscillator model considered in Lecture 4, the requirement
$\phi = 0$ played the role of such gauge fixing constraint. The constraints
imposed in QCD may have a lot of different forms. Various constraints
(choices of gauge) are used for different purposes. We will discuss here
Lorentz-invariant gauges and begin with Landau (or Lorentz) gauge.

$$\partial^\mu A_\mu^a = 0 . \tag{6.18}$$

Two remarks are in order. First, the gauge condition (6.18) does not
have a Hamiltonian form in a sense that it involves the Lagrange multipliers
$A_0^a(x)$ in addition to the dynamical variables $A_i^a(x)$. This results in some
minor methodical complications, which can be successfully coped with (see
e.g. Ref. [4]). For the path integral in Lagrangian form, the Landau gauge
(6.18) is as good as the Coulomb gauge $\partial_i A_i^a = 0$ as long as it picks out a
representative from each gauge orbit.

The second complication is more serious. In the non-Abelian case, the
latter property is actually not fulfilled. Unlike the gauge $\phi = 0$ in the
oscillator model, neither Landau gauge, nor Coulomb gauge are sufficient
to slice nicely the space of all field configurations into gauge orbits, with
each orbit being counted only once. We will see in the next lecture that
there are different gauge fields satisfying the condition (6.18) and related

by a gauge transformation: Gribov copies. It turns out, however, that the gauge transformation which relates the trivial perturbative vacuum $A_\mu^a = 0$ and its Gribov copy is large in the sense that the distance in function space between the copies is large. And this means that the copies may be relevant for the nonperturbative dynamics, but that they are irrelevant as far as perturbative expansions are concerned. Indeed, the loop expansion for the amplitudes is nothing but a semi-classical expansion near the perturbative vacuum. On the other hand, the configuration space in the vicinity of the perturbative vacuum *is* sliced into gauge orbits by the condition (6.18) without problems, and the presence of the copies is not felt.

Let us try to write a modified path integral implementing the constraint (6.18) as a δ function:

$$Z = \int \prod_{xa\mu} dA_\mu^a(x) \prod_{xa} \delta[\partial^\lambda A_\lambda^a(x)] \exp\left\{ \frac{-i}{4g^2} \int d^4x F_{\mu\nu}^a F^{a,\mu\nu} \right\} \qquad (6.19)$$

(we may forget about θ in perturbation theory). This is what is usually done in QED. However, this is *wrong* in the non-Abelian case.

If taking the expression (6.19) at its face value and building up the corresponding Feynman rules, one can see that the theory thus constructed makes no sense: it is not unitary. We will discuss this in details in the next Lecture. The reason for the trouble is that, in non-Abelian theory, the path integral (6.19) does not actually follow from the original path integral (4.25) which *is* our starting point.

Faddeev and Popov were the first to realize how to write down a correct path integral with gauge-fixing condition imposed in the non-Abelian case. They invented two ingenious tricks. The first trick is to insert unity

$$1 = \prod_x \int \mathcal{D}\Omega(x) \delta\left[R(A^\Omega) \right] \det\left\| \frac{\delta R(A^\Omega)}{\delta\Omega} \right\| \qquad (6.20)$$

in the integrand (4.25), where $R(A)$ is a gauge fixing function. The Landau gauge choice corresponds to $R(A) = \prod_{ax} \partial^\mu A_\mu^a$. The integral in (6.20) is done over all gauge transformations with the Haar measure $\mathcal{D}\Omega(x)$. As the original expression (4.25) was gauge invariant, the expression

$$\int \mathcal{D}\Omega \, \delta R(A^\Omega) \det\left\| \frac{\delta\left[R(A^\Omega)\right]}{\delta\Omega} \right\| \mathcal{D}A \, e^{iS[A]} \qquad (6.21)$$

is also gauge invariant. [In the above formula $\mathcal{D}\Omega \equiv \prod_x \mathcal{D}\Omega(x)$, $\mathcal{D}A \equiv$

$\prod_{x a \mu} dA^a_\mu(x)$.] Gauge invariance allows us to change the variables $A^\Omega \to A$ in the above integral and rewrite it as

$$\int \mathcal{D}\Omega \int \mathcal{D}A \; \delta[R(A)] \det \left\| \frac{\delta R(A^\chi)}{\delta \chi} \right\|_{\chi=0} e^{iS[A]}$$

$$= \int \mathcal{D}A \; \delta[R(A)] \det \|M^{ab}(A)\| \; e^{iS[A]} \qquad (6.22)$$

with

$$M^{ab}(A) = \partial^2 \delta^{ab} + f^{acb} \partial^\mu A^c_\mu \equiv \partial^\mu \mathcal{D}_\mu . \qquad (6.23)$$

The integral over $\mathcal{D}\Omega$ on the right side is lifted as the integrand does not depend on Ω anymore and $\int \mathcal{D}\Omega$ just brings about an irrelevant infinite constant.

The meaning of the performed transformations is the following: the integral (6.21) is done over all gauge fields as the integral (4.25) was.[§] For each A, δ function in Eq. (6.21) picks out a gauge transformation $\Omega(x)$ such that the gauge transformed field satisfies the gauge fixing constraint $R(A^\Omega) = 0$. Thereby, the whole range of integration in the path integral is split into a set of gauge orbits. As the contribution of each element of a given orbit to the path integral is the same (it is the determinant factor which takes care of it), we can suppress the integral over the orbit $d\Omega$ and write the integral in a way that only one element of the orbit satisfying the gauge fixing constraint $R(A^\Omega) = 0$ is picked out.

The "perturbative path integral" (6.22) has a nice property that the integrand does not have gauge zero modes anymore. It can be evaluated in the saddle approximation and expressed as a perturbative series in the coupling constant g. Still, the presence of the pre-exponential determinant factor brings about complications in practical calculations. And here comes the second Faddeev and Popov trick. Let us express the determinant in Eq. (6.22) as a path integral over extra Grassmann variables $c^a(x)$ and $\bar{c}^a(x)$:

$$\det \|M^{ab}(A)\| = \int \mathcal{D}c \mathcal{D}\bar{c} \; \exp \left\{ i \int d^4x \; \bar{c}^a[-\partial^2 \delta^{ac} - f^{abc} \partial^\mu A^b_\mu]c^c \right\}$$

$$(6.24)$$

[§]To be quite exact, these integrals are equivalent only in the perturbative framework. We disregard Gribov copies, which means that $A^a_\mu(x)$ is assumed to be small and $\Omega(x)$ close to unity. The key formula (6.20) was written under this assumption.

(the operator ∂^μ acts on everything to the right of it), which is true up to an irrelevant constant factor due to the property (3.41). The product of the determinant (6.24) and the exponential $e^{iS[A]}$ in the path integral can be written as $e^{i\tilde{S}[A]}$, where $\tilde{S}$ is a modified action involving, in addition to gauge fields, also scalar fermion *ghost* fields $c^a(x)$:

$$\tilde{S} = \int d^4x \left[-\frac{1}{4g^2}(F^a_{\mu\nu})^2 - \bar{c}^a(\partial^\mu \mathcal{D}^{ac}_\mu)c^c \right] . \tag{6.25}$$

We still have the δ function $\delta(\partial^\mu A_\mu)$ in the integrand, which is not too convenient. But this is handled in the same way as in QED. One can introduce a family of gauges

$$\partial^\mu A^a_\mu = \omega^a(x) \tag{6.26}$$

and integrate over all $\omega^a(x)$ with the weight

$$\exp\left\{ -i\frac{1}{2\xi g^2} \int d^4x\, \omega^a(x)\omega^a(x) \right\} .$$

This procedure is equivalent to inserting in the path integral the weighing factor

$$\exp\left\{ \frac{-i}{2\xi g^2} \int d^4x (\partial^\mu A^a_\mu)^2 \right\} \tag{6.27}$$

and defines a general Lorentz-invariant ξ-gauge. Landau gauge corresponds to the limit $\xi \to 0$. The determinant factor appears for each gauge in the family (6.26) by the same token as it did for Landau gauge. It does not depend on $\omega^a(x)$. The exponent in (6.27) is added to the Lagrangian, and we finally obtain

$$\mathcal{L}_{FP} = -\frac{1}{4}(\partial_\mu A^a_\nu - \partial_\nu A^a_\mu + gf^{abc}A^b_\mu A^c_\nu)^2 - \frac{1}{2\xi}(\partial^\mu A^a_\mu)^2$$
$$- \bar{c}^a(\partial^2 \delta^{ac} + gf^{abc}\partial^\mu A^b_\mu)c^c , \tag{6.28}$$

where we have performed a rescaling $A \to gA$, which is convenient for perturbative analysis. The ghosts are not real physical particles. If they were, the theory would make no sense: the Hamiltonian involving scalar fermionic fields would not have a ground state and unitarity would be lost. Therefore, one is not allowed to consider ghosts in the physical $|in\rangle$ and $\langle out|$

asymptotic states.¶ However, ghosts proliferate in loops. Loop integrals over ghosts fields produce the perturbative expansion of the Faddeev–Popov determinant. We will see later that ghost loops are not a luxury but a necessity. One *has* to take them into account to provide for unitarity of the amplitudes.

Before going further, let us make an important comment concerning the gauge invariance of the path integral. The integral

$$\int \mathcal{D}c \, \mathcal{D}\bar{c} \, \mathcal{D}A \, \exp\left\{ i \int d^4x \mathcal{L}_{\text{FP}} \right\} \tag{6.29}$$

comes from the gauge invariant path integral (4.25) and should be gauge invariant. This invariance is not seen at the level of the integrand. If we transform the gauge fields according to Eq. (1.10), both the gauge fixing term and the ghost–ghost–gluon interaction term in Eq. (6.28) are varied nontrivially. One can be convinced, however, that the Lagrangian (6.28) is invariant under a *global* symmetry transformation which acts both on the gauge and the ghost fields:

$$\delta A_\mu^a = \epsilon (\mathcal{D}_\mu c)^a ,$$
$$\delta c^a = -\frac{1}{2}\epsilon g f^{abc} c^b c^c ,$$
$$\delta \bar{c}^a = -\epsilon \frac{1}{\xi} \partial^\mu A_\mu^a , \tag{6.30}$$

where ϵ is an anticommuting Grassmann parameter. The invariance of (6.28) under (6.30) is a remnant of the gauge invariance after the gauge is fixed. It is called the BRST-symmetry (after Becchi, Rouet, Stora, and Tyutin). The symmetry (6.30) will be used in the next two lectures when deriving generalized Ward identities and arguing that QCD is a renormalizable theory.

Also, the full QCD gauge-fixed Lagrangian,

$$\mathcal{L}_{\text{g.f.}}^{\text{QCD}} = \mathcal{L}_{\text{FP}} + \sum_f \bar{\psi}_f (i\slashed{D} - m_f)\psi_f , \tag{6.31}$$

¶Strictly speaking, the physical QCD asymptotic states do not involve quarks and gluons either due to confinement — only colorless hadron states are physical. But confinement is a nonperturbative effect. In perturbation theory at any finite order, quarks and transverse gluons can and should be treated as asymptotic particles.

is invariant under a combination of the transformation (6.30) and the transformation

$$\delta\psi_f = igt^a(\epsilon c^a)\psi_f ,$$
$$\bar{\delta}\psi_f = -ig\bar{\psi}_f t^a(\epsilon c^a) . \tag{6.32}$$

Note that Eq. (6.32) and the first line in Eq. (6.30) have the meaning of a gauge transformation with the parameter $\epsilon c^a(x)$. Note also that, as the transformations (6.30) for $c^a(x)$ and $\bar{c}^a(x)$ have nothing to do with each other, $c^a(x)$ and $\bar{c}^a(x)$ should not be thought of as being complex conjugates of each other. They are completely independent variables. One can assume, for example, that ϵ and $\bar{c}^a(x)$ are real (invariant under involution) and $c^a(x)$ are purely imaginary (changing sign under involution) Grassmann variables. It is probably more proper to use different notations for the ghost and antighost fields and write $b^a(x)$ instead of $\bar{c}^a(x)$. We will actually do it in the next lecture when discussing the canonical BRST formalism. But we will keep the standard notation for a while, bearing in mind that $c^a(x)$ and $\bar{c}^a(x)$ are different independent variables.

    With the path integral

$$\int \mathcal{D}c\,\mathcal{D}\bar{c}\,\mathcal{D}A \prod_f \mathcal{D}\bar{\psi}_f\,\mathcal{D}\psi_f \exp\left\{i\int d^4x\left[\mathcal{L}_{\text{FP}} + \sum_f \bar{\psi}_f(i\not{\mathcal{D}} - m_f)\psi_f\right]\right\} \tag{6.33}$$

in hand, we can derive the Feynman rules in the same way as was outlined for $\lambda\phi^4$ theory. If one wishes, one could also introduce the unphysical ghosts creation and annihilation operators and derive the Feynman rules using the operator language and keeping in mind that ghosts never appear as asymptotic states. The result is

$b\nu \, \rotatebox{0}{\wwww} \, a\mu$ p

gluon propagator: $D_{\mu\nu}^{ab}(p) \;=\; \dfrac{-i\delta^{ab}}{p^2 + i0}\left[\eta_{\mu\nu} - \dfrac{(1-\xi)p_\mu p_\nu}{p^2 + i0}\right]$

3-gluon vertex: $\Gamma_{\mu\nu\lambda}^{abc}(p,q,r) \;=\; -gf^{abc}[(p-q)_\lambda \eta_{\mu\nu} +$
$(q-r)_\mu \eta_{\nu\lambda} + (r-p)_\nu \eta_{\mu\lambda}]$

4-gluon vertex: $\Gamma_{\mu\nu\lambda\sigma}^{abcd} \;=\; -ig^2 f^{abe} f^{cde}(\eta_{\mu\lambda}\eta_{\nu\sigma} - \eta_{\mu\sigma}\eta_{\nu\lambda})$
$-ig^2 f^{ace} f^{bde}(\eta_{\mu\nu}\eta_{\sigma\lambda} - \eta_{\mu\sigma}\eta_{\nu\lambda})$
$-ig^2 f^{ade} f^{bce}(\eta_{\mu\nu}\eta_{\sigma\lambda} - \eta_{\mu\lambda}\eta_{\nu\sigma})$

$b \cdots\cdots\overset{p}{\longleftarrow}\cdots\cdots a$

ghost propagator: $C^{ab}(p) \;=\; \dfrac{i\delta^{ab}}{p^2 + i0}$

$\bar{c}cA$ – vertex: $\Gamma_\mu^{abc}(p) \;=\; gf^{abc}p_\mu$

quark propagator: $G_j^i(p) \;=\; \dfrac{i}{\not{p} - m + i0}\delta_j^i$

quark–gluon vertex: $(\Gamma_\mu^a)_i^j \;=\; ig\gamma_\mu(t^a)_i^j \, .$ (6.34)

The 3-gluon and 4-gluon vertices come from the expansion of the first term in the Lagrangian (6.28). We see that the 3-gluon vertex involves momenta coming from the derivatives $\partial_\mu A_\nu^a$. We choose the convention that all the momenta p, q, and r are outgoing so that $p + q + r = 0$. Also, the ghost–ghost–gluon vertex involves the ghost momentum. Again, the convention is that p_μ is the outgoing ghost momentum. All the vertices depend on the color indices [adjoint indices a, b, c, d for gluon and ghost lines and fundamental (antifundamental) indices i, j for quark and antiquark lines] in a nontrivial way. There are no external ghosts. External gluon lines are represented by the transverse polarization vectors $e_\mu^{(T)}(k) = e_\mu^{(\alpha)}(k)$, $\alpha = 1, 2$, satisfying the properties $(e_\mu^{(\alpha)})^2 = -1$, $e_\mu^{(\alpha)}(k)k^\mu = 0$. This is quite parallel to QED with the only difference that here $e_\mu(k)$ carries also the color index a which we will not display explicitly. External quarks and antiquarks are represented by bispinors u_i and $\bar{u}^j$ carrying fundamental or antifundamental color indices. An algebraic expression corresponding to a QCD Feynman graph and constructed by the rules (6.34) contributes to iM_{fi} where M_{fi} is a scattering amplitude.

We have constructed the diagram technique in QCD. What is it good for? As was mentioned before, the "physical amplitudes" involving quarks and transverse gluons in the asymptotic states are not physical in QCD due to confinement. Still, such amplitudes (not the amplitudes themselves, but rather properly defined cross sections) are related to physical observables and the perturbative calculation of the former is *not* a meaningless exercise. We will touch upon some phenomenological applications in the following lectures.

Lecture 7

When the Gauge is Fixed ...

In this lecture, we will discuss some fine issues associated with the gauge-fixing procedure and the properties of perturbative amplitudes in QCD which are related to the non-Abelian nature of gauge fields.

7.1 Gribov Copies.

As was already mentioned, Landau gauge condition (6.18) does not completely remove gauge freedom, there are Gribov ambiguities or else Gribov copies: different fields satisfying the condition $\partial^\mu A_\mu = 0$ and related by a gauge transformation. The same is true for a lot of other gauge choices, in particular for Coulomb gauge $\partial_i A_i = 0$. In pure Yang–Mills theory for the vacuum fields with $E_i = 0$, the color charge density $\sim [A_i, E_i]$ is zero. The condition $\partial_i A_i = 0$ also implies $A_0 = 0$, and a vacuum Gribov copy in Coulomb gauge is simultaneously a Gribov copy in Landau gauge.

Following the original derivation of Gribov (see also the review [16]), we will construct here the vacuum copies in Coulomb gauge in $SU(2)$ theory. In other words, we will build up a static field $A_i(x)$ with zero field strength $F_{ij}(x) = 0$ satisfying the Coulomb gauge constraint $\partial_i A_i = 0$.

We start by noting that the condition $F_{ij} = 0$ implies that the gauge potential $A_i(x)$ is given by a gauge transformation of the perturbative vacuum $A_i(x) = 0$. Therefore, it has the (pure gauge) form

$$A_i(x) = -i[\partial_i \Omega(x)]\Omega^\dagger(x) . \tag{7.1}$$

Let us look for $\Omega(x)$ in the following spherically symmetric ansatz,

$$\Omega(x) = \exp\left\{\frac{if(r)}{2}n_a\sigma_a\right\}, \tag{7.2}$$

with $r = \sqrt{x_i x_i}$, $n_i = x_i/r$. Substituting (7.2) in (7.1), we obtain

$$A_i = \frac{n_i n_a \sigma_a}{2}f'(r) + \frac{\sigma_a(\delta_{ai} - n_a n_i)}{2r}\sin[f(r)]$$
$$+ \frac{1 - \cos[f(r)]}{2r}\epsilon_{ibc}n_b\sigma_c. \tag{7.3}$$

The condition $\partial_i A_i = 0$ implies then that the function $f(r)$ satisfies the following differential equation

$$\frac{d}{dr}\left(r^2\frac{df}{dr}\right) - 2\sin f = 0. \tag{7.4}$$

We have to solve this on the interval from $r = 0$ to $r = \infty$ with the initial condition $f(0) = 0$ [otherwise the gauge potential (7.3) would be singular at $r = 0$]. The derivative $f'(0)$ is not zero and sets up the overall scale of the solution [i.e. the region of space where the solution is essentially changed].

It is convenient to make the substitution $r = e^s$, after which Eq. (7.4) acquires the form

$$\ddot{f} + \dot{f} - 2\sin f = 0, \tag{7.5}$$

where the dot means d/ds. It can now be interpreted as the equation of motion for a particle moving in the potential $V(f) = 2\cos f$ with the friction term $\propto \dot{f}$. The Gribov solution corresponds to the particle starting from the point of unstable equilibrium at $f = 0$ in the distant "past" ($s = -\infty$ or $r = 0$) with zero velocity ($\dot{f} = rf' = 0$ for $r = 0$) and gliding downhill to the right or to the left (depending on whether $f'(0)$ is positive or negative) and then after some oscillations coming to rest at the minimum of the potential $f = \pm\pi$ at distant future. The solution can be obtained numerically, and the reader is welcome to do it with his/her Maple or Mathematica program.

For very large $r \gg \rho = |f'(0)|^{-1}$, the gauge transformation function acquires the form

$$\Omega_{\text{copy}}(x) \overset{r\to\infty}{\longrightarrow} \pm in_a\sigma_a, \tag{7.6}$$

i.e. $\Omega_{\text{copy}}(\infty)$ is not uniquely defined. This is the major difference with the instanton gauge transformation (5.22) with $\Omega_*(\infty) = -1$. If one forgets for a moment about unimportant oscillations and assumes that $f(r)$ is a monotonic function going from 0 at $r = 0$ to $\pm\pi$ at $r = \infty$, we see that the mapping (7.2) of R^3 onto $SU(2)$ covers only a half of the latter. That means, in particular, that the Chern–Simons number (5.23) of the configuration (7.2) is equal to $\pm 1/2$. This can be verified by direct calculation. Substituting (7.2), (7.3) in Eq. (5.23), one obtains after some algebra

$$N_{\text{c.s.}} = \frac{1}{2\pi} \int_0^\infty \frac{\partial}{\partial r}(f - \sin f)\,dr = \pm\frac{1}{2} \tag{7.7}$$

depending on whether $f(\infty) = \pi$ or $f(\infty) = -\pi$ (and *not* depending on whether the function $f(r)$ is monotonic or not).

The configuration (7.3) is separated from the perturbative vacuum by a high barrier. As was already mentioned in Lecture 5, a semi-classical Euclidean configuration describing the tunneling through this barrier and associated with the meron solution (5.28), with the Pontryagin index (2.6), equal to $1/2$ has an *infinite* action. Seemingly, this means that merons do not contribute to functional integrals describing physical quantities.

The action integral with the meron field strength (5.29) diverges, however, only logarithmically and the irrelevance of merons is not so clear if one considers a gas or some "interacting medium" of merons and takes into account an enthropy factor. There are also some indications (coming from the analysis of supersymmetric gauge theories and of pure $SU(N)$ Yang–Mills theory in the large N limit) that, probably not just merons, but some other *delocalized* configurations with fractional net topological charge may provide essential contributions to the partition function. Also, if the theory is regularized on the lattice with finite size and spacing, this provides a cutoff for the action integral, and meron configurations might become important...

Gribov copies should also be accurately taken into account when doing (now rather fashionable) lattice calculations with partially fixed gauge. One cannot exclude the fact that the effects brought about by the Gribov copies survive also in the continuum limit. Indeed, Gribov copies are not just reduced to merons: also nontrivial (nonvacuum) fields have copies which are not directly related to merons. Thus, a (questionable) premise that merons are not relevant does not lead to a (probably wrong) conclusion that Gribov copies are irrelevant too.

We have to make two more remarks here. First, there are some non-covariant gauges (like the fixed-point gauge $x^\mu A_\mu = 0$, to be discussed in

more details in Lecture 12) where Gribov copies do not appear whatsoever. Second, when people study nonperturbative dynamics of QCD, they
do not in most cases fix the gauge and, therefore, do not have to cope with
the Gribov ambiguities problem. The primary motivation to fix the gauge
in the path integral is to develop perturbation theory and to derive the
corresponding Feynman rules. But whatever nontrivial nonperturbative
dynamics might or might not be associated with the Gribov copies, they
are absolutely irrelevant as far as QCD perturbation theory is concerned.

7.2 Ward Identities.

After a brief foray into nonperturbative physics, we return onto the firm
ground of perturbation theory.

When the path integral for the partition function of a theory enjoys a
symmetry, this fact leads to an infinite series of certain relations between
Green's functions of the theory. These are called the generalized Ward
identities or just *Ward identities* (WI). The gauge symmetry of the path
integral in QCD also entails a lot of different WI, some of which are very
important both when analyzing the formal structure of the theory and
when applying the perturbative techniques for studying particular QCD
processes.

What *are* the Green's functions in a gauge theory? The question is
not so trivial because a Green's function like $\langle A_\mu^a(x) A_\nu^b(y)\rangle_0$ vanishes in
QCD if the averaging in the path integral is done over all field configu-
rations. Indeed, the Green's function $\langle A_\mu^a(x) A_\nu^b(y)\rangle_0$ is not invariant under gauge transformations of fields. When going along all gauge orbits,
positive and negative contributions in the path integral cancel each other
and the net result is zero. This applies to *any* gauge-dependent operator. Thus, only the averages of some gauge-independent operators (like
the correlators $\langle j_\mu(x) j_\nu(y)\rangle_0$ of electromagnetic and other vector currents
$j_\mu = \sum_{fg} c_{fg} \bar\psi_f \gamma_\mu \psi_g$ or the correlator $\langle (F_{\mu\nu}^a)^2(x)\,(F_{\mu\nu}^a)^2(y)\rangle_0$, etc.) have a
meaning in the full theory treated at the nonperturbative level.

In this lecture, however, we are interested in the perturbative Green's
functions defined as the averages over small fluctuations around the perturbative classical vacuum $A_\mu^a = 0$ in the framework of a particular gauge-
fixing procedure. Actually, the notion of "perturbative vacuum" is also not
absolutely trivial and requires comments. In Coulomb or Landau gauge,

the perturbative vacuum is just the configuration $A_\mu^a(x) = 0$. When the gauge fixing condition (6.26) is chosen, we have to solve the equation $\partial^\mu[-i\partial_\mu\Omega(x)\Omega^\dagger(x)] = \omega(x)$ and pick out a solution which goes over to $\Omega(x) = 1$ in the limit $\omega(x) \to 0$. For small $\omega(x)$ (we need only small $\omega(x)$ as the weight (6.27) involves a large factor g^{-2} in the exponent), this can be done perturbatively...

The Faddeev–Popov formalism allows one to forget about all these complications. In general ξ-gauge, perturbative Green's functions are defined as averages of the corresponding operators over the fluctuations in the vicinity of $A_\mu^a(x) = 0$, with the weight in Eq. (6.33). The Lagrangian (6.31) is not gauge-invariant, but it is invariant with respect to the BRST transformations (6.30) and (6.32). The easiest way to derive a set of WI is to employ the BRST symmetry.* Consider the generating functional

$$
Z[J_\mu^a, \bar\eta^a, \eta^a, \bar\zeta_f, \zeta_f] = \int \mathcal{D}A\mathcal{D}\bar c\mathcal{D}c \prod_f \mathcal{D}\bar\psi_f\mathcal{D}\psi_f \times
$$

$$
\exp\left\{i\int d^4x \left[\mathcal{L}_{\text{g.f.}}^{\text{QCD}} + J^{\mu a}A_\mu^a + \bar c^a\eta^a + \bar\eta^a c^a + \sum_f(\bar\zeta_f\psi_f + \bar\psi_f\zeta_f)\right]\right\}. \quad (7.8)
$$

The Grassmann sources $\zeta_f(x)$ and $\bar\zeta_f(x)$ are Dirac conjugates of each other just as the quark and antiquark fields are. But the sources $\eta^a(x)$ and $\bar\eta^a(x)$ are completely mutually independent just as the ghost and antighost fields are.

The functional integral (7.8) is invariant under the change of variables (6.30) and (6.32). Now, $\mathcal{L}_{\text{g.f.}}^{\text{QCD}}$ is invariant under this transformation. The measure is invariant too: the Jacobian of the transformation (6.30) can be written symbolically as

$$
J = \det \begin{vmatrix} \delta_\mu^\nu\delta^{ab} & -\epsilon(\mathcal{D}_\mu)^{ab} & 0 \\ 0 & \delta^{ab} + \epsilon g f^{abc}c^c & 0 \\ -\frac{\epsilon}{\xi}\delta^{ab}\partial^\nu & 0 & 1 \end{vmatrix} = 1, \quad (7.9)
$$

which is also true if the transformations (6.32) for the fermion fields are taken into account.

But the terms involving the sources in the integrand are not invariant.

---

*One could also derive the same set of WI exploiting the gauge invariance of the partition function written in the form (6.22), as was originally done by Slavnov and Taylor.

To leading order in ϵ, we have

$$-iZ^{-1}[J_\mu^a, \eta^a, \bar\eta^a, \bar\zeta_f, \zeta_f]\delta_\epsilon Z[J_\mu^a, \bar\eta^a, \eta^a, \bar\zeta_f, \zeta_f] \overset{\text{def}}{=} \epsilon\Phi$$

$$= \epsilon\left\langle\left[\int d^4x \left[J^{a\mu}\mathcal{D}_\mu c^a - \frac{1}{\xi}(\partial^\mu A_\mu^a)\eta^a + \frac{g}{2}\bar\eta^a f^{abc}c^b c^c\right.\right.\right.$$

$$\left.\left.\left. + ig\sum_f(\bar\psi_f c^a t^a \zeta_f - \bar\zeta_f t^a c^a \psi_f)\right]\right]\right\rangle_{J,\eta,\bar\eta,\zeta_f,\bar\zeta_f} = 0 \ . \quad (7.10)$$

[the averaging is done with the weight (7.8)]. The WI are obtained taking variations of Eq. (7.10) with respect to the sources and setting $J = \bar\eta = \eta = \bar\zeta_f = \zeta_f = 0$ afterwards. The simplest nontrivial interesting WI is obtained by considering the identity

$$i\frac{\delta^2}{\delta J^{\mu a}(x)\delta\eta^b(y)}\Phi[\text{sources}]\bigg|_{\text{sources}=0} =$$

$$\left\langle\frac{1}{\xi}A_\mu^a(x)\partial^\nu A_\nu^b(y) - \mathcal{D}_\mu c^a(x)\bar c^b(y)\right\rangle_0 = 0 \ . \quad (7.11)$$

This is a certain relation between the gluon Green's function $D_{\mu\nu}^{ab}(x-y) = \langle A_\mu^a(x)A_\nu^b(y)\rangle_0$, the ghost Green's function $\langle c^a(x)\bar c^b(y)\rangle_0$ and the 3-point Green's function $\sim \langle A_\mu^c(x)c^a(x)\bar c^b(y)\rangle_0$.

Alternatively, one can define the ghost Green's function in a given gluon background

$$\langle c^a(x)\bar c^b(y)\rangle_A = C_A^{ab}(x,y) \ , \quad (7.12)$$

satisfying the equation

$$\partial^\mu \mathcal{D}_\mu^{ac} C_A^{cb}(x,y) = -i\delta^{ab}\delta(x-y) \ , \quad (7.13)$$

and rewrite Eq. (7.11) as

$$\frac{1}{\xi}\frac{\partial}{\partial y_\nu}D_{\mu\nu}^{ab}(x-y) = \langle \mathcal{D}_\mu^{ac} C_A^{cb}(x,y)\rangle_0 \ . \quad (7.14)$$

The promised simple and interesting relation is obtained when Eq. (7.14) is acted upon by the operator $\partial/\partial x_\mu$. We obtain

$$\frac{\partial^2}{\partial x_\mu \partial y_\nu}D_{\mu\nu}^{ab}(x-y) = \xi\langle\partial^\mu \mathcal{D}_\mu^{ac} C_A^{cb}(x,y)\rangle = -i\xi\delta^{ab}\delta(x-y) \ , \quad (7.15)$$

which simply means that the longitudinal part of the exact gluon Green's function is exactly the same as for the tree level Green's function [given by the first line in Eq. (6.34)]. In other words, the longitudinal part of the gluon Green's function is not renormalized.

Another beautiful WI is obtained from the following fourth variation of the functional (7.10):

$$
i\frac{\delta^4}{\delta J^{\mu a}(x)\delta J^{\nu b}(y)\delta J^{\sigma c}(z)\delta\eta^d(u)}\Phi[\text{sources}]\bigg|_{\text{sources}=0} = 0 =
$$
$$
-\frac{1}{\xi}\left\langle A_\mu^a(x)A_\nu^b(y)A_\sigma^c(z)\partial^\lambda A_\lambda^d(u)\right\rangle_0 + \left\langle \mathcal{D}_\mu c^a(x)A_\nu^b(y)A_\sigma^c(z)\bar{c}^d(u)\right\rangle_0
$$
$$
+ \left\langle A_\mu^a(x)\mathcal{D}_\nu c^b(y)A_\sigma^c(z)\bar{c}^d(u)\right\rangle_0 + \left\langle A_\mu^a(x)A_\nu^b(y)\mathcal{D}_\sigma c^c(z)\bar{c}^d(u)\right\rangle_0 .
$$
$$(7.16)$$

Acting on this with the operator $\partial^3/(\partial x_\mu\partial y_\nu\partial z_\sigma)$, we manage to get rid of the ghost Green's function. Bearing in mind (7.13) and (7.15), we obtain

$$
\left\langle \partial^\mu A_\mu^a(x)\,\partial^\nu A_\nu^b(y)\,\partial^\sigma A_\sigma^c(z)\,\partial^\lambda A_\lambda^d(u)\right\rangle_0 =
$$
$$
-\xi^2\left[\delta^{ab}\delta^{cd}\delta(x-y)\delta(z-u) + \delta^{ac}\delta^{bd}\delta(x-z)\delta(y-u)\right.
$$
$$
\left. + \delta^{ad}\delta^{bc}\delta(x-u)\delta(y-z)\right] ,
$$
$$(7.17)$$

which means that the connected part of the correlator of four longitudinal gluons vanishes. Going over into momentum space and introducing the exact 1-particle irreducible vertices $\Gamma^{(4)}_{\mu\nu\sigma\lambda}(p,k,q,r)$ and $\Gamma^{(3)}_{\mu\nu\sigma}(p,k,q)$, we finally obtain an identity

$$
p^\mu k^\nu q^\sigma r^\lambda\left[\Gamma^{(4)}_{\mu\nu\sigma\lambda}(p,k,q,r) + \right.
$$
$$
\Gamma^{(3)}_{\mu\nu\alpha}(p,k,-p-k)D_{\alpha\beta}(p+k)\Gamma^{(3)}_{\beta\sigma\lambda}(p+k,q,r) +
$$
$$
\Gamma^{(3)}_{\mu\sigma\alpha}(p,q,-p-q)D_{\alpha\beta}(p+q)\Gamma^{(3)}_{\beta\nu\lambda}(p+q,k,r) +
$$
$$
\left. \Gamma^{(3)}_{\mu\lambda\alpha}(p,r,-p-r)D_{\alpha\beta}(p+r)\Gamma^{(3)}_{\beta\nu\sigma}(p+r,k,q)\right] = 0 , \quad(7.18)
$$

relating the gluon propagator and the proper 3-gluon and 4-gluon vertices.

On the other hand, there is no simple analog of the classical Ward–Takahashi identity relating the vertex $\Gamma^{\bar{e}e\gamma}_\mu$ and the electron propagator in QED. Taking the variation $\delta^3\Phi/\delta\bar{\zeta}_f(x)\delta\zeta_f(y)\delta\eta^a(z)$ (no summation over f)

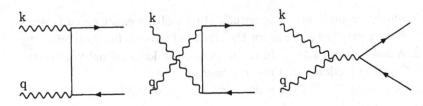

Fig. 7.1   The process $\bar{q}q \to 2g$.

and applying the operator $\partial^2/(\partial z_\sigma)^2$, we can derive

$$q^\mu G(p+q)\Gamma^a_\mu(p+q,p;q)G(p) =$$

$$gt^a[G(p+q) - G(p)] + \begin{array}{l} \text{something ugly} \\ \text{involving ghosts} \end{array} . \qquad (7.19)$$

7.3   Ghosts and Unitarity.

The sum of all ghost loops reproduces the nontrivial Faddeev–Popov determinant. If not taking the ghost loops into account, physical amplitudes would not be gauge-invariant. Let us illustrate the necessity of ghosts by studying a simple pedagogical example.

Consider the process of the annihilation of a quark–antiquark pair into two gluons. Compared to the analogous process $e^+e^- \to 2\gamma$ in QED, it involves an extra graph with the 3-gluon vertex (see Fig. 7.1). The amplitude of the process can be calculated according to the Feynman rules (6.34) and has the form $M^{\mu\nu}e^{(\alpha)}_\mu(k)e^{(\beta)}_\nu(q)$, where $\alpha, \beta = 1, 2$ mark two possible physical polarizations of gluons.

Let us first recall the situation in QED. For the process $e^+e^- \to 2\gamma$, one could write

$$iM^{\mu\nu}(k,q) = (-ie_0)^2 \int d^4x d^4y e^{ikx+iqy} \langle 0|T\{j^\mu(x)j^\nu(y)\}|e^+e^-\rangle ,$$
$$(7.20)$$

where $j^\mu(x) = \bar{e}(x)\gamma^\mu e(x)$ is the electromagnetic current and e_0 is the electron charge. In QED the electromagnetic current is conserved: $\partial_\mu j^\mu = 0$, which implies $k_\mu M^{\mu\nu}(k,q) = q_\nu M^{\mu\nu}(k,q) = 0$. In the non-Abelian case, the situation is different. Indeed, the current $j^{a\mu} = \bar{q}\gamma^\mu t^a q$ has the property

$$\mathcal{D}_\mu j^{a\mu} = (\partial_\mu \delta^{ac} + gf^{abc}A^b_\mu)j^{c\mu} = 0 \qquad (7.21)$$

with the covariant derivative. This relation does not describe any local conservation law.[†] Besides, $M^{\mu\nu}$ cannot be presented in the form (7.20) due to the third graph in Fig. 7.1.

Therefore, there is no reason for $k_\mu M^{\mu\nu}(k, q)$ to be zero, and it is not. However, if we multiply $k_\mu M^{\mu\nu}$ by the transverse polarization vector of the second gluon, we obtain zero again:

$$k_\mu M^{\mu\nu}(k, q) e_\nu^{(T)}(q) = 0 . \tag{7.22}$$

Indeed, an explicit calculation (do it!) reveals that

$$k_\mu M^{ab,\,\mu\nu}(k, q) \sim (q^\nu q^\rho - \eta^{\nu\rho} q^2) \bar{u} \gamma_\rho t^c u f^{abc} . \tag{7.23}$$

The term $\propto q^2$ is zero when the second gluon is on mass shell, and the first term gives zero when multiplied by the transverse polarization vector $e_\nu^{(T)}(q)$. Note that the contribution of only the first two graphs in Fig. 7.1 in the amplitude would not be transverse even in the limited sense (7.22); this property holds only when, in addition, the third graph is taken into account.

Generally, it is true that for any amplitude involving an incoming or outgoing gluon, with the momentum k and thereby representable in the form $e_\mu^{(T)}(k) M^\mu(k)$, where M^μ involves only transverse polarization vectors of all other external gluons, the relation

$$M^\mu(k) k_\mu = 0 \tag{7.24}$$

holds. It is one of the QCD Ward identities, a rigorous proof of which will be given at the end of this lecture. The physical meaning of the property (7.22) [and its generalization (7.24)] is the following. Besides two transverse polarizations, one could also consider two unphysical polarizations, the scalar polarization and the spatial longitudinal polarization:

$$e^{(s)} = (1, 0) , \qquad e^{(l)} = \frac{1}{|\mathbf{k}|}(0, \mathbf{k}) . \tag{7.25}$$

[†]An analogy with general relativity can be drawn: as is well-known, a locally conserved energy-momentum tensor does not exist in curved space. The canonical energy-momentum tensor $T^{\mu\nu}$ satisfies only the property $T^{\mu\nu}{}_{;\mu} = 0$ with covariant derivatives involving Christoffel symbols.

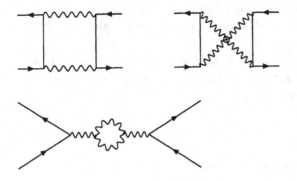

Fig. 7.2   The process $\bar{q}q \to \bar{q}q$.

Or, alternatively, two light cone polarizations

$$e_\mu^{(+)}(k) = \frac{1}{\sqrt{2}|k|}k_\mu , \qquad e_\mu^{(-)}(k) = \frac{1}{\sqrt{2}|k|}(k_0, -k) . \qquad (7.26)$$

The quantity $M^{\mu\nu}k_\mu e_\nu^{(T)}(q) \propto M^{\mu\nu}e_\mu^{(+)}(k)e_\nu^{(T)}(q)$ can be interpreted as the amplitude of the process of the production of the transverse gluon with momentum q and the gluon with momentum k carrying an unphysical polarization $e_\mu^{(+)}(k)$. This is something which we do not want: only tranverse gluons are physical particles and nothing else can be produced in collision of quark and antiquark physical (at the perturbative level) states. The Ward identity (7.22) ensures that the amplitude of the process $\bar{q}q \to g^{(+)}g^{(T)}$ is zero. One can be convinced that the amplitude of the process $\bar{q}q \to g^{(-)}g^{(T)}$ is zero too.

This is not yet the end of the story. Consider the process $\bar{q}q \to g^{(+)}g^{(-)}$. Multiplying the tensor (7.23) by the vector $e_\nu^{(-)}(q) \sim (q_0, -q)$, we find with dismay that the amplitude of the process with the creation of *two* unphysical polarizations is nonzero. One could try to find a way out of the paradox postulating that one is just not allowed to consider amplitudes of such unphysical processes. That is OK as far as tree amplitudes are concerned, but the trouble strikes back when loops are taken into account.

Consider the one loop contribution to the elastic zero angle scattering amplitude $\bar{q}q \to \bar{q}q$ with a two-gluon intermediate state. The corresponding graphs are drawn in Fig. 7.2. Let us calculate their imaginary parts using the Cutkosky rules, so that the internal gluon lines are put onto mass shell. If the graphs are calculated in Feynman gauge, the residue of the

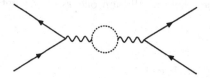

Fig. 7.3   Ghost loop contribution to $M_{\bar{q}q \to \bar{q}q}$.

propagators at the pole is $\eta_{\mu\nu}$, which can be decomposed as

$$\eta_{\mu\nu} = e_\mu^{(+)} e_\nu^{(-)} + e_\mu^{(-)} e_\nu^{(+)} - \sum_{\alpha=1,2} e_\mu^{(\alpha)} e_\nu^{(\alpha)} \qquad (7.27)$$

(we choose the basis where $e_\mu^{(\alpha)}$ are real). Then the imaginary part of the elastic forward scattering amplitude is given by the sum of the cross section of the physical process $\bar{q}q \to 2g^{(T)}$ *and* the cross section of the unphysical process $\bar{q}q \to g^{(+)} g^{(-)}$. We have either to abandon the restriction that only physical polarizations are created (and thereby gauge invariance) or the optical theorem and thereby unitarity.

This paradox is cured by the ghosts. Consider another contribution to the amplitude drawn in Fig. 7.3, involving the ghost loop. Its imaginary part due to the two-ghost "intermediate state" is nonzero. It is related to the "cross section" of the process $\bar{q}q \to \bar{c}c$. An explicit calculation shows that this "cross section" is negative and cancels exactly the positive "cross section" of another unphysical process $\bar{q}q \to g^{(+)} g^{(-)}$. Thereby, the two contributions which cancel each other can be disregarded. Only the physical cross section with transverse gluons in the final state contributes to the imaginary part, and gauge invariance and unitarity can be reconciled again.[‡]

Actually, a cancellation of this kind also occurs in QED, if one chooses as a basis for unphysical polarization not the light cone vectors (7.26) (in which case unphysical contributions to the cross section are zero right from the beginning), but the scalar and spatial longitudunal vectors (7.25). In QED the "cross sections" $\sigma_{e^+e^- \to \gamma^{(s)} \gamma^{(s)}}$ and $\sigma_{e^+e^- \to \gamma^{(l)} \gamma^{(l)}}$ are not zero. Only the net contribution of all this trash to $\mathrm{Im}\,[M_{e^+e^- \to e^+e^-}]$ is zero. In

---

[‡]This was the argument which led Feynman to the suggestion about the existence of "fictitious scalar particles", i.e. the ghosts, back in 1963. An elaboration of this dispersive relations + unitarity method allowed de Witt to build up the ghost Feynman rules independently of Faddeev and Popov.

the non-Abelian case, the cancellation occurs only when ghosts are taken
into account.

7.4   BRST Quantization.

An "analytic experiment" of the previous section displayed unitarity of the
S-matrix restricted on the physical Hilbert space involving only quark and
transverse gluon asymptotic states. To prove that it is *always* the case,
some extra work should be done. We will present a proof of "semi-Greek"
quality. A more detailed and accurate one can be found in the original
paper [17].

Let us first recall how it is proven in the Abelian case. The Lagrangian
of scalar QED in Feynman gauge (where the analysis is the most simple)
reads

$$\mathcal{L} = -\frac{1}{4}F_{\mu\nu}F^{\mu\nu} + |\mathcal{D}_\mu\phi|^2 - \frac{1}{2}(\partial^\mu A_\mu)^2 . \tag{7.28}$$

Treating A_0 as a dynamic variable, its canonical momentum is $\pi_0 = -\partial^\mu A_\mu$.
The canonical hamiltonian is

$$\mathcal{H} = \frac{1}{2}E_i E_i + \frac{1}{2}B_i B_i + \bar{\pi}\pi + |\mathcal{D}_i\phi|^2 - \frac{\pi_0^2}{2} - \pi_0(\partial_i A_i) , \tag{7.29}$$

where E_i, B_i are the electric and magnetic fields, and π is the canonical
momentum of the charged scalar field. The *Gupta–Bleuler* quantization
procedure consists in imposing the standard equal-time commutation relations,

$$[\pi_\Phi(x),\ \Phi(y)] = -i\delta(x - y) , \tag{7.30}$$

for all variables Φ including A_0. Expanding

$$A_\mu(x) = \sum_k \frac{1}{\sqrt{2\omega}}[a_\mu(k)e^{-ikx} + a_\mu^\dagger(k)e^{ikx}]$$

and substituting it in (7.30) together with the expressions for $\pi_i \equiv E_i$ and
π_0, we deduce

$$[a_\mu(k), a_\nu^\dagger(q)] = -\eta_{\mu\nu}\delta(k - q) . \tag{7.31}$$

In particular, $[a_0(k), a_0^\dagger(q)] = -\delta(k - q)$. The negative sign tells us that the norm of the scalar gluon state $a_0^\dagger|vac\rangle$ is negative, and we are dealing here with a system with *indefinite metric*.

The physical subspace $\mathfrak{h}_{\text{phys}}$ of the large Hilbert space $\mathfrak{H}$ is defined in two steps. We first impose the subsidiary condition

$$(\partial^\mu A_\mu)^\dagger|\text{phys}\rangle = 0, \qquad (7.32)$$

where $(\partial^\mu A_\mu)^\dagger$ is the positive frequency part of $\partial^\mu A_\mu$ involving only annihilation operators. The Hilbert space $\mathfrak{h}$ satisfying Eq. (7.32) involves the states with positive norm (asymptotically: transverse photons and charged scalar particles) and also unphysical states $\sim \left[a_0^\dagger(k) + a_i^\dagger \frac{k_i}{|k|}\right]^n |vac\rangle$, which have zero norm. In other words, $\mathfrak{h}$ is the space with positive semi-definite metric.

Let $\mathfrak{h}_0$ be the subspace of $\mathfrak{h}$ consiting of the states with zero norm. Note that it is orthogonal to $\mathfrak{h}$, i.e., for any two states $|\phi\rangle \in \mathfrak{h}_0$ and $|\psi\rangle \in \mathfrak{h}$, the inner product $\langle\phi|\psi\rangle$ vanishes. Indeed, positive semi-definiteness of metric in $\mathfrak{h}$ allows one to write the Cauchy–Bunyakovsky inequality,

$$\|\langle\phi|\psi\rangle\| \leq \|\langle\phi|\phi\rangle\|^{1/2}\|\langle\psi|\psi\rangle\|^{1/2} = 0. \qquad (7.33)$$

Vanishing of all matrix elements involving the states of $\mathfrak{h}_0$ means that the states $|\psi\rangle$ and $|\psi + \phi\rangle$ are equivalent for all physical purposes. Thus, the physical subspace is the quotient space $\mathfrak{h}_{\text{phys}} = \mathfrak{h}/\mathfrak{h}_0$. It involves asymptotically only transverse photons and is equipped with positive definite metric.

The Hamiltonian (7.29) as well as its nonlinear part H^{int} are Hermitian. This means that the scattering matrix $T \exp\left\{-i \int_{-\infty}^{\infty} H^{\text{int}} dt\right\}$ is unitary in the large Hilbert space $\mathfrak{H}$ with indefinite metric. But the operator $(\partial^\mu A_\mu)^\dagger$ commutes with the Hamiltonian. Therefore, $\mathfrak{h}$ and by the same token $\mathfrak{h}_{\text{phys}}$ are invariant under time evolution, and hermiticity of the Hamiltonian in $\mathfrak{H}$ implies that a Hermitian Hamiltonian H_{phys} acting on the subspace $\mathfrak{h}_{\text{phys}}$ can be defined. The physical S-matrix $T \exp\left\{-i \int_{-\infty}^{\infty} H_{\text{phys}}^{\text{int}} dt\right\}$ is unitary.

The proof of unitarity of S_{phys} in the non-Abelian case goes along the same lines, but is somewhat trickier. Let us first rewrite the Faddeev–Popov Lagrangian (6.28) in the form

$$\mathcal{L} = -\frac{1}{4}F_{\mu\nu}^a F^{\mu\nu\,a} - \partial^\mu B^a A_\mu^a + \frac{\xi}{2}(B^a)^2 + \partial^\mu b^a (\mathcal{D}_\mu c)^a, \qquad (7.34)$$

where the extra variable B^a satisfying the equations of motion $\xi B^a + \partial^\mu A_\mu^a = 0$ is introduced. Now, $b^a \equiv \bar{c}^a$ is Hermitian and c^a is anti-Hermitian. Speaking of A_0^a, it is no longer an independent dynamic variable as in Eq. (7.28), but plays the role of the canonical momentum of the variable B^a. The canonical Hamiltonian is

$$\mathcal{H} = \frac{1}{2}(\pi_i^a)^2 + \frac{1}{4}F_{ij}^a F_{ij}^a + \pi_B^a (\mathcal{D}_i \pi_i)^a + B^a \partial_i A_i^a$$

$$- \frac{\xi}{2}(B^a)^2 - \pi_c^a \pi_b^a + g f^{abc} \pi_B^a c^b \pi_c^c + \partial_i b^a \mathcal{D}_i c^a , \qquad (7.35)$$

where

$$\pi_i^a = F_{0i}^a , \quad \pi_b^a = (\mathcal{D}_0 c)^a , \quad \pi_c^a = -\partial_0 b^a , \quad \pi_B^a = -A_0^a . \qquad (7.36)$$

The Lagrangian (7.34) is invariant with respect to the BRST transformations

$$\delta_\epsilon A_\mu^a = \epsilon (\mathcal{D}_\mu c)^a ,$$

$$\delta_\epsilon c^a = -\frac{1}{2}\epsilon g f^{abc} c^b c^c ,$$

$$\delta_\epsilon b^a = \epsilon B^a ,$$

$$\delta_\epsilon B^a = 0 . \qquad (7.37)$$

A direct application of the Noether's theorem [see Eq. (12.1)] allows one to write down a canonical conserved BRST charge:

$$Q = \int d^3x \left[\pi_i^a (\mathcal{D}_i c)^a - \frac{g}{2} f^{abc} \pi_c^a c^b c^c + \pi_b^a B^a \right] . \qquad (7.38)$$

Canonical quantization consists in imposing the (anti)commutation relations

$$[\pi_{\Phi_I}(x), \Phi_J(y)] = -i\delta(x - y)\delta_{IJ} . \qquad (7.39)$$

Here, $[\pi_{\Phi_I}(x), \Phi_J(y)]$ is the commutator when used with the bosonic variables $\Phi_I^{bos} = A_i^a, B^a$, and is the anticommutator for the fermionic variables $\Phi_I^{ferm} = b^a, c^a$.

A very important property of the quantum BRST charge (7.38) is its *nilpotency* $Q^2 = 0$. This is easily seen by observing that $\delta_\kappa(\delta_\epsilon \Phi_I) = 0$, for all Φ_I, but can also be checked directly using the canonical commutators (7.39).[§] Note also that Q is Hermitian.

The large Hilbert space $\mathfrak{H}$ where the Hamiltonian (7.35) is defined involves states with negative norm. Asymptotic scalar gluon states $[a_0^a(k)]^\dagger|vac\rangle$ have negative norm by the same token as in Abelian case. Scalar gluons can be represented as linear combinations of longitudinal gluons $k_i[a_i^a(k)]^\dagger|vac\rangle$ and "B-states" $[B^a(k)]^\dagger|vac\rangle$. The latter are nothing but the states $g^{(+)}$ of the previous section [see Eq. (7.26)]. Their norm vanishes.

Let us discuss the ghost sector. Expand

$$
\begin{aligned}
b^a(x) &= \sum_k \frac{1}{\sqrt{2\omega_k}}[b_k^a e^{-ikx} + (b_k^a)^\dagger e^{ikx}] , \\
c^a(x) &= \sum_k \frac{1}{\sqrt{2\omega_k}}[-c_k^a e^{-ikx} + (c_k^a)^\dagger e^{ikx}] .
\end{aligned}
\tag{7.40}
$$

The canonical equal-time anticommutators

$$
\begin{aligned}
\{\partial_0 c^a(x), b^b(y)\}_+ &= \{-\partial_0 b^a(x), c^b(y)\}_+ = -i\delta^{ab}\delta(x-y) , \\
\{b,b\}_+ &= \{c,c\}_+ = \{\partial_0 b, b\}_+ = \{\partial_0 c, c\}_+ = 0
\end{aligned}
\tag{7.41}
$$

imply that

$$
\{(b_k^a)^\dagger, c_q^b\}_+ = \{(c_k^a)^\dagger, b_q^b\}_+ = -\delta^{ab}\delta_{kq} ,
\tag{7.42}
$$

and all other anticommutators including $\{c,c^\dagger\}_+$ and $\{b,b^\dagger\}_+$ vanish. This means that the ghost and antighost states

$$
|ghost_k\rangle = (c_k^a)^\dagger|vac\rangle , \quad |antighost_k\rangle = (b_k^a)^\dagger|vac\rangle
$$

have zero norm, but

$$
\langle antighost_k^a|ghost_q^b\rangle = \langle ghost_k^a|antighost_q^b\rangle = -\delta^{ab}\delta_{kq} ,
\tag{7.43}
$$

and the state $|ghost_k\rangle + |antighost_k\rangle$ has negative norm. Six asymptotic states of a given color and momentum $|g_{1,2}^{(Trans)}\rangle$, $|g^{(Long)}\rangle$, $|B\rangle$, $|ghost\rangle$,

[§]Nilpotency of Q *off mass shell* is the main advantage of the *Nakanishi–Lautrup* formalism involving an extra variable B^a. It is possible to work in the framework of the original Lagrangian (6.28), but, in this case, Q^2 does not vanish identically, but only on mass shell by virtue of the operator equations of motion.

and $|antighost\rangle$ can be classified with respect to the action of the BRST charge (7.38), where the coupling constant g can be set to zero. We have

$$Q|g_{1,2}^{(T)}\rangle = 0 \, ,$$
$$Q|B\rangle = Q|ghost\rangle = 0 \, ,$$
$$Q|antighost\rangle = |B\rangle \, ,$$
$$Q|g^{(L)}\rangle = |ghost\rangle \, . \tag{7.44}$$

We have obtained two true BRST singlets $|g_{1,2}^{(T)}\rangle$ and two doublets $\{|g^{(L)}\rangle$, $|ghost\rangle\}$ and $\{|antighost\rangle, |B\rangle\}$. As the states $|ghost\rangle$ and $|antighost\rangle$ have vanishing norm, but nonvanishing inner product, two BRST doublets are naturally combined into a *quartet*.

You might already have guessed what our next step will be. We first define the subspace $\mathfrak{h}$, obeying $Q|\mathfrak{h}\rangle = 0$. The states in $\mathfrak{h}$ have either positive norm as $|g^{(T)}\rangle$ or zero norm as $|B\rangle$ and $|ghost\rangle$. The latter property also follows from the fact that $|B\rangle$ and $|ghost\rangle$ are obtained by acting on some state $|\psi\rangle$ with the Hermitian nilpotent operator Q: $\||Q|\psi\rangle\|^2 = \langle\psi|Q^2|\psi\rangle = 0$.

As it was the case in Abelian theory and by the same reason, the states with zero norm (forming the subspace $\mathfrak{h}_0$) are orthogonal to all the states in $\mathfrak{h}$ and are irrelevant. The physical space $\mathfrak{h}_{phys}$ represents the quotient space $\mathfrak{h}/\mathfrak{h}_0$.[¶] The Hamiltonian (7.35) is Hermitian and hence the S-matrix in the large Hilbert space $\mathfrak{H}$ is unitary. The BRST charge commutes with the Hamiltonian, therefore $\mathfrak{h}$, and $\mathfrak{h}_{phys}$ are invariant under time evolution, and so the physical S-matrix is unitary *quod erat demonstrandum*.

Note that the subsidiary condition $Q|\mathfrak{h}_{phys}\rangle = 0$ *does* not imply the Gauss law condition $(\mathcal{D}_i\pi_i)^a |\mathfrak{h}_{phys}\rangle = 0$. The latter simply is not true in the BRST formalism. But the only thing we really need is the structure of $\mathfrak{h}_{phys}$ in the asymptotic limit $t \to \pm\infty$, where interactions become irrelevant. In the limit $g \to 0$ ghosts decouple and, as is seen from Eq. (7.38), $Q|\Psi\rangle = 0$ leads to $\partial_i\pi_i^a|\Psi\rangle = 0$. Factorizing over the zero norm states is tantamount to disregarding the dependence of Ψ on the ghost variables and on $B^a(x)$. We see once again that S_{phys} defined above relates only physical asymptotic

---

[¶]A mathematician knowing well the theory of differential forms would say that physical states are "BRST-closed" (i.e. $Q|phys\rangle = 0$), but not "BRST-exact" (i.e. $|phys\rangle \neq Q|something\rangle$). Indeed, the operator Q is rather analogous to the operator of external differentiation of forms d, which is also nilpotent $d(d\omega) = 0$. Closed forms obey $d\omega = 0$, exact forms are represented as $\omega = d\chi$, and the quotient space closed/exact is related to cohomology group ...

states containing transverse gluons.

The BRST quantization method outlined in this section allows one to prove many other important theorems. We can, for example, derive a set of Ward identities for physical amplitudes without cost. The simplest such identity is Eq. (7.24). Using the reduction formula (6.2), the equations of motion $\xi B + \partial^\mu A_\mu = 0$, and the relation $B = i\{Q, b\}$, we obtain

$$
\begin{aligned}
M^\mu(k)k_\mu \quad &\sim \quad \Box_x \langle phys_1 | \partial^\mu A_\mu(x) | phys_2 \rangle \ \sim \ \Box_x \langle phys_1 | B | phys_2 \rangle \\
&\sim \quad \Box_x \langle phys_1 | \{Q, b\}_+ | phys_2 \rangle \ = \ 0 \, ,
\end{aligned}
\tag{7.45}
$$

due to the property $Q | \mathfrak{h}_{phys} \rangle = 0$.

It is also not difficult to prove the generalized Ward identity

$$
M^{\mu_1 \cdots \mu_n} k^1_{\mu_1} \cdots k^n_{\mu_n} \ = \ 0 \, ,
\tag{7.46}
$$

where $M^{\mu_1 \cdots \mu_n}$ is an amplitude involving an arbitrary number of extra external transverse gluons. Proceeding along the same lines, we are left with the matrix element

$$
\langle phys_1 | \{Q, b(x_1)\} \cdots \{Q, b(x_n)\} | phys_2 \rangle \, ,
$$

which is zero due to the BRST-closed nature of $|phys_{1,2}\rangle$ and nilpotency of Q.

Lecture 8

Regularization and Renormalization

8.1 Different Regularization Schemes.

Euclidean Feynman integrals for loop graphs diverge at large momenta k_E and the Minkowski integrals diverge at large virtualities $|k^2|$. To handle this divergence, we should introduce an ultraviolet cutoff, to *regularize* the theory in the ultraviolet. We should do this in a gauge-invariant way, if not, huge gauge-non-invariant terms would appear in the amplitudes. For example, a simple momentum cutoff (which breaks gauge invariance) gives a quadratically divergent gluon mass (and a quadratically divergent photon mass in QED).

There are several gauge-invariant regularization procedures. The most "politically correct" one is probably *lattice regularization*. As we have seen, the path integral symbol can be given a meaning which is valid also beyond perturbation theory if discretizing spacetime and trading the Lagrangian (1.17) of continuous Yang–Mills theory for the gauge-invariant Wilson lattice action entering the path integral (4.34).* The finite lattice spacing $a = \Lambda_{\mathrm{UV}}^{-1}$ serves as an ultraviolet cutoff. One *can* calculate loops in Yang–Mills theory with the lattice ultraviolet cutoff, but it is not very convenient. In particular, fixing the gauge for the lattice action brings about some (purely technical) problems.

From a "philosophical" point of view, the second best method of regularization is *Slavnov's* scheme, which combines the *Pauli–Villars* and the higher-derivative regularization method. The Pauli–Villars method is habitually used in QED. Its simplified form consists in subtracting from each

---

*In Lecture 13, we will learn how to do it when quarks are included.

QED diagram containing electron–positron loops a similar diagram with loops of some extra fermion with a very large mass M. Heavy fields are irrelevant when the loop momenta are of the order of the physical external momenta $p_{\text{char}} \ll M$. The graphs with electron loops and with heavy Pauli–Villars fermion loops separately diverge at large momenta, but their difference is finite and depends on the heavy mass M, which plays the role of the ultraviolet regulator.

The procedure just described works for photon–photon scattering box graphs and for the triangle graph associated with the axial anomaly,[†] but it does not work for the 1-loop graph for the photon polarization operator, which involves not only logarithmic, but also quadratic divergences. To get rid of them both, we need to introduce at least two different fermion fields. To be more precise, we modify the functional integral of QED by substituting

$$\det \| - i\slashed{\partial} + m \| \; \longrightarrow \; \det \| - i\slashed{\partial} + m \| \prod_{i=1}^{N} \left(\det \| - i\slashed{\partial} + M_i \| \right)^{c_i} \; , \quad (8.1)$$

$M_i \gg m$. This procedure is, obviously, gauge-invariant. It regularizes all fermion loops if the conditions

$$1 + \sum_{i=1}^{N} c_i \;=\; 0 \, ,$$

$$m^2 + \sum_{i=1}^{N} c_i M_i^2 \;=\; 0 \qquad\qquad (8.2)$$

are fulfilled (see **Problem 1** below). We see that, provided not all the M_i are equal, a set $\{c_i\}$ satisfying this exists for $N \geq 2$.

The modification (8.1) of the measure regularizes the electron loops in QED and quark loops in QCD, but not the loops involving gauge particles. To handle these, we modify the Lagrangian by adding a higher derivative

---

[†]Anomaly will be discussed in detail in Lecture 12 and, in the diagrammatic approach, in Lecture 14, where an explicit calculation of the "anomalous diangle" in the Schwinger model using Pauli–Villars regularization is presented as a **Problem**.

term

$$\mathcal{L}_{\text{Maxwell}} \rightarrow -\frac{1}{4}\left[\{F_{\mu\nu}F^{\mu\nu} + \frac{1}{\Lambda^4}\partial^2 F_{\mu\nu}\partial^2 F^{\mu\nu}\right],$$

$$\mathcal{L}_{\text{YM}} \rightarrow -\frac{1}{2}\left[\text{Tr}\{F_{\mu\nu}F^{\mu\nu}\} + \frac{1}{\Lambda^4}\text{Tr}\{\mathcal{D}^2 F_{\mu\nu}\mathcal{D}^2 F^{\mu\nu}\}\right]. \quad (8.3)$$

With this new term in the QED (resp. QCD) action, the photon (resp. gluon) propagator is modified according to

$$\frac{1}{k^2} \rightarrow \frac{1}{k^2 + k^6/\Lambda^4}, \quad (8.4)$$

and so most of the integrals converge in the ultraviolet. Those which do not can be handled in a Pauli–Villars-like way; see the book [4] for more details. The higher-derivative and lattice regularization schemes have a common drawback: a lot of new vertices appear, which makes explicit calculations complicated. The most artificial and physically least transparent, yet technically most convenient method is *dimensional regularization*. This is the most popular approach used in practical calculations. It consists in changing the dimension of spacetime:

$$S_{YM} \rightarrow -\frac{1}{2}\int d^d x \, \text{Tr}\{F_{\mu\nu}F^{\mu\nu}\}, \quad (8.5)$$

where $\mu = 1, \ldots, d$. The canonical dimension of the field A is modified: $[A] = m^{d/2-1}$. Also, the constant g (in $F \sim \partial A + gA^2$) acquires a dimension $[g] = m^{2-d/2}$, so that the regularized action (8.5) is dimensionless.

Consider an integral which one often meets in one-loop calculations in four-dimensional theory:

$$\int \frac{d^4 k_E}{(k_E^2 + M^2)^2}, \quad (8.6)$$

where M^2 is an expression involving external momenta, Feynman parameters, and/or fermion masses. The integral diverges logarithmically at large momenta. With the regularization (8.5), we consider the same integral done over $d^d k_E$, where d is an arbitrary parameter which, to begin with, is an integer. If $d < 4$, the integral is convergent and can be done explicitly. Using the property $d^d k_E = k_E^{d-1} dk_E V_{d-1}$, where

$$V_{d-1} = \frac{2\pi^{d/2}}{\Gamma(d/2)}$$

is the volume of the $(d-1)$-dimensional unit sphere, one can derive

$$\int \frac{d^d k_E}{(k_E^2 + M^2)^n} = \frac{\pi^{d/2}\Gamma(n - d/2)}{(M^2)^{n-d/2}\Gamma(n)} \tag{8.7}$$

[the integral is done with the change of variables $x = M^2/(k_E^2 + M^2)$, which reduces it to the Euler integral $B(\alpha, \beta)$]. We next analytically continue this formula derived for an integer dimension d to arbitrary real values of d. We will be interested in those values of d just a little bit less than 4: $d = 4 - 2\epsilon$, $\epsilon \ll 1$. For $n = 2$, the expression (8.7) develops a singularity in the limit $\epsilon \to 0$ due to the property

$$\Gamma(\epsilon) \sim \frac{1}{\epsilon} - \gamma + O(\epsilon) \tag{8.8}$$

(γ is the Euler constant), so that the logarithmic singularity in the original four-dimensional integral (8.6) displays itself now as a pole $\sim 1/\epsilon$.

The reader might be confused by the fact that, instead of the dimensionful regularization parameter $\Lambda_{UV} \gg$ *everything* appearing in other regularization schemes, we are dealing here with a dimensionless parameter ϵ. However, this is an illusion. As was mentioned above, at $d = 4 - 2\epsilon$, the coupling constant carries dimension and can be written as

$$g = \mu^\epsilon g_0 , \tag{8.9}$$

where g_0 is the value of the coupling in the limit $\epsilon \to 0$ and μ is a parameter carrying the dimensions of mass. Therefore, dimensional regularization involves actually two constants: ϵ and μ. The integrals like the one in Eq. (8.7) are typically multiplied by g^2, so that we have

$$I^{(d)} \propto \frac{\mu^{2\epsilon}}{\epsilon} + \ldots \sim \frac{1}{\epsilon} + 2\ln\mu + \ldots = \ln\left(\mu^2 e^{1/\epsilon}\right) + \ldots , \tag{8.10}$$

where the dots stand for the terms not depending on ϵ and μ in the limit $\epsilon \to 0$. We see that everything depends on the parameter $\Lambda_{UV}^{\text{dim. reg.}} = \mu \exp\{1/2\epsilon\}$, which becomes very large in the limit $\epsilon \to 0$ and plays the role of an ultraviolet regulator.

In practical calculations, not only logarithmic, but also quadratically divergent integrals $\sim \int d^4 k_E/(k_E^2 + M^2)$ appear [cf. Eq. (6.16)]. A distinguishing feature of the dimensional regularization method is that quadratic divergences do not show up: the integral (8.7) with $n = 1$ does not involve terms $\propto \left(\Lambda_{UV}^{\text{dim. reg.}}\right)^2$, but only the terms

$$\propto \Gamma(-1+\epsilon) \sim -\frac{1}{\epsilon} \propto \ln\left(\Lambda_{UV}^{\text{dim. reg.}}\right) .$$

In some cases such as theories involving scalar particles, this may be considered a nuissance, but for QED and QCD it is a nice expected feature of a gauge-invariant regularization procedure. The only places where quadratic divergences might have appeared in these theories are the 1-loop graphs for the photon (gluon) mass, but these should vanish as a consequence of gauge invariance (see **Problems 1, 2** at the end of the section).

We have learned how to calculate scalar integrals like (8.7), but in practical calculations, the integrand may involve tensor structures depending on loop and external momenta. For fermion loops, γ-matrices and their traces are also present. A tensor like $\eta_{\mu\nu}$ does not have a direct meaning in a space of fractional dimension. However, bearing in mind that eventually all such tensors are going to be contracted with other tensors and/or with polarization vectors of external gluons, it suffices to define formally $\eta^{\mu\nu}p_\mu q_\nu = p \cdot q$) and

$$\eta_{\mu\nu}\eta^{\mu\nu} = d = 4 - 2\epsilon . \tag{8.11}$$

It is often necessary to keep the term of order ϵ, because the structure like $\eta_{\mu\nu}\eta^{\mu\nu}$ may be multiplied by a divergent integral $\sim 1/\epsilon$ and constant terms should be kept to provide for a correct gauge-invariant answer.

Tensor integrals are calculated with the rules

$$\int k_\mu f(k)\, d^d k = 0 ,$$

$$\int k_\mu k_\nu f(k)\, d^d k = \frac{\eta_{\mu\nu}}{d} \int k^2 f(k)\, d^d k , \tag{8.12}$$

etc. γ-"matrices" in d dimensions are defined formally as objects obeying the Clifford algebra relations

$$\gamma_\mu \gamma_\nu + \gamma_\nu \gamma_\mu = 2\eta_{\mu\nu}\mathbb{1} , \tag{8.13}$$

where $\mathbb{1}$ is the unit matrix of "dimension" $2^{d/2}$ so that $\text{Tr}\{\mathbb{1}\} = 2^{d/2}$. Bearing (8.13) and (8.11) in mind, the usual relations of the Dirac matrices algebra are somewhat modified, e.g.

$$\gamma_\mu \gamma^\nu \gamma^\mu = -2(1 - \epsilon)\gamma^\nu ,$$
$$\gamma_\mu \gamma^\nu \gamma^\rho \gamma^\mu = 4\eta^{\nu\rho} - 2\epsilon\gamma^\nu\gamma^\rho . \tag{8.14}$$

The advantage of dimensional regularization is its simplicity. A disadvantage is that it operates not with the functional integral as a whole, but with individual Feynman graphs. Also, though it works well for QCD, it does not work for chiral gauge theories like the standard electroweak model because the notion of chirality and, at the technical level, the matrix γ^5 can be defined only in 4 dimensions.

Besides dimensional regularization, another convenient practical way to calculate Feynman graphs order by order is associated with the use of unitarity and dispersion relations. We will illustrate it in Lecture 9.

Problem 1. Using Pauli–Villars regularization, prove that the 1-loop contribution to the photon mass does indeed vanish.

Solution. We have

$$m_\gamma^2 = \Pi_\mu^\mu(0) \propto \int \frac{\text{Tr}\{\gamma_\mu(\not{k} + m)\gamma^\mu(\not{k} + m)\}}{(k^2 - m^2)^2} d^4k . \tag{8.15}$$

Calculating the trace, performing the Wick rotation, and using the recipe (8.1), we obtain to leading order

$$m_\gamma^2 \propto \int \frac{k_E^2 + 2m^2}{(k_E^2 + m^2)^2} d^4k_E + \sum_{i=1}^{N} c_i \int \frac{k_E^2 + 2M_i^2}{(k_E^2 + M_i^2)^2} d^4k_E =$$

$$\int \frac{d^4k_E}{k_E^2}\left[1 + \sum_{i=1}^{N} c_i\right] - m^4 \int \frac{d^4k_E}{k_E^2(k_E^2 + m^2)^2} - \sum_{i=1}^{N} c_i M_i^4 \int \frac{d^4k_E}{k_E^2(k_E^2 + M_i^2)^2}$$

which vanishes due to (8.2).

Problem 2. The same, but using dimensional regularization.

Solution. Substituting d^dk for d^4k in Eq. (8.15) and using the properties (8.13), (8.14), we obtain (after Wick rotation)

$$m_\gamma^2 \propto \int \frac{m^2(2 - \epsilon) + k_E^2(1 - \epsilon)}{(k_E^2 + m^2)^2} d^dk_E = m^2 \int \frac{d^dk_E}{(k_E^2 + m^2)^2}$$

$$+ (1 - \epsilon) \int \frac{d^dk_E}{k_E^2 + m^2} \propto \left[1 - \epsilon - m^2\frac{\partial}{\partial m^2}\right] m^{2(1-\epsilon)} = 0 .$$

Problem 3. Calculate the 1-loop polarization operator of massless quarks in the Feynman gauge $\xi = 1$ using dimensional regularization.

Solution. We have to calculate the integral

$$-i\Sigma(p) = (ig)^2 c_F(2-d) \int \frac{d^d k}{(2\pi)^d} \frac{\not{p}-\not{k}}{k^2(p-k)^2}$$

$[c_F = t^a t^a$, the factor $2 - d$ appears due to (8.14)]. Introducing as usual the Feynman parameter x:

$$\frac{1}{k^2(p-k)^2} = \int_0^1 \frac{dx}{[(1-x)k^2 + x(p-k)^2]^2} \, ,$$

shifting the integration variable $k - xp \to k$ (it is not dangerous here as the integral converges for $d < 4$), and disregarding the term $\propto \not{k}$ due to (8.12), we obtain

$$\Sigma(p) = 2ig^2 c_F(1-\epsilon)\not{p} \int_0^1 dx(1-x) \int \frac{d^d k}{(2\pi)^d} \frac{1}{[k^2 + p^2 x(1-x)]^2} \, .$$

Performing the Wick rotation, using (8.7), bearing in mind (8.8), expressing g^2 as in Eq. (8.9), expanding

$$\int_0^1 x^{-\epsilon}(1-x)^{1-\epsilon}dx = \frac{1}{2} + \epsilon + o(\epsilon) \, ,$$

and rotating back to Minkowski space, we finally obtain

$$\Sigma(p) = \frac{g_0^2}{(4\pi)^2} c_F \not{p} \left(-\frac{1}{\epsilon} + \gamma - 1 + \ln \frac{-p^2}{4\pi\mu^2} \right) + O(\epsilon) \, . \tag{8.16}$$

8.2 Renormalized Theory as an Effective Theory. Slavnov–Taylor Identities.

QCD like QED is a renormalizable theory. This means that, after the divergences are regularized, the *only* net effect of all the troublesome divergent contributions to different physical amplitudes consists in redefining the fundamental constants of the theory: the charge and mass of the electron in QED and the coupling constant g and the quark masses in QCD. In other words, all the physical quantities are expressed in terms of renormalized parameters g^{ren}, m_q^{ren} and do not involve any ultraviolet divergences anymore. All such divergences are absorbed into the renormalized constants

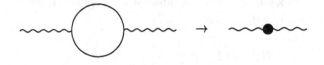

Fig. 8.1   Appearance of counterterms.

which depend on the ultraviolet cutoff Λ_0, the bare coupling constants g_0 and m_0, and an arbitrary chosen scale μ. It is convenient to choose μ of the order of the characteristic energy scale of the process of interest.

The assertion of the renormalizability of QCD can be proven as an exact rigorous theorem. We will not do it here, but rather provide a heuristic physical explanation. Ultraviolet divergences come from large loop momenta. When loop momenta are much larger than characteristic external momenta, the latter are not really important. It is instructive to consider a one-loop correction to, say, the photon propagator in the coordinate representation. Large momenta correspond to small distances in the loop. In that case, the loop "looses its structure" and can be treated as a point (see Fig. 8.1). This point brings about a new quadratic contribution to the Lagrangian which, by gauge and Lorentz invariance, is bound to have the structure $\sim (\partial_\mu A_\nu - \partial_\nu A_\mu)^2$. This *counterterm* is added to the similar structure in the tree Lagrangian, so that the effective value of e^2 is changed.

Renormalizability is closely related to the general notion of *effective Lagrangians*, which is not limited to field theory, but is also heavily used in ordinary quantum and classical mechanics. Suppose we have a system whose Hamiltonian involves two essentially different dimensionful parameters m and M associated with some "light" and "heavy" degrees of freedom ϕ_{light} and ϕ_{heavy}. Suppose also that we are interested in the spectrum and other characteristics of the system at the energies of order $\mathcal{E}_{\text{char}} \sim m \ll M$. It is then true that these low-energy properties can be studied not with the full Hamiltonian $H(\phi_{\text{light}}, \phi_{\text{heavy}})$, but with an effective Hamiltonian $H^{\text{eff}}(\phi_{\text{light}})$ depending on the light degrees of freedom only. One can further build up a systematic expansion in the dimensionless small parameter m/M for $H^{\text{eff}}(\phi_{\text{light}})$ and analyze low-energy dynamics in the effective Hamiltonian framework order by order.

A classical example of such an effective Hamiltonian is the Born–Oppenheimer Hamiltonian for a two-atom molecule. In that case, heavy degrees of freedom are the positions and momenta of atomic electrons, while the light degrees of freedom are the positions and momenta of atomic nuclei. Though the electrons are, of course, lighter than the nuclei, the electron degrees of freedom are called "heavy" because the characteristic energies related to electronic excitations have atomic scale, while the characteristic energy due to oscillations of the nuclei around their equilibrium position is much lower. The effective Born–Oppenheimer Hamiltonian is simply an oscillator, whose rigidity depends on how the energy of the lowest electronic term depends on the distance between the nuclei.

Another well-known example is the effective Lagrangian (or Hamiltonian) of quasi-static and quasi-homogeneous electromagnetic fields in QED. For simplicity, let the field be just static, homogeneous, and purely magnetic. In that case, ϕ_{light} is the magnetic field strength B and ϕ_{heavy} are the electron and positron field degrees of freedom. The effective Hamiltonian

$$\mathcal{H}^{\text{eff}} = \frac{1}{2}B^2 - \frac{e^4}{360\pi^2 m^4}|B|^4 + O(|B|^6) \tag{8.17}$$

involves the tree term $B^2/2$ and the corrections, which are all suppressed if the dimensionless parameter $b = e|B|/m^2$ is small.

The renormalization procedure is nothing but the construction of the effective Lagrangian in the spirit of Born and Oppenheimer, where the light degrees of freedom are those field modes with momenta in the region of physical interest and the heavy degrees of freedom are the modes with momenta of order Λ_0. This approach and interpretation is due to Wilson.

To be precise, one should establish some separation scale μ and treat all the modes with momenta $p < \mu$ as light variables and the modes with momenta $p > \mu$ as heavy variables. To be still more precise, the separation of scales should be done in a gauge-invariant way. Again, lattice regularization provides (theoretically!) the best way of such a separation. Consider pure Yang–Mills theory (we still have to learn what *is* lattice gauge theory with fermions). Suppose we defined the theory on the lattice with spacing $a = \Lambda_0^{-1}$. To construct the effective Lagrangian defined at the scale μ, we introduce a lattice which is Λ_0/μ times more coarse than the original one. The effective action depending on the link variables U^{new} for this new lattice is defined as

$$\exp\left\{-S^{\text{eff}}[U^{\text{new}}]\right\} =$$

$$\int \prod_{\substack{\text{old} \\ \text{links}}} dU^{\text{old}} \delta \left[U^{\text{new}} \left(\prod_{\substack{\text{along a} \\ \text{new link}}} U^{\text{old}} \right)^{\dagger} - 1 \right] \exp\left\{-S[U^{\text{old}}]\right\} . \quad (8.18)$$

$S^{\text{eff}}[U^{\text{new}}]$ should be gauge-invariant [in the lattice sense, see Eq. (4.35)]. Assume now that it is essentially local (probably, one can even prove it). Then $S^{\text{eff}}[U^{\text{new}}]$ should have the same functional form as $S[U^{\text{old}}]$ given in Eq. (4.34) + the terms which lead to the structures of higher dimension like $\text{Tr}\left\{(\mathcal{D}_\alpha F_{\mu\nu})^2\right\}$ in continuum theory.

Returning now to the continuum notation, we can deduce that the Wilsonian effective Lagrangian has the form

$$\mathcal{L}_\mu^{\text{eff}} = -\frac{1}{2g^2} \text{Tr}\left\{[F_{\mu\nu}(\mu)]^2\right\} + \sum_i c_i O_i , \quad (8.19)$$

where O_i are local gauge-invariant operators of dimension 6 and higher. The argument μ for $F_{\mu\nu}$ means that they came from U^{new} corrresponding to the modes with $p < \mu$ in the naïve Wilson procedure. The coefficients g^2 and c_i depend on g_0^2, Λ_0^2, and on μ. The coefficients c_i carry negative dimension and should be proportional to inverse powers of parameters μ and Λ_0. As the procedure (8.18) corresponds to integrating over large loop momenta, c_i should depend essentially on the ultraviolet cutoff, i.e. be of order of Λ_0^{-2}, Λ_0^{-4}, etc., and the effect these counterterms exert on the physics at momenta $p \sim \mu$ is negligible (that is why the operators of higher dimension are called *irrelevant* operators). So for all practical purposes, the effective Lagrangian can be taken in the form (8.19) keeping only the first term, which has the same functional form as the original Lagrangian. Theories enjoying this property are called *renormalizable*, and we have just "proven" that Yang–Mills theory belongs to this class.[‡]

In the theory involving quarks, there are several gauge-invariant structures of dimension 4: $\text{Tr}\left\{(F_{\mu\nu})^2\right\}$ and $\bar{\psi}_f[i\not{D} - m_f]\psi_f$ for each quark flavor.

[‡]We find this heuristic reasoning rather conclusive, but it does not have the status of exact proof, of course. The standard proof of renormalizability of QCD is based on the dimensional regularization or on the higher-derivative regularization methods and relies rather heavily on the analysis of particular Feynman graphs.

Rescaling if necessary the fermion fields, we can render the coefficient of $i\bar{\psi}\not{D}\psi$ in the effective Lagrangian equal to 1. Hence,

$$\mathcal{L}_{\mu}^{\text{eff}} = -\frac{1}{2g^2(\mu)}\text{Tr}\left\{[F_{\mu\nu}(\mu)]^2\right\} + \sum_f \bar{\psi}_f(\mu)[i\not{D} - m_f(\mu)]\psi_f(\mu)$$

$$+ \quad \text{irrelevant terms} \quad . \qquad (8.20)$$

It has the same form as the tree Lagrangian, only the parameters — the coupling constant and quark masses — depend on the scale μ.

If everything is done correctly, the result (8.20) should be effectively reproduced also in practical perturbative calculations which do not appeal to the lattice formulation. There is a way to calculate the Wilsonian effective Lagrangian directly using the background field method. We will use, however, a more standard approach where renormalization factors of the different structures in the Lagrangian are calculated separately, and the full effective Lagrangian is obtained in the second step. Let us see how it works.

Expanding the Lagrangian of Yang–Mills perturbation theory (6.31) in the powers of the fields produces quadratic, cubic, and quartic terms, which give rise to the gluon, ghost, and quark propagators, and various bare vertices. Let us now take into account loop corrections and consider the exact propagators and exact vertices of our theory. Following the philosophy outlined above, we will be interested only in the contributions coming from the virtualities of internal lines that exceed a Wilsonian separation scale μ. We will assume that μ is much larger than the external momenta p^{ext} and quark masses m_q. A nontrivial and very important statement is that under such conditions the exact propagators and vertices have the same color and Lorentz structure as the bare ones. All other structures are suppressed as powers of p^{ext}/μ. Thus, we can write

$$
\begin{aligned}
[\mathcal{D}_{\alpha\nu}^{ab}(p)]^{\text{trans.}}\big|_\mu &\approx Z_A(\Lambda_0^2/\mu^2,\, g_0^2, \xi_0)\, [\mathcal{D}_{\alpha\nu}^{ab\,(\text{bare})}(p)]^{\text{trans}}\,, \\
\xi(\mu) &\approx Z_\xi(\Lambda_0^2/\mu^2,\, g_0^2, \xi_0)\, \xi_0\,, \\
C^{ab}(p)\big|_\mu &\approx Z_c(\Lambda_0^2/\mu^2,\, g_0^2, \xi_0)\, C^{ab\,(\text{bare})}(p)\,, \\
\mathcal{G}_q(p)\big|_\mu &\approx Z_q(\Lambda_0^2/\mu^2,\, g_0^2, \xi_0)\, \frac{i}{\not{p} - m^q(p)}\,, \\
m^q(\mu) &\approx Z_m(\Lambda_0^2/\mu^2,\, g_0^2, \xi_0)\, m_0^q\,, \\
\Gamma_{\alpha\nu\lambda}^{abc}(p,q,r)\big|_\mu &\approx Z_{3A}^{-1}(\Lambda_0^2/\mu^2,\, g_0^2, \xi_0)\, \Gamma_{\alpha\nu\lambda}^{abc\,(\text{bare})}(p,q,r)\,, \\
\Gamma_{\alpha\nu\lambda\sigma}^{abcd}\big|_\mu &\approx Z_{4A}^{-1}(\Lambda_0^2/\mu^2,\, g_0^2, \xi_0)\, \Gamma_{\alpha\nu\lambda\sigma}^{abcd\,(\text{bare})}\,, \\
\Gamma_\alpha^{abc}(p)\big|_\mu &\approx Z_{\bar{c}cA}^{-1}(\Lambda_0^2/\mu^2,\, g_0^2, \xi_0)\, \Gamma_\alpha^{abc\,(\text{bare})}(p)\,, \\
\Gamma_\alpha^{a}\big|_\mu &\approx Z_{\bar{q}qA}^{-1}(\Lambda_0^2/\mu^2,\, g_0^2, \xi_0)\, \Gamma_\alpha^{a\,(\text{bare})}\,,
\end{aligned}
$$

$$(8.21)$$

where we have defined $[\mathcal{D}_{\alpha\nu}^{ab}(p)]^{\text{trans.}} \propto (p^2\eta_{\alpha\nu} - p_\alpha p_\nu)$ and decomposed

$$
[\mathcal{D}_{\alpha\nu}^{ab}(p)]\big|_\mu = [\mathcal{D}_{\alpha\nu}^{ab}(p)]^{\text{trans.}}\big|_\mu - i\delta^{ab}\xi(\mu)\frac{p_\alpha p_\nu}{(p^2 + i0)^2}\,.
$$

The bare propagators and vertices are written in Eq. (6.34). The renormalization factors Z are just some numbers. Under the conditions specified above, they only depend on the bare coupling g_0, bare gauge parameter ξ_0, and the dimensionless ratio Λ_0/μ.

Note that the transverse part of the gluon Green's function and the longitudinal one involving the gauge parameter ξ are renormalized with their own factors. Actually, we derived earlier the WI (7.15), which implies that the longitudinal part is not renormalized at all, so that $Z_\xi = 1$. Two different spinor structures in the quark Green's function $G^{-1}(k)$ are also renormalized with their own factors, their ratio giving the mass renormalization. All the terms in the expression (6.34) for the 3-gluon and 4-gluon vertices are multiplied by one and the same renormalization factor, however. This is guaranteed by the symmetry considerations.

The Green's functions (8.21) "build up" an effective Lagrangian

$$\mathcal{L}^{\text{eff}} = -\frac{1}{4Z_A}(\partial_\mu A_\nu^a - \partial_\nu A_\mu^a)^2 - \frac{g_0}{Z_{3A}}f^{abc}(\partial_\mu A_\nu^a)A^{b\mu}A^{c\nu}$$

$$-\frac{g_0^2}{4Z_{4A}}f^{abe}f^{cde}A_\mu^a A_\nu^b A^{\mu c}A^{\nu d} - \frac{1}{2\xi_0}(\partial^\mu A_\mu^a)^2 + \frac{1}{Z_c}\partial_\mu \bar{c}^a \partial^\mu c^a$$

$$+\frac{g_0}{Z_{\bar{c}cA}}f^{abc}\partial^\mu \bar{c}^a A_\mu^b c^c + \frac{i}{Z_q}\bar{\psi}\not{\partial}\psi - \frac{m_0 Z_m}{Z_q}\bar{\psi}\psi + \frac{g_0}{Z_{\bar{q}qA}}\bar{\psi}\!\!\not{A}\psi \quad (8.22)$$

(with this expression in hand, the renormalized propagators and vertices (8.21) follow from the tree Feynman rules).

To bring the kinetic terms to the standard form, it is convenient to redefine,[§]

$$A \to Z_A^{1/2}A, \quad c \to Z_c^{1/2}c, \quad \psi \to Z_q^{1/2}\psi.$$

Gauge invariance [or rather the BRST invariance (6.30) (6.32) which takes over the role of gauge invariance for the gauge-fixed Lagrangian (6.31)] requires now that the redefined effective Lagrangian would coincide in form with the original one up to renormalization of the constants g, ξ. This requirement is rather rigid. It tells us that the renormalization factors cannot be arbitrary, but satisfy the following *Slavnov–Taylor* identities

$$\frac{Z_A^3}{Z_{3A}^2} = \frac{Z_A^2}{Z_{4A}} = \frac{Z_A Z_c^2}{Z_{\bar{c}cA}^2} = \frac{Z_A Z_q^2}{Z_{\bar{q}qA}^2} \equiv Z_{\text{inv}}. \quad (8.23)$$

Their meaning is that the strength of coupling $Z_{\text{inv}}g_0^2$ extracted from the 3-gluon, 4-gluon, ghost–ghost–gluon, and quark–quark–gluon vertices coincides even after the renormalization is performed. Any such vertex can be used to extract the renormalization factor for the effective charge given by the ratio (8.23).

The relations (8.23) can also be derived directly from the properly chosen Ward identities, and this is the way they were derived originally. For example, the relation $1/Z_{4A} = Z_A/Z_{3A}^2$ follows from Eq. (7.18).

Let us discuss in some more details how this general renormalization procedure is implemented in practice. In the dimensional regularization framework the separation scale μ is associated with the parameter μ introduced in Eq. (8.9). Taking into account only the internal virtualities

---

[§]Note that such a redefinition results in the renormalization of the effective gauge parameter: $\xi_0 \to \xi_0/Z_A$.

exceeding μ^2 modifies Green's functions. For example, the one-loop contribution to $\Sigma(p)|_\mu$ is

$$\Sigma(p)|_\mu = \frac{g_0^2}{(4\pi)^2} c_F \not{p} \left[-\frac{1}{\epsilon} + \gamma - 1 - \ln(4\pi) \right] , \qquad (8.24)$$

and the term $\sim \ln(-p^2/\mu^2)$ entering Eq. (8.16) disappears.

Sometimes, it is convenient to choose another *renormalization scheme* and identify μ with the characteristic external virtuality, in which case no restriction on the range of loop integration is imposed. This nicely works for QED, where the renormalization of charge boils down to the renormalization of the photon propagator. In QCD one can also consider the gluon, quark, and ghost propagators at Euclidean virtuality[¶] $p^2 = -\mu^2$ and define the renormalization factors $Z_A, Z_\xi, Z_c, Z_q, Z_m$ in the same way as in Eq. (8.21), but with the full exact propagators on the left side.

Let us try to do the same for the exact vertices. To begin with, choose the symmetric Euclidean normalization points[‖]

$$p^2 = q^2 = r^2 = -\mu^2 \quad \text{for } 3A, \ \bar{c}cA, \ \text{and } \bar{q}qA \text{ vertices} ,$$

$$\left\{ \begin{array}{l} p^2 = q^2 = r^2 = k^2 = -\mu^2 \\ p \cdot q = q \cdot r = r \cdot k = k \cdot p = p \cdot r = q \cdot k = \mu^3/3 \end{array} \right. \quad \text{for } 4A \text{ vertex.} \quad (8.25)$$

And here we meet a problem. Exact off-shell vertices involve many different tensor structures. The symmetric choice (8.25) somewhat reduces their number, but still a generic 3-gluon vertex admits three distinct structures

$$\begin{aligned} \Gamma_{\mu\nu\lambda}^{abc}(p,q,r) = \ & -g f^{abc}[(p-q)_\lambda \eta_{\mu\nu} + (q-r)_\mu \eta_{\nu\lambda} \\ & + (r-p)_\nu \eta_{\mu\lambda}] Z_{3A}^{-1}(\mu^2) + g f^{abc}(p-q)_\lambda (q-r)_\mu (r-p)_\nu \, B(\mu^2) \\ & + g f^{abc}[r_\mu p_\nu q_\lambda - r_\nu p_\lambda q_\mu] C(\mu^2) , \qquad (8.26) \end{aligned}$$

where we have defined Z_{3A}^{-1} as the coefficient of $\Gamma_{\mu\nu\lambda}^{abc}(p,q,r)^{\text{bare}}$ in the decomposition (8.26). On top of this, for $N_c \geq 3$ there are structures involving the color tensor d^{abc}. Extra terms in the vertex is an obvious nuissance. They make the notion of the renormalization factor Z_{3A}^{-1} ambiguous [18].

[¶] Green's functions at Euclidean momenta are more convenient to analyze as they do not involve imaginary parts.

[‖] We hasten to comment that choosing the symmetric kinematics is a pure convention. It makes calculations with several loops easier, but other conventions leading eventually to the same physical results are possible.

Indeed, the choice of basis other than in Eq. (8.26) would modify the value of the coefficient Z_{3A}^{-1}.

This problem is actually not specific for QCD. We would meet it also in QED, if trying to define the vertex renormalization factor in the Euclidean kinematics (8.25). The off-shell $\bar{e}e\gamma$ vertex in QED and the off-shell $\bar{q}qA$ vertex in QCD involve at least three extra tensor structures (more than that if we keep the terms proportional to fermion mass).

If some particular way of defining the vertex renormalization factors is chosen, with almost any such choice, the Slavnov–Taylor identities (8.23) do not exactly hold anymore.

The appearance of new vertices and breakdown of the Slavnov–Taylor identities in this "off-shell momentum–space subtraction" or just *MOM* renormalization scheme is not so surprising. When internal momenta of the same order as external ones are allowed, we are leaving the firm ground of the Born–Oppenheimer approach and are not guaranteed anymore that the effective Lagrangian has a local form (8.22) with the universal coupling $Z_{\text{inv}}g_0^2$. Still, the MOM renormalization procedure has a certain meaning because the *main* contribution to the divergent in the ultraviolet loop integrals always comes from large internal momenta. In particular, at one-loop level, the coefficents $B(\mu^2)$ and $C(\mu^2)$ of the new structures in Eq. (8.26) do not involve ultraviolet logarithms and are suppressed compared to Z_{3A}^{-1}. For many applications these extra structures are irrelevant.

Also, the Slavnov–Taylor identities are not completely destroyed. Although the effective charges

$$g_{\bar{q}qA}^2(\mu) \;=\; \frac{Z_A Z_q^2}{Z_{\bar{q}qA}^2}g_0^2\,, \quad g_{3A}^2(\mu) \;=\; \frac{Z_A^3}{Z_{3A}^2}g_0^2\,, \tag{8.27}$$

and the charges defined from the $\bar{c}cA$ and $4A$ vertices are now all different, one can claim that they are expressed as analytic functions of each other:

$$g_{\bar{q}qA}^2(\mu) \;=\; F[g_{3A}^2(\mu)] \;=\; g_{\bar{q}qA}^2(\mu) + A g_{\bar{q}qA}^4(\mu) + \cdots\,, \tag{8.28}$$

etc. Furthermore, it turns out that the coefficients A are rather small, the functions $F(g^2)$ are rather close to 1, and the ambiguity in the definition of the invariant charge is rather weak.

The advantage of the MOM scheme is that it is more physical. Coupling constants depend not on somewhat artificial Wilsonian separation scale μ, but directly on the characteristic external momenta, i.e. on the characteristic energy scale of a physical process in interest. We will return to the discussion of different renormalization schemes at the end of the next lecture.

Running Coupling Constant

To begin with, let us find the renormalization of the coupling constant at the one-loop level.

9.1 One-loop Calculations.

We will choose the $\bar{q}qA$-vertex for this purpose. Technically, the calculations for the ghost–ghost–gluon vertex are a little bit easier (ghosts are scalars while quarks are fermions involving extra spinor indices). The calculation with quarks is a little bit more instructive, however, because it is parallel to a similar calculation in QED and both the similarities and differences between Abelian and non-Abelian theories are seen more clearly in this way. So we define the renormalized charge as

$$g^2(\mu) = Z_{\text{inv}} g_0^2 = \frac{Z_A Z_q^2}{Z_{\bar{q}qA}^2} g_0^2 . \tag{9.1}$$

We will calculate the renormalization factors entering Eq. (9.1) via exact one-loop Euclidean propagators and vertices in the framework of the MOM prescription. (At the one-loop level there is no difference between different renormalization schemes and there is no point to distinguish between the characteristic external momentum and the scale-separation parameter.) We start by calculating the renormalization factor for the gluon propagator. The relevant graphs are drawn in Fig. 9.1. Now, Z_A, as well as Z_q and $Z_{\bar{q}qA}$, is not as such a physical quantity and depends on the gauge. We will work in Feynman gauge $\xi = 1$.

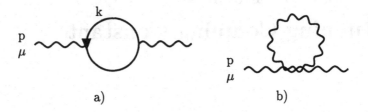

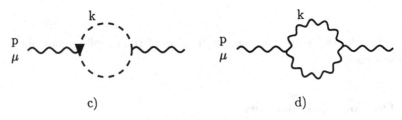

Fig. 9.1   Gluon polarization operator.

First, we note that, as we have already mentioned, the longitudinal part of the gluon propagator is not renormalized:

$$p^\mu \mathcal{D}_{\mu\nu}(p) \;=\; p^\mu D_{\mu\nu}(p) \; . \tag{9.2}$$

From this and from the Dyson equation

$$\mathcal{D}_{\mu\nu}(p) \;=\; D_{\mu\nu}(p) + D_{\mu\alpha}(p)\left[i\mathcal{P}^{\alpha\beta}(p)\right]\mathcal{D}_{\beta\nu}(p) \; , \tag{9.3}$$

involving the 1-particle-irreducible gluon polarization operator $\mathcal{P}_{\mu\nu}(p)$, it follows that $\mathcal{P}_{\mu\nu}(p)$ is transverse:

$$p^\mu \mathcal{P}_{\mu\nu}(p) = 0 \; . \tag{9.4}$$

In QED the property (9.4) follows trivially from current conservation. As was discussed earlier, the colored current is not conserved in QCD, but the property (9.4) holds nevertheless. Lorentz invariance and transversality dictate

$$\mathcal{P}^{ab}_{\mu\nu}(p) = \delta^{ab}\left(\eta_{\mu\nu} - \frac{p_\mu p_\nu}{p^2}\right)\mathcal{P}(p^2) \; .$$

So like in QED, it suffices to calculate $\mathcal{P}^{aa}_{\mu\mu}(p) = 3(N^2 - 1)\mathcal{P}(p^2)$, which simplifies calculations. For $\mathcal{P}^{ab}_{\mu\nu}(p)$ to be non-singular, $\mathcal{P}(0)$ (the gluon mass) should be zero and it is. With dimensional regularization, it is seen immediately. The graph with fermion loop gives the same expression as in QED and gives zero for $\mathcal{P}(0)$ for the same reason (see **Problem 2** in the previous lecture). Also, the contributions to $\mathcal{P}(0)$ from all other graphs with massless particles in the loop are proportional, if any, to

$$\int \frac{d^4k}{k^2} \, ,$$

which, according to the rule (8.7), gives zero identically as soon as the dimension of spacetime is changed $4 \to 4 - 2\epsilon$.

We will calculate $\mathcal{P}(p^2)$ in the case of massless quarks (we are interested now only in the renormalization factor, which, as was mentioned, does not depend on mass when the latter is small compared to characteristic momenta) in the Euclidean region $p^2 = -\mu^2 < 0$ via its imaginary part at positive $p^2 = s$ using the dispersion relation along the same lines as it is done in the book [8] for QED. This method is much more physical than dimensional regularisation (though its direct implication is associated with certain technical difficulties for complicated graphs in higher orders). The fact that the gluon mass is zero allows us to write for $\mathcal{P}(p^2)$ the dispersive relation with one substraction:

$$\mathcal{P}(-\mu^2) = \frac{-\mu^2}{\pi} \int_0^{\Lambda_0^2} \frac{\mathrm{Im}\,\mathcal{P}(s')\,ds'}{s'(s' + \mu^2 - i0)} \, . \tag{9.5}$$

We will calculate the imaginary part as a half of the discontinuity of the polarization operator at the cut. The discontinuity is determined with the help of the Cutkosky rules when the fermions, gluons, and ghosts in the corresponding loops are put on mass shell and their propagators are substituted by δ functions:

$$\frac{1}{k^2 + i0} \to -2\pi i\delta(k^2) \, .$$

Consider first the diagram with a quark loop. The integrand in the corresponding Feynman integral involves the factor

$$\begin{aligned} I_{\text{quark}} &= (i^2)_{\text{propagators}} \times [(ig)^2]_{\text{vertices}} \\ &\times (-1)_{\text{ferm.loop}} \times \mathrm{Tr}\{t^a t^b\} \times \mathrm{Tr}\{\gamma_\mu \not{k}\gamma^\mu(\not{k} - \not{p})\} \, . \end{aligned} \tag{9.6}$$

It is the same as in QED up to the factor $\text{Tr}\{t^a t^b\} = \frac{1}{2}\delta^{ab}$. We should also multiply it by the number of quark flavors N_f. Calculating the trace and noting that, as this expression is multiplied by $(2\pi i)^2\delta(k^2)\delta[(k-p)^2]$, we can safely put $k^2 \equiv 0$ and $k \cdot p \equiv p^2/2$, so we obtain

$$\text{Im } \mathcal{P}_{\mu\mu}^{ab} \propto 4N_f g^2[k^2 - k \cdot p]\delta^{ab} = -2N_f g^2 p^2 \delta^{ab}. \tag{9.7}$$

Restoring all the relevant numerical factors or just multiplying the known QED result by $N_f/2$, we obtain

$$\text{Im } \mathcal{P}^q(s) = -\frac{g^2 s N_f}{24\pi}. \tag{9.8}$$

Let us now consider three other graphs in Fig. 9.1. The easiest is the graph with the 4-gluon vertex: its imaginary part is just zero. However, the graphs with the triple gluon vertices and with the ghost loop contribute. First, consider the ghost loop contribution. Calculating the imaginary part of the corresponding graph with Cutkosky rules, we obtain the analog of Eqs. (9.6, 9.7):

$$\begin{aligned} I_{\text{ghost}} &= (i^2)_{\text{propagators}} \times (-1)_{\text{ferm.loop}} \times g^2 f^{dac} f^{cbd} k_\mu (k-p)^\mu \\ &= -Ng^2[k^2 - k \cdot p]\delta^{ab} \equiv \frac{g^2 N p^2}{2}\delta^{ab}. \end{aligned} \tag{9.9}$$

Comparing it with Eq. (9.7) and the latter with Eq. (9.8), we derive for the fourth term of the proportion

$$\text{Im } \mathcal{P}^c(s) = \frac{g^2 N s}{96\pi}. \tag{9.10}$$

The imaginary part of the ghost loop contribution has the opposite sign, compared to the quark loop. This is very natural in the light of the discussion in Lecture 7. In fact, the Cutkosky trick we are using amounts to calculating the unphysical amplitude $\langle \text{virtual gluon}(p^2)|\text{virtual gluon}(p^2)\rangle$ by unitarity, saturating it, in the first case, by physical quark–antiquark states and, in the second case, by unphysical ghost states. The graph in Fig. 9.1c is related to the graph in Fig. 7.3 for the physical process $\bar{q}q \to \bar{q}q$. As was mentioned, the cross section for the production of unphysical ghost degrees of freedom is *negative*.

Finally, let us calculate the graph with the gluon loop in Fig. 9.1d.

Proceeding in the same way as before, we obtain for the integrand

$$
\begin{aligned}
I_{\text{gluon}} &= \left(\frac{1}{2}\right)_{\text{symmetry}} \times [(-i)^2]_{\text{propagators}} \\
&\quad \times g^2 f^{acd} f^{bdc} [(k+p)_\beta \eta_{\mu\alpha} + (p - 2k)_\mu \eta_{\alpha\beta} + (k - 2p)_\alpha \eta_{\mu\beta}]^2 \\
&= 9g^2 N (p^2 - k \cdot p + k^2) \delta^{ab} \equiv \frac{9g^2 N p^2}{2} \delta^{ab} ,
\end{aligned}
\tag{9.11}
$$

which gives

$$
\text{Im } \mathcal{P}^c(s) = \frac{3g^2 N s}{32\pi} .
\tag{9.12}
$$

We see that the gluon loop contribution is 9 times larger than that from the ghost loop and has the same sign. The latter comes from the unphysical gluon polarizations which, according to Eq. (7.27), appear in the residue $\eta_{\mu\nu}$ of the gluon propagator in Feynman gauge.

Note that the contributions of unphysical gluon polarizations and ghosts do not cancel each other, as was the case for the inclusive cross section $\text{Im } M_{\bar{q}q \to \bar{q}q} \propto \sigma_{\bar{q}q \to 2g} + \sigma_{\bar{q}q \to \bar{c}c}$ considered in the previous section. There is nothing wrong here: the decay of the virtual gluon is not a physical process, and there is no reason for such a cancellation to occur. Actually, the calculation we have just done is nothing but the calculation of two of the graphs contributing to $\text{Im } M_{\bar{q}q \to \bar{q}q}$: the graph in Fig. 7.3 and the last graph in Fig. 7.2; the cancellation occurs only if taking two other graphs in Fig. 7.2 into account.

Adding all the pieces together, we obtain

$$
\text{Im } \mathcal{P}(s) = g^2 s \left[\frac{5N}{48\pi} - \frac{N_f}{24\pi} \right] .
\tag{9.13}
$$

Substituting it in the dispersive relation (9.5), we finally derive

$$
\begin{aligned}
Z_A &= \frac{\mathcal{D}(-\mu^2)}{D(-\mu^2)} = \frac{1}{1 + \mathcal{P}(-\mu^2)/\mu^2} \\
&= 1 + \frac{g_0^2}{48\pi^2} [5N - 2N_f] \ln \frac{\Lambda_0^2}{\mu^2} + O(g_0^4) .
\end{aligned}
\tag{9.14}
$$

If calculating the same graphs in a generic ξ-gauge, the result is

$$
Z_A = 1 + \frac{g_0^2}{48\pi^2} \left[\frac{13 - 3\xi}{2} N - 2N_f \right] \ln \frac{\Lambda_0^2}{\mu^2} + O(g_0^4) .
\tag{9.15}
$$

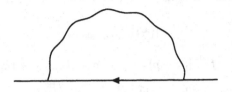

Fig. 9.2   Quark polarization operator.

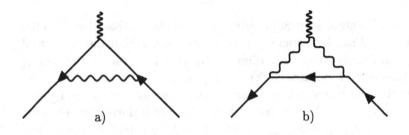

Fig. 9.3 $\bar{q}qA$ vertex.

Our next task is to find the renormalization factors for the quark propagator and for the $\bar{q}qA$ vertex. The former can actually be fixed without calculation: the corresponding graph depicted in Fig. 9.2 has exactly the same structure as in QED up to the color factor (A.18), the Casimir eigenvalue in the fundamental representation. Taking the known QED result (for $\xi = 1$, we have reproduced it in the previous lecture while solving **Problem 3**), we can immediately write

$$Z_q = 1 - \frac{g_0^2 c_F \xi}{16\pi^2} \ln \frac{\Lambda_0^2}{\mu^2} + O(g_0^4) \,. \qquad (9.16)$$

Consider now renormalization of the vertex. At the 1-loop level, two graphs depicted in Fig. 9.3 contribute. The first graph has the same structure as in QED, again. The QED result should be multiplied by the color factor $c_F - N/2$, which is obtained from the equality (A.19). The second diagram involving the 3-gluon vertex should be calculated anew. We will not do it here and just quote the result for the sum of the two graphs:

$$Z_{\bar{q}qA} = 1 - \frac{g_0^2}{16\pi^2} \left[c_F \xi + \frac{3+\xi}{4} N \right] \ln \frac{\Lambda_0^2}{\mu^2} + O(g_0^4) . \quad (9.17)$$

Note that in QED $Z_e = Z_{\bar{e}e\gamma}$ which follows from the Ward–Takahashi identity $q^\mu \Gamma_\mu(p+q, p; q) = e[G^{-1}(p+q) - G^{-1}(p)]$. In QCD the WI (7.19) is "contaminated" by ghosts, there is no reason for the corresponding Z-factors to coincide, so they do not, and their ratio contributes to the effective charge renormalization.

Substituting the Z-factors (9.14, 9.16, 9.17) in Eq. (9.1), we obtain the final result

$$g^2(\mu) = Z_{\text{inv}} g_0^2 = \left[1 + \frac{g_0^2}{16\pi^2} \left(\frac{11N}{3} - \frac{2N_f}{3} \right) \ln \frac{\Lambda_0^2}{\mu^2} + O(g_0^4) \right] g_0^2 . \quad (9.18)$$

The result (9.18) is only valid to leading nontrivial order in g_0^2 and does not alone allow one to draw a far-reaching conclusion. Pretty soon we will derive an improved expression taking into account the leading contributions to all orders in the coupling constant. A remarkable fact can, however, be observed right away: if the number of flavors is not too large,* the coefficient of g_0^2 in the square bracket has the opposite sign compared to QED [the QED result can be obtained from Eq. (9.18) if setting $N \to 0$ and $N_f/2 \to 1$]. As a result, the effective charge *grows* when μ goes down. In QED it, on the contrary, decreases.

Let us recall the reason why the sign of the term $\sim e_0^2 \ln \frac{\Lambda_0^2}{\mu^2}$ in the renormalization factor Z_{inv} in QED was negative (cf. [8], pp. 465–468). First, in QED $Z_e = Z_{\bar{e}e\gamma}$ and the charge renormalization comes exclusively from the renormalization of the photon propagator. The renormalization factor of the propagator is $Z_\gamma = 1 - \Pi(-\mu^2)/\mu^2$, where

$$i\Pi_{\mu\nu} = \left(\eta_{\mu\nu} - \frac{p_\mu p_\nu}{p^2} \right) \Pi(p^2) = iP_{\mu\nu} + [iP_{\mu\alpha}] D^{\alpha\beta} [iP_{\beta\nu}] + \cdots \quad (9.19)$$

is related to the correlator of two electromagnetic currents

$$\Pi_{\mu\nu}(p) = ie_0^2 \int e^{ip \cdot x} \langle T\{j_\mu(x) j_\nu(0)\}\rangle_0 \, d^4x . \quad (9.20)$$

---

*In the real world, $N_f = 6$ or even effectively less if a characteristic scale of interest μ is less than the heavy quark masses.

Now, $\Pi(p^2)$ is related to its imaginary part by the dispersive relation (9.5). The imaginary part of the photon polarization operator (9.20) taken with negative sign is called the photon *spectral density*: $\rho(s) \overset{\text{def}}{=} -\text{Im}\,\Pi^{\text{QED}}(s)$. It can be expressed as

$$\rho(s) \propto -\sum_n \langle 0|j_\mu|n\rangle\langle n|j^\mu|0\rangle \delta(P_n^2 - s) , \qquad (9.21)$$

where the sum runs over all physical intermediate states. In QED the current is conserved, which dictates $P_n^\mu \langle 0|j_\mu|n\rangle = 0$. Going in the frame where the virtual photon is at rest, we see that the vectors $\langle 0|j_\mu|n\rangle$ are purely spacelike, from which we derive the fact that $\rho(s)$ is strictly positive to all orders of perturbation theory; it is an exact theorem of QED.[†] Therefore in QED,

$$Z_{\text{inv}} = Z_\gamma = 1 - \frac{1}{\pi}\int_0^{\Lambda_0^2} \frac{\rho(s)ds}{s(s+\mu^2)} < 1 , \qquad (9.22)$$

and the physical charge is necessarily smaller than the bare one.

Due to lack of current conservation on one hand and, on the other hand, impossibility to introduce some extra quarks with small color charge (in non-Abelian case the latter is quantized and the ordinary quarks have the minimal color charge corresponding to the fundamental representation), this kind of reasoning does not apply to QCD, and $-\text{Im}\,\Pi(s)$ [or $-\text{Im}\,\mathcal{P}(s)$ coinciding with $-\text{Im}\,\Pi(s)$ in the lowest order] does not need to be positive. In fact, we have seen that it is not, due to contribution of unphysical polarizations when calculating it with Cutkosky rules. Besides, $Z_q \neq Z_{\bar{q}qA}$, which affects charge renormalization. Nothing prevents Z_{inv} to be greater than 1 and, indeed, we have seen that it *is* greater than 1 at the one-loop level.

[†]Another way of reasoning is noticing that $-\text{Im}\,\Pi^{\text{QED}}(s)$, the decay probability for the virtual photon, can be related to a physical cross section. Consider some extra charged particles E carrying very small charge $e' \ll e$. To leading order in e'^2, the inclusive cross section $\sigma_{\bar{E}E\to\text{all}}(s)$ (with *all* orders in e^2 taken into account) is proportional to $-\text{Im}\,\Pi^{\text{QED}}(s)$. Needless to say, a physical cross section is always positive.

9.2 Renormalization Group. Asymptotic Freedom and Infrared Slavery.

The result (9.18) can be trusted when the correction $\propto g_0^2$ is small. Note, however, that even if g_0^2 is small, the correction can be of order 1 or larger due to the presence of the large logarithm

$$L = \ln \frac{\Lambda_0^2}{\mu^2}$$

as a factor. If $g_0^2 L \sim 1$, higher loop corrections are of the same order as the 1-loop corrections and we are in a position to take them into account. Fortunately, this can be done. There are two ways to do it. The first way is to single out accurately the *leading logarithmic contributions* $\sim (g_0^2)^n L^n$ (they are indeed leading because each loop involves *at most* one ultraviolet logarithm) in the relevant graphs to all orders in perturbation theory and sum up all such terms.

In QED it is relatively easy. It turns out that the only source of the leading logarithmic corrections to the photon polarization operator $\mathcal{P}(p^2)$ is the 1-loop graph. Consider a higher-loop graph contributing to the polarization operator $\mathcal{P}(p^2)$. Note first of all that, for a correction to be leading logarithmic, characteristic virtualities in the corresponding Feynman integral should be ordered

$$\Lambda_0^2 \gg |k_n^2| \gg \cdots \gg |k_1^2| \gg |p^2| = \mu^2 . \tag{9.23}$$

We have a "matryoshka" hierarchy of loops, one within another. The largest virtuality $|k_n^2|$ is associated with the innermost loop and the virtuality $|k_1^2|$ with the outer one. The Feynman integral for the external loop can be written down explicitly:

$$\mathcal{P}(p^2) =$$
$$-\frac{i}{3} \int \frac{d^4 k_1}{(2\pi)^4} \text{Tr} \left\{ \Gamma_\mu(p + k_1, k_1; p) \mathcal{G}(p + k_1) \Gamma^\mu(k_1, p + k_1; -p) \mathcal{G}(k_1) \right\} \tag{9.24}$$

(with $\Gamma_\mu^{(0)} = -ie_0 \gamma_\mu$). This is almost the standard Schwinger–Dyson equation, but *both* vertices entering Eq. (9.24) are exact. The prescription (9.24) reflects the fact that we want $|k_1^2|$ to be much smaller than any virtuality in internal loops for the exact propagators and vertices. This takes a correct account of overlapping divergences.

Leading logarithmic corrections in $\mathcal{P}(p^2)$ should be associated with the leading logarithmic corrections in $\mathcal{G}(k)$ and Γ_μ. The QED Ward identity $Z_e = Z_{\bar{e}e\gamma}$ is responsible for the cancellation of the latter. Moreover, we can choose the Landau gauge $\xi = 0$, where the leading logarithmic corrections both in $\mathcal{G}(k)$ and Γ_μ are absent. At the 1-loop level, this is seen from the direct calculation [see Eq. (9.16)]. Also, by not so difficult reasoning, which uses, again, the ordering of the virtualities (9.23) (See **Problem 2** at the end of this lecture), it can be shown to be true to all orders. Thus, in higher order corrections to $\mathcal{P}(p^2)$, at least one logarithm is lost and, for $n > 1$, the n-loop contribution to $\mathcal{P}^{\text{QED}}(-\mu^2)$ is of order $\mathcal{P}_n^{\text{QED}}(-\mu^2) \sim (e_0^2)^n L^{n-1}$, which is subleading. Thereby, to leading order, it suffices to consider only the one-loop graph in $\mathcal{P}(-\mu^2)$ and to sum all these one-loop bubbles with the Dyson equation. As a result, we have a geometric progression,

$$e^2(\mu) \stackrel{\text{def}}{=} -i\mu^2 D(-\mu^2) = \frac{e_0^2}{1 + \frac{e_0^2}{12\pi^2}\ln\frac{\Lambda_0^2}{\mu^2}} . \qquad (9.25)$$

A similar program can in principle be carried out for QCD, but it is much more difficult and I even do not know whether somebody managed to do it explicitly. People usually use here another very powerful method known as *renormalization group*.

Let me first quote the result. To leading order, it is very simple and is expressed, again, as a sum of the geometric progression,

$$g^2(\mu) = \frac{g_0^2}{1 - \frac{g_0^2 b_0}{16\pi^2}\ln\frac{\Lambda_0^2}{\mu^2}} , \qquad (9.26)$$

where

$$b_0 = \frac{11N}{3} - \frac{2}{3}N_f . \qquad (9.27)$$

Let us now derive it. Renormalizability of the theory means that all the physical results should not depend on the bare charge g_0^2 and the ultraviolet cutoff Λ_0, but only on the effective charge $g^2(\mu)$. Let us consider the effective charges $g^2(\mu)$ and $g^2(\mu')$ at two different scales and let us first assume that $\mu' \gg \mu$. The effective charges are related to g_0^2 and Λ_0^2 according to

$$g^2(\mu) = Z_{\text{inv}}\left(\ln\frac{\Lambda_0^2}{\mu^2},\, g_0^2\right) g_0^2 ,$$

$$g^2(\mu') = Z_{\text{inv}}\left(\ln\frac{\Lambda_0^2}{\mu'^2},\, g_0^2\right) g_0^2 , \qquad (9.28)$$

with one and the same invariant function Z_{inv}.

Suppose we are interested in a process with a characteristic energy scale μ and hence, eventually, in $g^2(\mu)$. A crucial observation is that we can treat the parameters μ' and $g^2(\mu')$ exactly on the same footing as Λ_0 and g_0^2. In the Wilsonian effective action spirit, we proceed here in several steps. On the 0th step, we define our theory with the coupling constant g_0^2 at the scale Λ_0. On the first step, we integrate over the modes with momenta greater than μ' and derive thereby the renormalized effective action on the scale μ' which, as far as the modes with momenta less than μ' are concerned, plays exactly the same role as the bare one. We can perform now the second step and integrate over the modes with momenta less than μ', but greater than μ to obtain the effective action at the scale μ and the effective charge $g^2(\mu)$. Nothing is changed except the upper and low scale from and where we are going. That means that the charge $g^2(\mu)$ will be related to $g^2(\mu')$ as

$$g^2(\mu) = Z_{\text{inv}}\left(\ln\frac{\mu'^2}{\mu^2},\, g^2(\mu')\right) g^2(\mu') \qquad (9.29)$$

with *the same* function $Z_{\text{inv}}(L, g^2)$ as in Eq. (9.28). Substituting here the effective charges (9.28), we see that Z_{inv} satisfy a functional equation

$$Z_{\text{inv}}\left(\ln\frac{\Lambda_0^2}{\mu^2},\, g_0^2\right) = Z_{\text{inv}}\left(\ln\frac{\mu'^2}{\mu^2},\, g^2(\mu')\right) Z_{\text{inv}}\left(\ln\frac{\Lambda_0^2}{\mu'^2},\, g_0^2\right) . \quad (9.30)$$

This equation defines the renormalization group (or the group of multiplicative renormalizations), which just consists in tuning the effective charge, when scale is changed, with the factor Z_{inv}, which depends on the *ratio* of two scales and on the charge defined at the upper scale, but no *explicit* scale dependence appears. The elements of the group are the functions $Z_{\text{inv}}(L, g^2)$ and the product of two such functions gives a third one.

The functional equation (9.30) is the main magic trick. It provides stringent restrictions on the form of the function $Z_{\text{inv}}(L, g^2)$. The best way to handle Eq. (9.30) is to differentiate it over μ and to set $\mu = \mu'$ afterwards.

This amounts to representing one of the group elements in Eq. (9.30) in the infinitesimal Lie form. It is convenient to differentiate with respect to $\ln \mu$ and not Eq. (9.30), but rather directly Eq. (9.29). We obtain the differential equation for the evolution of effective charge:

$$\mu \frac{dg^2(\mu)}{d\mu} = \beta[g^2(\mu)] , \tag{9.31}$$

where

$$\beta(g^2) = -2g^2 \left. \frac{\partial}{\partial L} Z_{\text{inv}}(L, g^2) \right|_{L=0} \tag{9.32}$$

is called *Gell-Mann–Low function* or just the β function. Now, $\beta(g^2)$ can be expanded into a series in g^2. The first term of this series can be inferred from the one-loop result (9.18):

$$\beta(g^2) = -\frac{b_0 g^4}{8\pi^2} + O(g^6) , \tag{9.33}$$

with b_0 defined in Eq. (9.27). When only the leading term in $\beta(g^2)$ is taken into account, equation (9.31) can be easily integrated. For

$$\alpha_s = \frac{g^2}{4\pi} ,$$

we obtain

$$\alpha_s(\mu^2) = \frac{2\pi}{b_0 \ln \frac{\mu}{\Lambda_{\text{QCD}}}} , \tag{9.34}$$

where Λ_{QCD} is the integration constant. One can be easily convinced that Eq. (9.34) coincides actually with Eq. (9.26) if one identifies

$$\Lambda_{\text{QCD}} \equiv \Lambda_0 \exp \left\{ -\frac{8\pi^2}{b_0 g_0^2} \right\} . \tag{9.35}$$

That means that we have summed up a geometric progression, indeed. The form (9.34) of the result is much more illuminating, however. First of all, it does not involve the dependence of unphysical parameters Λ_0 and g_0, but rather on their combination Λ_{QCD} which is a true physical parameter of QCD. A miracle has happened. The original Yang–Mills Lagrangian involves a *dimensionless* coupling g^2. We see, however, that a *real* physical parameter of the theory is Λ_{QCD}, which carries dimension. This phenomenon is called *dimensional transmutation*. Together with the physical quark masses, Λ_{QCD} sets a scale for all relevant dimensionful quantities in QCD: in particular, to all hadron masses.[‡]

Both from Eqs. (9.31, 9.33) (telling us that the derivative of the effective charge with respect to scale is negative) and from their solution (9.34), it follows that the effective charge falls off when the characteristic energy grows. It means that the larger the energy (the smaller are the distances), the smaller the coupling constant and hence the more trustable is perturbation theory. This behavior is called *asymptotic freedom*. It is just opposite to what we had in QED: the growth of charge at small distances, so that perturbative expansion eventually breaks down, however small an initial large-distance charge was.

The latter property discovered first in QED by Landau, Abrikosov, and Khalatnikov and known as *zero charge* problem is rather troublesome and means in fact, that QED is probably not a self-consistent theory. Really, to define quantum field theory, we should attribute a meaning to the path integral symbol. That can be done by putting a theory on the lattice and introducing thereby an ultraviolet cutoff. Naturally, we want the results not to depend on the ultraviolet cutoff in the limit when it is sent to infinity. However, in QED and in many other field theories where the coupling grows at small distances ($\lambda\phi^4$ theory, Yukawa theory, etc.), we cannot do it. If we tend $\Lambda_0 \to \infty$ while keeping g_0 fixed, the effective charge (9.25) at any physical energy scale would go to zero. Of course, the result (9.25) was obtained only in perturbation theory and does not allow us to make definite conclusions concerning the nonperturbative regime. But no serious reason why this trouble should be cured in the full theory is seen, and probably it is not. For $\lambda\phi^4$ theory, the fact that the trouble *is* there and that the physical coupling constant goes to zero in the limit $\Lambda_0 \to \infty$ was confirmed by numerical lattice calculations.

On the contrary, QCD is very nice in this respect. The continuum limit when the cutoff is sent to infinity and the bare coupling to zero exists and presents no difficulties. Everything depends only on the combination (9.35).

There is, however, another side of the coin. According to (9.34), when the physical scale μ goes down, the effective charge *rises*. Eventually, at $\mu \sim \Lambda_{QCD}$, it becomes of order 1 and perturbative calculations in terms

---

[‡]The appearance of scale in a theory which did not have it at the tree level can be described as breaking of the conformal symmetry of the tree action by quantum effects. We will return to the discussion of this question in Lecture 12.

of quarks and gluons lose any sense. This growth of charge is sometimes called *infrared slavery*. Indeed, we know from experiment that, at large distance scales, there is no trace of perturbative quarks or gluons whatsoever. Confinement occurs and, instead of quarks, we have hadrons with the characteristic energy scale $\sim \Lambda_{QCD}$. We cannot prove confinement now, but obviously, infrared slavery represents a necessary condition of it. If the coupling constant stayed small at large distances, the asymptotic states would be quarks and gluons as given by perturbative analysis.

Asymptotic freedom and the infrared slavery are not, however, related so rigidly. To understand it, remember that the result (9.34) corresponds to the summation of the leading logarithmic terms $\propto \alpha_s^n L^n$ in the Green's function. The equation (9.31) also allows one to sum up next-to-leading logarithms $\propto \alpha_s^n L^{n-1}$ and $\propto \alpha_s^n L^{n-2}$ etc. at almost no cost: we only need to know the higher-order terms in the expansion of the β function. According to Eq. (9.32), the term $\sim g^6$ in $\beta(g^2)$ and hence all the subleading terms $\propto \alpha_s^n L^{n-1}$ in Z_{inv} can be determined if the coefficients of the two-loop overlapping ultraviolet logarithm $\alpha_s^2 L$ in the Green's functions are determined. The term $\sim g^8$ in $\beta(g^2)$ gives all the subleading terms $\alpha_s^n L^{n-2}$ and is obtained from the 3-loop calculation of the terms $\sim \alpha_s^3 L$ etc.

We quote here the result of the two-loop calculations (with dimensional regularization — see the discussion below):

$$\beta(g^2) = -\frac{b_0 g^4}{8\pi^2} - \frac{b_1 g^6}{128\pi^4} , \qquad (9.36)$$

where

$$b_1 = \frac{34}{3} N^2 - \left(\frac{13}{3} N - \frac{1}{N} \right) N_f . \qquad (9.37)$$

Now look: when $N = 3$ and $N_f = 6$, the first term in Eq. (9.36) is negative and so is the second term. However, the coefficient b_0 given by Eq. (9.27) falls off if N_f increases: quarks bring about the conventional screening of charge, like in QED. If their effect overshoots the antiscreening effect due to gluons and ghosts, which happens at $N_f > 16$, asymptotic freedom is lost. Consider an imaginary world with $N = 3$ and $N_f = 16$, then the first term in Eq. (9.36) is still negative, but very small. The second term is now positive and is not particularly small. When we start to evolve the equation (9.31) from *very* small distances (with *very* small coupling constant) into the infrared, the second term in (9.36) is, at first, unessential, and the effective charge grows. Sooner or later, the coupling grows to a point when the second term balances the first one. Due to the chosen boundary value $N_f = 16$ with an artificially small coefficient b_0, this balancing occurs at the point when the coupling constant is rather small,

$$\alpha_s^* = \frac{2\pi}{151},$$ (9.38)

and the third and higher terms in the β function can be ignored.

The coupling constant freezes at the *fixed point* (9.38) and does not grow with distance anymore. In this theory, we do have asymptotic freedom, but do not have infrared slavery and confinement.

Problem 1. *a)* Show that the explicit solution of the differential equation (9.31) can be written in the form

$$\mu' = \mu \exp \left\{ \int_{g^2(\mu)}^{g^2(\mu')} \frac{dt}{\beta(t)} \right\}$$ (9.39)

(the *Gell-Mann–Low equation*).

b) From this, deduce that, to any order in the coupling constant, Λ_{QCD}, defined as the scale where the effective charge $g^2(\Lambda_{QCD})$ becomes singular, is related to Λ_0 and g_0^2 as

$$\Lambda_{QCD} = \Lambda_0 \exp \left\{ \int_{g_0^2}^{\infty} \frac{dt}{\beta(t)} \right\}.$$ (9.40)

Solution is straightforward.

9.3 Observables. Ambiguities. Anomalous Dimensions.

We defined the effective charge $g^2(\mu)$ as the coefficient in the effective Lagrangian (8.20) describing the dynamics at the characteristic energy scale $E \sim \mu$. This is, however, a mathematical definition. A physical definition requires pointing out an *observable* which would allow one to *measure* $g^2(\mu)$ in a real physical experiment. An important remark is that, though it can be done, it cannot be done in a unique way. There are many different definitions of $g^2(\mu)$, different *renormalization schemes*, which are more or less equivalent from the theoretical and practical viewpoint, all of them leading to identical results for physically observable quantities.

The situation is nontrivial even in QED. Speaking of the low-energy constant α_{phys}, it is defined relatively unambiguously. A natural definition is

$$\alpha_{\text{phys}} = \lim_{\omega \to 0} \left[\frac{3m_e^2}{8\pi} \sigma^{\text{Compt}}(\omega) \right]^{1/2} , \qquad (9.41)$$

where σ^{Compt} is the cross section of the Compton scattering $\gamma e \to \gamma e$ and ω is the photon frequency. The point is that, in the limit $\omega \to 0$, all higher-order contributions to the Compton amplitude are suppressed as powers of ω/m and do not "contaminate" the definition (9.41). *The same* α_{phys} is measured in the Josephson effect, etc. The reason for such a benign behaviour is that we are dealing here with the case where the Born–Oppenheimer philosophy applies in a straightforward way. We have two distinct scales: ω and m. Knowing the expansion of the set of scattering amplitudes in ω/m, one can in principle construct the Born–Oppenheimer expansion for the effective Lagrangian, describing interactions of soft photons in the presence of a scattering center.[§]

However, consider the process of Compton scattering (or any other QED process) at high frequency $\omega \gg m$. Then $\sigma^{\text{Compt}}(\omega)$ can be represented as a series over the low-energy coupling constant α_{phys} with the coefficient involving large logarithms $\ln(\omega/m)$. The renormalizability of QED implies that $\sigma^{\text{Compt}}(\omega)$ can also be expressed into powers of running charge $\alpha(\omega)$, defined as in Eq. (9.25) with $\mu \equiv \omega$, without any logarithms in the coefficients. But the same is true for $\alpha(2\omega)$! One can *define* $\tilde{\alpha}(\omega) = \alpha(2\omega)$ and express $\sigma^{\text{Compt}}(\omega)$ as a series over $\tilde{\alpha}(\omega)$. This new series is as good as the old one. $\tilde{\alpha}(\omega) \neq \alpha(\omega)$ and the coefficients in the two series are different, but the final answer for $\sigma^{\text{Compt}}(\omega)$ is of course *scheme-independent*.

The underlying reason for this complication is that here, in constrast to the low frequency case, we cannot separate high energy modes from the low energy modes in such a neat and clean manner. In the cases when we do not really have a gap between two regions, the Wilsonian procedure of defining the effective Lagrangian involves an inherent ambiguity. That is, one can, of course, suggest some particular precise definition of $\mathcal{L}_{\text{eff}}$ like the one in Eq. (8.18), but other definitions can be suggested.

[§]This Lagrangian does not have, however, a very simple and nice form. In contrast to the Euler–Heisenberg Lagrangian (8.17), it is nonlocal and not translationally invariant (cf. the problem of polaron [19]).

One of the ambiguities involves the change of scale $\mu \to c\mu$ and the corresponding redefinition $\alpha_c(\mu) = \alpha(c\mu)$. It is comparatively "benign" in a sense that the β function defined in Eq. (9.31) is not sensitive to a change of scale:

$$
\begin{aligned}
\frac{1}{4\pi}\beta^{\text{new}}(4\pi\alpha_c) &= \mu\frac{d\alpha_c(\mu)}{d\mu} = c\mu\frac{d\alpha(c\mu)}{d(c\mu)} \\
&= \frac{1}{4\pi}\beta^{\text{old}}[4\pi\alpha(c\mu)] = \frac{1}{4\pi}\beta^{\text{old}}(4\pi\alpha_c) \, .
\end{aligned}
\tag{9.42}
$$

However, other redefinitions are possible. In fact, any analytic redefinition

$$
\tilde{\alpha}(\mu) = F[\alpha(\mu)] = \alpha(\mu) + C_2\alpha^2(\mu) + \cdots
\tag{9.43}
$$

with finite C_n is as good as the original one [cf. Eq. (8.28)]. With an arbitrary change of convention (9.43), the β function is no longer invariant, though the first two terms of the expansion of $\beta(\alpha)$ in α *are*.¶ The invariance is lost, however, in the next order in α.

In QCD the situation is more difficult than in QED. Due to infrared slavery and confinement, we cannot explore perturbatively the low-energy limit of QCD and the notion of "physical low-energy color charge" does not exist. The only meaningful definition of "physical charge" refers to the region of high momenta, where the situation is essentially the same as in QED: many different definitions of $g^2(\mu)$ are possible, with the β function being sensitive to a particular definition of the charge. In some schemes, $\beta(g^2)$ is even gauge-dependent. Also, the fundamental parameter Λ_{QCD} depends on the scheme. [This is seen e.g. from Eq. (9.40); a simple analysis shows that a change in the coefficient of the 3-rd term in $\beta(g^2)$, while keeping Λ_0 fixed, multiplies Λ_{QCD} by some factor which does not depend on the ultraviolet scale.] ‖

¶To understand that, write

$$
\tilde{\beta}(\tilde{\alpha}) = F'(\alpha)\beta(\alpha) \, .
$$

Substituting here $\beta(\alpha)$ in the form (9.36) and expressing $\alpha = \tilde{\alpha} - C_2\tilde{\alpha}^2 + \cdots$, we see that the coefficients of the terms $\sim \tilde{\alpha}^2$ and $\sim \tilde{\alpha}^3$ in $\tilde{\beta}(\tilde{\alpha})$ are, indeed, the same as the corresponding coefficients in $\beta(\tilde{\alpha})$.

‖In standard QCD only analytic redefinitions of the effective charge are usually considered. But nonanalytic redefinitions, which can modify not only the 3-rd, but also the 2-nd term in the β function, are admissible as well. Using a proper such definition, one can kill *all* the coefficients in the β function except the leading one. In supersymmetric gauge theories, this is not a mere trick, but a sensible way to renormalize the theory [20] (see some further discussion of this issue in Lecture 12).

We have to live with this complication, but the reader should not be scared too much. As is also the case in QED, final results for physical quantities *are* scheme-independent. Also, the ambiguity shows up only at subleading orders, and $g^2(\mu)$ can be called a kind of "semi-physical" quantity because its sensitivity to the renormalization scheme is not so strong after all. In particular, the statement that in some particular process the effective charge is small is scheme-invariant: it just means that the corresponding cross section can be represented as a reasonably well convergent perturbative series.

Consider, as an example (a very important one), the process $e^+e^- \rightarrow$ hadrons. Its cross section is related to the contribution of hadron loops in the imaginary part of the photon polarization operator $\Pi(s)$. As was mentioned already in the Introduction, a fundamental theorem of QCD tells us that, when the energy is high enough and the invariant charge $\alpha_s(s)$ is small, $\Pi(s)$ can be calculated by evaluating the contribution of the quark and gluon loops:

$$\sigma^{\text{observed}}(e^+e^- \rightarrow \text{hadrons}) \approx \sigma^{\text{theor}}(e^+e^- \rightarrow \text{quarks and gluons}). \quad (9.44)$$

This is a manifestation of so called *quark–hadron duality*. The property (9.44) and its generalizations will be exploited on many occasions in the following lectures.

The right side of Eq. (9.44) represents a series in α_s. One of the ways to write it is

$$R(s) = \frac{\sigma(e^+e^- \rightarrow \text{hadrons})}{\sigma(e^+e^- \rightarrow \mu^+\mu^-)}$$

$$= N_c \sum_f Q_f^2 \left[1 + \frac{3}{4}c_F\frac{\alpha_s(\mu)}{\pi} + A\left(\frac{\alpha_s(\mu)}{\pi}\right)^2 + O(\alpha_s^3)\right], \quad (9.45)$$

where Q_f are the electric charges of the quarks, assumed to be massless,** and

$$A = \frac{c_F}{16}\left[\frac{c_F}{2} + \frac{1}{N_c} + \left(\frac{11}{2} - 4\zeta(3) - \ln\frac{s}{\mu^2}\right)(11N_c - 2N_f)\right], \quad (9.46)$$

with $\zeta(3) = 1.202\ldots$. The coefficient of α_s in Eq. (9.45) (the number 3/4) was determined by Schwinger a long time ago. It amounts to calculating

---

**Practically, that means that, for LEP energy scales, the sum runs over 5 flavors while the top quark with the mass ≈ 170 GeV is not taken into account.

the imaginary part of the two-loop graphs contributing to the photon polarization operator (the quark loops with a gluon rung). The masslessness of quarks renders the calculation not too difficult and the reader is welcome to try to repeat it. The calculation of the coefficient A is already rather involved (and the calculation of the coefficient of α_s^3 requires the use of a computer).

A and all higher coefficients are scheme-dependent. The value (9.46) refers to the so-called $\overline{\text{MS}}$ (*modified minimal subtraction*) scheme, the most popular one in practical calculations. Let us first explain what the MS scheme (without "bar") is. It is based on the dimensional regularization method and just consists in crossing out all the poles $\propto 1/\epsilon$, $\propto 1/\epsilon^2$, etc. in the unrenormalized Green's functions. For example, the unrenormalized one-loop fermion polarization operator (8.16) (calculated in Feynman gauge) goes over to the following renormalized one:

$$\Sigma_{1\,\text{loop}}^{\text{MS}}(p, \xi = 1) = \frac{g_{\text{MS}}^2(\mu)}{16\pi^2} c_F \slashed{p} \left(\gamma - 1 + \ln \frac{-p^2}{4\pi\mu^2} \right). \tag{9.47}$$

The $\overline{\text{MS}}$ prescription is $1/\epsilon - \gamma + \ln(4\pi) \to 0$, rather than just $1/\epsilon \to 0$. It is more convenient than the simple-minded MS scheme, because the expressions for renormalized Green's functions obtained this way are simpler. For example,

$$\Sigma_{1\,\text{loop}}^{\overline{\text{MS}}}(p, \xi = 1) = \frac{g_{\overline{\text{MS}}}^2(\mu)}{16\pi^2} c_F \slashed{p} \ln \frac{-p^2}{e\mu^2}. \tag{9.48}$$

The relation $g_{\text{MS}}^2(\mu e^{\gamma/2}/\sqrt{4\pi}) = g_{\overline{\text{MS}}}^2(\mu)$ holds, which means that

$$\Lambda_{\text{QCD}}(\overline{\text{MS}}) \equiv \Lambda_{\overline{\text{MS}}} = e^{-\gamma/2}\sqrt{4\pi}\Lambda_{\text{MS}}.$$

These subtraction recipes correspond to the clear-cut Wilsonian procedure with separation scale $\mu\sqrt{4\pi}e^{(1-\gamma)/2}$ for the MS scheme and $\mu\sqrt{e}$ for the $\overline{\text{MS}}$ scheme [cf. Eq. (8.24)].

As everything associated with dimensional regularization, the $\overline{\text{MS}}$ scheme is theoretically clear and practically convenient, but is not very physical. The MOM renormalization scheme, described at the end of the previous lecture, is less convenient and less elegant from the theoretical viewpoint, but it is more physical. As a result, the coefficients of perturbative expansion of different physical quantities in $g_{\text{MOM}}^2(E_{\text{char}})$ tend to be a little bit smaller. If the effective charge is defined according to Eq. (8.27) on the basis of 3-gluon vertex, the relation $\Lambda_{\text{MOM}} \approx 2.16 \dots \Lambda_{\overline{MS}}$ holds.

The coefficient A of the term $\propto \alpha_s^2(\mu)$ in $R(s)$ depends on an arbitrarily chosen normalization point μ, but the physical quantity $R(s)$ is μ-independent as is explicitly seen (at the level $\propto g^2$ and $\propto g^4$) if substituting the 1–loop result (9.18) for $g^2(\mu)$ into Eq. (9.45).

Up to now, we have only discussed renormalization of the invariant charge. The factor $Z_{\text{inv}}(L, g_0^2)$ was defined in Eq. (8.23) via certain combinations of the renormalization factors of some particular Green's functions. We can also pose a problem of summing up the leading, subleading, etc. logarithmic ultraviolet contributions in particular renormalization factors like Z_q, Z_{3A}, etc. Even though they are not gauge-invariant and do not have a *direct* physical meaning, their values in a particular gauge can be related (in a gauge-dependent way) to certain gauge-independent physically observable quantities.

Again, the method of renormalization group is the most adequate and convenient one. Consider, for example, the propagator of a quark defined at (Wilsonian separation) scale μ. Like earlier, we assume the quarks to be massless and their virtuality to be small, $|p^2| \ll \mu^2$. According to the definition,[††] $\mathcal{G}(p) = Z_q\left(\Lambda_0/\mu, g_0^2\right)(i/\not{p})$. On the other hand, it can be calculated in the framework of the effective theory defined on some intermediate scale μ', $\mu < \mu' < \Lambda_0$. We have

$$
\begin{aligned}
\mathcal{G}(p) &= Z_q\left(\ln\frac{\mu'^2}{\mu^2}, \, g^2(\mu'^2)\right) \mathcal{G}_{\mu'}^{\text{eff}}(p) \\
&= Z_q\left(\ln\frac{\mu'^2}{\mu^2}, \, g^2(\mu'^2)\right) Z_q\left(\ln\frac{\Lambda_0^2}{\mu'^2}, \, g_0^2\right)\left(\frac{i}{\not{p}}\right) .
\end{aligned}
\tag{9.49}
$$

It follows that the renormalization factor Z_q satisfies the functional equations

$$
Z_q\left(\ln\frac{\Lambda_0^2}{\mu^2}, \, g_0^2\right) = Z_q\left(\ln\frac{\mu'^2}{\mu^2}, \, g^2(\mu'^2)\right) Z_q\left(\ln\frac{\Lambda_0^2}{\mu'^2}, \, g_0^2\right) .
\tag{9.50}
$$

It has the same form as the equation (9.30) for Z_{inv} with the important difference that it involves now *two* different functions: Z_q and the ratio of effective charges Z_{inv}. A convenient way to handle Eq. (9.50) is to

---

[††]The Green's functions and their renormalization factors depend also on the gauge parameter ξ. This can be taken into account (see e.g. the book [6]), but the final result (9.54) is not changed.

differentiate it with respect to $\ln \mu'$ and set $\mu' = \Lambda_0$ afterwards. Bearing Eq. (9.31) in mind, we obtain the *Ovsyannikov–Callan–Symanzik* (OCS) equation [21],

$$\left[\Lambda_0 \frac{\partial}{\partial \Lambda_0} + \beta(g_0^2) \frac{\partial}{\partial g_0^2} - \gamma_q(g_0^2) \right] Z_q \left(\ln \frac{\Lambda_0^2}{\mu^2}, g_0^2 \right) = 0 , \qquad (9.51)$$

where

$$\gamma_q(g^2) = 2 \frac{\partial}{\partial L} Z_q(L, g^2) \Big|_{L=0} \qquad (9.52)$$

is called the *anomalous dimension* of the quark Green's function. This name comes from the observation that *if* $\beta(g^2)$ would vanish (the invariant charge would not run), the solution of Eq. (9.51) would be just

$$Z_q = \left(\frac{\Lambda_0}{\mu} \right)^{\gamma(g^2)} \qquad (9.53)$$

and the Green's function $\mathcal{G}(p)$ would acquire the factor $c^{1+\gamma(g^2)}$ under the scale transformation $x \to cx$, $p \to c^{-1}p$. The canonical dimension of the Green's function $d_{\text{can}} = 1$ would be modified due to perturbative corrections.

The solution of Eq. (9.51) with initial conditions $Z_q(0, g^2) = 1$ can also be found in the general case with arbitrary $\beta(g^2) \neq 0$. It has the form [‡‡]

$$Z_q \left(\ln \frac{\Lambda_0^2}{\mu^2}, g_0^2 \right) = \exp \left\{ - \int_{g_0^2}^{g^2(\mu^2)} \frac{\gamma(t)}{\beta(t)} dt \right\} . \qquad (9.54)$$

To lowest order,

$$\beta(t) = -\frac{b_0 t^2}{8\pi^2}, \qquad \gamma_q(t) = \frac{\gamma_q^{(0)} t}{8\pi^2} , \qquad (9.55)$$

and we obtain

$$Z_q \left(\ln \frac{\Lambda_0^2}{\mu^2}, g_0^2 \right) = \left[\frac{g^2(\mu^2)}{g_0^2} \right]^{\gamma_q^{(0)}/b_0} . \qquad (9.56)$$

The OCS equations and their solutions (9.54), (9.56) have exactly the same form for all other renormalization factors of the 2-point Green's functions and of the exact vertices in the symmetric normalization point (8.25). One

[‡‡]In the limit $\beta(g^2) \to 0$, $g^2(\mu^2) - g_0^2 \approx \beta(g_0^2) \ln \frac{\mu}{\Lambda_0}$, and we reproduce the result (9.53).

need not be restricted by the vertices present in the QCD Lagrangian. The problem of finding the renormalization of, say, pseudoscalar color singlet quark vertex is perfectly well posed, has a solution, and is important for some physical applications.

The last comment is that $Z_{\text{inv}}(L, g_0^2)$ also satisfies the partial differential equation (9.51). Comparing Eq. (9.32) with Eq. (9.52), we see that in this case

$$\gamma_{\text{inv}}(g^2) = -\frac{\beta(g^2)}{g^2}.$$

Substituting it into Eq. (9.54), we find $Z_{\text{inv}}(L, g_0^2) = g^2(\mu^2)/g_0^2$ which is, of course, true, but does not give any new information.

Problem 2. Reproduce the leading order result (9.56) for the renormalization factor $Z_e(L, e_0^2)$ in QED by direct summation of the relevant leading logarithmic graphs.

Solution. The 1-loop graph like in Fig. 9.2 for the electron polarization operator $\Sigma(p)$ provides a logarithmically divergent factor

$$\Sigma^{(1)}(p) = -\frac{e_0^2 \xi_0}{16\pi^2} \not{p} \int_{\mu^2}^{\Lambda_0^2} \frac{dK^2}{K^2} = -\frac{e_0^2 \xi_0}{16\pi^2} \not{p} \ln \frac{\Lambda_0^2}{\mu^2}, \qquad (9.57)$$

with $\mu^2 = -p^2 \gg m_e^2$. The logarithmically divergent part in the 1-loop correction to the $\bar{e}e\gamma$ vertex has the form

$$\Gamma_\mu^{(1)}(p', p; q) = \frac{e_0^2 \xi_0}{16\pi^2} \gamma_\mu \ln \frac{\Lambda_0^2}{\mu^2},$$

where $\mu^2 = \max(|p'^2|, |p^2|, |q^2|)$.

The leading logarithmic (LL) contribution to $\Sigma^{(n)}(p)$ comes from the kinematic region (9.23) with all virtualities ordered. That allows us to write for $\Sigma^{\text{LL}}(p)$ the integral equation

$$\Sigma^{\text{LL}}(p) =$$
$$ie_0^2 \int \frac{d^4 k_1}{(2\pi)^4} \Gamma_\mu^{\text{LL}}(p, p + k_1; -k_1) \mathcal{G}^{\text{LL}}(p + k_1) \Gamma_\nu^{\text{LL}}(p + k_1, p; k_1) \mathcal{D}_{\text{LL}}^{\mu\nu}(k_1).$$
$$(9.58)$$

Like in Eq. (9.24), the LL corrections to *both* vertices are taken into account, and it is assumed that $K_1^2 = |k_1^2|$ is much larger than the virtualities in the internal loops for $\mathcal{G}$, Γ, and $\mathcal{D}$. In the leading logarithmic order, the renormalization factor for Γ_μ depends on the ratio Λ_0^2/K_1^2, K_1^2 being the largest virtuality associated with the vertices in Eq. (9.58). Due to the Ward–Takahashi identity, it coincides with the renormalization factor Z_e^{-1}. Bearing in mind that $\Sigma(p) = \not{p}\left(1 - Z_e^{-1}\right)$, equation (9.58) is reduced to

$$Z_e^{-1}\left(\ln\frac{\Lambda_0^2}{\mu^2}, \, e_0^2\right) =$$

$$1 + \frac{e_0^2\xi_0}{16\pi^2}\int_{\mu^2}^{\Lambda_0^2}\frac{dK^2}{K^2}Z_e^{-1}\left(\ln\frac{\Lambda_0^2}{K^2}, \, e_0^2\right)Z_\gamma\left(\ln\frac{\Lambda_0^2}{K^2}, \, e_0^2\right). \qquad (9.59)$$

Acting on this integral equation by the operator $\mu\frac{\partial}{\partial\mu}$, we rewrite it in the differential form

$$\mu\frac{\partial}{\partial\mu}Z_e = \frac{e_0^2\xi_0}{8\pi^2}\frac{1}{1 - \frac{b_0e_0^2}{16\pi^2}\ln\frac{\Lambda_0^2}{\mu^2}}Z_e, \qquad (9.60)$$

where we substituted the LL expression (9.25) for Z_γ; $b_0^{\text{QED}} = -4/3$. The *ordinary* differential equation (9.60) is equivalent to the partial differential equation (9.51) to leading order. Bearing in mind that $\gamma_e^{(0)} = -\xi_0$, according to the definitions (9.52), (9.55), and the result (9.57), one can be directly convinced that the solution of Eq. (9.60) with the initial condition $Z_e(0, e_0^2) = 1$ coincides with Eq. (9.56). Note that when $\gamma_e^{(0)} = -\xi_0 = 0$, the LL corrections to $\mathcal{G}_e$ and $\Gamma_{\bar{e}e\gamma}$ are absent altogether: the fact we used earlier to derive (9.25) by diagrammatic methods.

Remark. Also, nonleading logarithmic corrections to $Z_e(L, e_0^2)$ could in principle be obtained by this method. One should, however, have in mind that the strict ordering of virtualities (9.23) does not hold anymore. For example, in the next-to-leading order, two of the virtualities can be comparable: $|k_i^2| \sim |k_{i+1}^2|$ for some i. To take into account this kinematic region, one should add the term coming from the two-loop skeleton graph in the integral equation (9.58). The renormalization group method is much simpler, indeed...

Lecture 10

Weathering Infrared Storms

As well as ultraviolet divergences, Feynman graphs for amplitudes involve also infrared divergences, coming from the region of soft virtual gauge bosons. The divergences appear in higher loop corrections to any elastic scattering amplitude when a charged particle in QED (or a color-charged particle in QCD) changes the direction of its motion.

In QED the resolution of this problem is well known and the physical picture behind it is well understood. When the infrared divergences in all loops are summed up, the elastic amplitude vanishes. This is very natural and can already be explained at the classical level: scattering of charged particles is always accompanied by the bremsstrahlung of soft photons and the probability *not* to emit such a photon is zero. But anyway, a pure elastic scattering is not a physical process. The apparatus always has a finite energy resolution $\Delta\epsilon$. We never know whether soft photons with energy $\leq \Delta\epsilon$ were emitted or not and we have to sum over the *probabilities* of all these processes. Such a sum is called a *physical cross section*. These physical cross sections are always finite: infrared divergences which show up in the probabilities of individual processes cancel out of the sum.

At the one-loop level, the mechanism of this cancellation in QED was understood by Bloch and Nordsieck back in 1937 and is explained in the textbooks [8; 9]. It was generalized to higher loops by Yennie, Frautchi, and Suura. Their proof was polished and presented in a nice transparent form by Weinberg. We follow here the approach of Weinberg and refer the reader to Chapter 13 of his book [3] for more details.

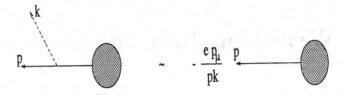

Fig. 10.1   Emission of a soft photon.

10.1   Bloch–Nordsieck Cancellation.

Consider the graph in Fig. 10.1 describing the emission of a soft (virtual
or real) photon with momentum k from an external electron line. The
amplitude is singular in the limit $k \to 0$:

$$M_\mu(p, k) \;=\; -e\bar{u}(p)\gamma_\mu \frac{\not{p} + \not{k} + m}{(p+k)^2 - m^2} X(p+k) \;=\; \eta \frac{e p_\mu}{p \cdot k} M(p) + O(1),$$

$$(10.1)$$

where $X(p+k)$ is the dashed blob in Fig. 10.1, $M(p) = \bar{u}(p)X(p)$ and
$\eta = -1$ in this case. When deriving Eq. (10.1), we used the identity $\gamma_\mu \not{p} = 2p_\mu - \not{p}\gamma_\mu$ and the equations of motion $\bar{u}(p)(\not{p} - m) = 0$. The graphs
describing the soft photon emission from an initial electron line and from
initial and final positron lines are also singular, and are similarly given by
Eq. (10.1) with $\eta = 1$ for an initial electron and a final positron, and $\eta = -1$
for an initial positron.

Consider now the amplitude of emission of n photons from an external
line. The $n!$ graphs which differ by the order in which the photons are emitted contribute to the amplitude. The most singular part of the amplitude
has the form

$$M_{\mu_1 \ldots \mu_n}(p; k_1, \ldots, k_n) =$$

$$e^n \eta^n p_{\mu_1} \cdots p_{\mu_n} \left[\frac{1}{(p \cdot k_1)(p \cdot k_1 + p \cdot k_2) \cdots (p \cdot k_1 + \ldots + p \cdot k_n)} \right.$$

$$\left. + \text{permutations} \right] M(p) + \ldots = \frac{e^n \eta^n p_{\mu_1} \cdots p_{\mu_n}}{(p \cdot k_1) \cdots (p \cdot k_n)} M(p) + \ldots,$$

$$(10.2)$$

where the dots stand for the less singular in the infrared terms. The result
(10.2), which can be proven by induction, is rather remarkable. It says that

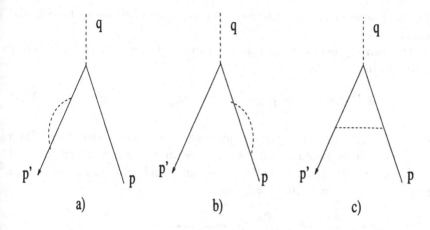

Fig. 10.2 Contributions to $M^{(1)}(p,p')$.

the amplitude of emission of several soft photons is factorized, which means that the photons are emitted independently from each other. In general, different soft photons can be emitted from different external lines, but the factorization still holds: each soft photon with momentum k generates the factor $e \sum_i \eta_i p_\mu^{(i)}/(p^{(i)} \cdot k)$, where the summation over all external lines with momenta $p^{(i)}$ is done.

As the simplest example where infrared divergences appear in the loops, consider the process of electron scattering in a static electromagnetic field. At the 1-loop level, the three graphs depicted in Fig. 10.2 contribute.

Note that we are considering now an *unrenormalized* amplitude and that is why the graphs of Figs. 10.2a,b describing corrections to external lines appear. An alternative [8; 9] would be to consider a renormalized amplitude. Then, only the graph in Fig. 10.2c should have been taken into account, but some extra infrared-divergent terms would be introduced by the renormalization factor,

$$Z_1 = 1 - \frac{\alpha}{2\pi}\left(\ln\frac{\Lambda_0}{m} + 2\ln\frac{\lambda}{m}\right), \qquad (10.3)$$

where λ is the infrared cutoff and the Feynman gauge, $\xi = 1$, is chosen.*

---

*Infrared divergences in renormalization factors did not show up in expressions like Eq. (9.16) in the previous lecture, because the external lines carried nonvanishing virtualities there and these virtualities served as an infrared cutoff. In the Wilsonian renormalization procedure outlined in Lecture 8, it is the separation scale μ that pro-

The final answer for the infrared-divergent part of the amplitude is the same.

One can express the sum of the three graphs in Fig. 10.2 in the following clever form:

$$M^{(1)}(p,p') = \frac{1}{2} \int \frac{d^4k}{(2\pi)^4} D^{\mu\nu}(k) M^{(0)}_{\mu\nu}(p,p';k,-k) , \qquad (10.4)$$

where $M^{(0)}_{\mu\nu}(p,p';k,-k)$ is the tree amplitude involving the emission of two photons with momenta k and $-k$. The factor $1/2$ is introduced to avoid double counting. Staying in Feynman gauge and using the factorization properties (10.1) and (10.2), we obtain

$$M^{(1)}(p,p') = \frac{ie^2}{2} \left[m^2 \int \frac{d^4k}{(2\pi)^4} \frac{1}{(p\cdot k)^2(k^2+i0)} \right.$$

$$+ m^2 \int \frac{d^4k}{(2\pi)^4} \frac{1}{(p'\cdot k)^2(k^2+i0)}$$

$$\left. - 2(p\cdot p') \int \frac{d^4k}{(2\pi)^4} \frac{1}{(p\cdot k)(p'\cdot k)(k^2+i0)} \right] M^{(0)}(p,p') . (10.5)$$

The most convenient way to do these integrals is to integrate first over dk_0, with closing the contour in the lower half-plane and picking up the contribution of the pole $k_0 = |k| - i0$. [†] We obtain

$$M^{(1)}(p,p') = \frac{e^2}{32\pi^3} \left[\int \frac{m^2 dk}{|k|^3(p_0 - |p|\cos\theta_{pk})^2} \right.$$

$$+ \int \frac{m^2 dk}{|k|^3(p'_0 - |p'|\cos\theta_{p'k})^2}$$

$$\left. - \int \frac{2(pp')dk}{|k|^3(p_0 - |p|\cos\theta_{pk})(p'_0 - |p'|\cos\theta_{p'k})} \right] M^{(0)}(p,p') . \quad (10.6)$$

vides an infrared cutoff.

[†]If the infinitesimal imaginary shifts in the fermion propagators, omitted in Eq. (10.5), were restored, one could be convinced that the fermion poles stay in this case outside the contour and do not contribute. They could provide an infrared-divergent contribution for, say, the scattering amplitude $ee \to ee$ in the graphs involving a soft photon line stretched between the two different final or two different initial charged particles. But this contribution involves an extra factor i and does not affect the correction $\propto \alpha$ to the Born cross section. When the contributions of this kind are summed up in all orders, they provide an infinite Coulomb phase for the amplitude which is seen also in the framework of nonrelativistic quantum scattering theory. See Weinberg's book for more details.

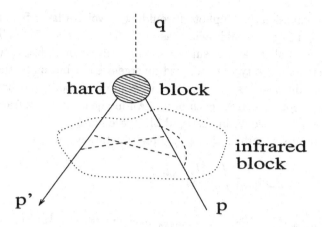

Fig. 10.3   An infrared contribution in $M(p, p')$. The hard block involves momenta $k > \mu$ and the infrared block momenta $k \le \mu$.

Integrating over angles and introducing an infrared cutoff $\lambda \ll m$ for the integral over $|\boldsymbol{k}|$, we obtain for the infrared-divergent part of the amplitude

$$M^{(1)}(p, p') = \frac{\alpha}{2\pi} \ln \frac{E_{\text{char}}}{\lambda} \left[2 - \frac{1}{\beta} \ln \frac{1 + \beta}{1 - \beta} \right] M^{(0)}(p, p') , \qquad (10.7)$$

where $\beta = \sqrt{1 - 4m^4/(Q^2 + 2m^2)^2}$ is the velocity of the outgoing electron in the rest frame of the incoming electron. E_{char} stands for the characteristic energy of the process. In our case, it is reasonable to choose it $E_{\text{char}} \sim \sqrt{Q^2}$, and we will stick to this choice in the next lecture when discussing the double logarithmic asymptotics of amplitudes. But, as far as the structure of infrared singularities is concerned, any other choice is also OK. We only claim that the coefficient of $\ln \lambda$ in $M^{(1)}(p, p')$ is determined correctly.

   The infrared-divergent part of the amplitude can also be determined to any order in α. To do it quite accurately, introduce a Wilsonian separation scale $\mu \ll E_{\text{char}},$[‡] and let us call the virtual photons in a complicated multiloop diagram with momenta $k \le \mu$ "soft" and those with momenta $k > \mu$ "hard". Note that, for a region of soft k_i to be relevant, the integral over $d^4 k_i$ should be sensitive to the infrared, which is only possible if the corresponding photon line is attached at both ends to external lines or the lines with small virtualities $\ll \mu E_{\text{char}}$. Otherwise, at least one of the ver-

---

[‡]The relation between μ and E_{char} is opposite compared to what we had in Lecture 8, where the Wilsonian procedure was used to renormalize the theory.

tices of the emission of the photon would not involve a large factor $\propto 1/k_i$ and the loop integral would behave as $\sim \int^\mu d^4k_i/k_i^3$ or $\sim \int^\mu d^4k_i/k_i^2$. This converges at small k_i and is suppressed as a power of μ/E_{char}, which we neglect. Therefore, a typical infrared-divergent contribution in the amplitude can be depicted as in Fig. 10.3, with infrared loops and hard loops being nicely separated from each other. All momenta in the infrared block are soft and all momenta in the hard block are hard.

An all-orders generalization of the representation (10.4) is

$$M(p,p') = \sum_{n=0}^{\infty} \frac{1}{2^n n!} \int \frac{d^4k_1}{(2\pi)^4} D^{\mu_1\nu_1}(k_1) \cdots$$

$$\int \frac{d^4k_n}{(2\pi)^4} D^{\mu_n\nu_n}(k_n) \, M_{\mu_1\nu_1\ldots\mu_n\nu_n}(p,p';k_1,-k_1,\ldots,k_n,-k_n) \;, \quad (10.8)$$

where $M_{\mu_1\ldots\nu_n}$ is the amplitude of emission of $2n$ photons with corresponding momenta. It may involve loops, but only hard ones. Again, the factor $1/2^n n!$ was introduced to count every topologically distinct graph for the amplitude only once. Only the most singular infrared part of $M_{\mu_1\ldots\nu_n}$ displayed in Eq. (10.2) should be taken into account in the integral (10.8). Indeed, the contribution of the next-to-leading term $\propto k^{-(2n-1)}$ is proportional to $\int^\mu d^{3n}k/k^{3n-1} \sim \mu/E_{\text{char}} \ll 1$ and the contribution of the other subleading terms is further suppressed.

In the approximation just described and justified, the integrals over different soft photon momenta in Eq. (10.8) factor out, and we finally obtain

$$M(p,p') = M^{\text{hard}}(p,p') \exp\left\{ \frac{\alpha}{2\pi} \ln\frac{\mu}{\lambda} \left[2 - \frac{1}{\beta} \ln\frac{1+\beta}{1-\beta} \right] \right\} \;. \quad (10.9)$$

As usual, $M^{\text{hard}}(p,p')$ depends on μ in such a way that the product of two factors on the right side of Eq. (10.9) is μ-independent. Eq. (10.9) can be generalized to any QED process with the substitution

$$2 - \frac{1}{\beta} \ln\frac{1+\beta}{1-\beta} \;\to\; \frac{1}{2} \sum_{nn'} \frac{\eta_n\eta_n'}{\beta_{nn'}} \ln\frac{1+\beta_{nn'}}{1-\beta_{nn'}} \;, \quad (10.10)$$

where the sum is over all pairs of external charged particles and

$$\beta_{nn'} = \sqrt{1 - \frac{m^4}{(p_n p_n')^2}} \quad (10.11)$$

is the relative velocity of the particles in the pair.

Let us continue our discussion of the electron scattering process. The elastic scattering cross section

$$d\sigma^{\text{el}} = d\sigma^{\text{hard}} \exp\left\{-\frac{\alpha A}{\pi} \ln \frac{\mu}{\lambda}\right\} \; , \quad A = \frac{1}{\beta} \ln \frac{1+\beta}{1-\beta} - 2 \; , \quad (10.12)$$

vanishes in the limit $\lambda \to 0$, as it should. Now, consider the inelastic processes involving emission of some number of real soft photons whose total energy does not exceed $\omega_m \ll E_{\text{char}}$. The amplitude of emission of n soft photons factorizes and can be written in the form

$$M(p, p'; k_1, \ldots, k_n)$$
$$= e^n \left[\frac{p \cdot \epsilon_1}{p \cdot k_1} - \frac{p' \cdot \epsilon_1}{p' \cdot k_1}\right] \cdots \left[\frac{p \cdot \epsilon_n}{p \cdot k_n} - \frac{p' \cdot \epsilon_n}{p' \cdot k_n}\right] M^{\text{el}}(p, p') \; , \quad (10.13)$$

where ϵ_n are the polarization vectors of the photon. The elastic amplitude $M^{\text{el}}(p, p')$ also involves here virtual infrared loops. Taking the square of Eq. (10.13), summing over polarizations, and integrating over the photon phase space, we obtain

$$d\sigma_n = d\sigma^{\text{el}} \frac{1}{n!} \left(\frac{\alpha A}{\pi}\right)^n \int_\lambda^\infty \frac{d\omega_1}{\omega_1} \cdots \int_\lambda^\infty \frac{d\omega_n}{\omega_n} \, \theta\left(\omega_m - \sum_{i=1}^n \omega_i\right), \quad (10.14)$$

where ω_i are the energies of individual photons and A is the result of the angular integration (*the same* as for the virtual infrared loops[§]). The phase space integrals corresponding to different photons have almost factored out. They would factor out completely were it not for the θ function implementing the restriction $\sum_i \omega_i \leq \omega_m$. A clever thing to do is to express it in the integral form

$$\theta(\omega_m - E) \overset{E \geq 0}{=} \frac{1}{\pi} \int_{-\infty}^\infty \frac{\sin(\omega_m u)}{u} e^{iEu} du \; , \quad (10.15)$$

after which the integrals over $d\omega_1, \ldots, d\omega_m$ are disentangled. The presence of the θ function allows us to choose ω_m for the upper limit of the integrals,

---

[§]It is seen immediately if the photon density matrix is chosen in the form $\rho_{\mu\nu} = -\eta_{\mu\nu}$: the structure of the *integrands* is then exactly the same as in Eq. (10.6). But calculations using the physical transverse density matrix $\rho_{ij}(k) = \delta_{ij} - k_i k_j / |k|^2$ give, of course, the same result.

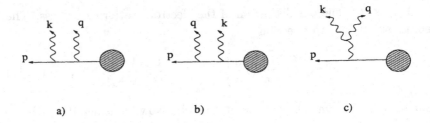

Fig. 10.4   Emission of two soft gluons.

and we obtain

$$d\sigma_n = d\sigma^{\text{el}} \frac{1}{n!} \left(\frac{\alpha A}{\pi} \right)^n \frac{1}{\pi} \int_{-\infty}^{\infty} d\sigma \frac{\sin \sigma}{\sigma} \left[\ln \frac{\omega_m}{\lambda} + \int_0^\sigma \frac{e^{ix} - 1}{x} dx \right]^n .$$

(10.16)

The physical cross section is defined as

$$d\sigma^{\text{phys}} = \sum_{n=0}^{\infty} d\sigma_n = d\sigma^{\text{el}} F(\alpha A) \exp \left\{ \frac{\alpha A}{\pi} \ln \frac{\omega_m}{\lambda} \right\},$$

(10.17)

where

$$F(\alpha A) = \frac{1}{\pi} \int_{-\infty}^{\infty} d\sigma \frac{\sin \sigma}{\sigma} \exp \left\{ \frac{\alpha A}{\pi} \int_0^\sigma \frac{e^{ix} - 1}{x} dx \right\}$$

(10.18)

is regular in the infrared. Substituting here $d\sigma^{\text{el}}$ from Eq. (10.12), we finally obtain

$$d\sigma^{\text{phys}}(\omega_m) = d\sigma^{\text{hard}}(\mu) F(\alpha A) \exp \left\{ -\frac{\alpha A}{\pi} \ln \frac{\mu}{\omega_m} \right\} .$$

(10.19)

The physical cross section thus defined is finite and does not depend on the infrared cutoff λ anymore. Neither does it depend on the artificial separation scale μ. On the other hand, it does depend on the kinematic cutoff ω_m. The lower ω_m is, and the better the energetic resolution of our device is , the smaller will be the fraction of those events involving emission of additional photons, which can be confused with the elastic scattering process and are included in the definition (10.17) of the physical cross section. As a result, $d\sigma^{\text{phys}}$ goes to zero in the limit $\omega_m \to 0$.

Problem. Find a gauge such that the loop corrections to the electron propagator would not involve infrared divergences.

Solution. Substituting

$$D_{\mu\nu} = \frac{i}{k^2 + i0} \left[g_{\mu\nu} - \frac{(1 - \xi)k_\mu k_\nu}{k^2 + i0} \right] \quad \text{and} \quad M_{\mu\nu} = \frac{e^2 p_\mu p_\nu}{(p \cdot k)^2} \frac{i}{\not{p} - m} + \ldots$$

into Eq. (10.4) and choosing $p_\mu = (m, \mathbf{0})$, we obtain

$$G^{(1)}(p) \propto \int \frac{d^4 k}{(2\pi)^4 k_0^2} \left[\frac{1}{k_0^2 - \mathbf{k}^2 + i0} - \frac{(1 - \xi)k_0^2}{(k_0^2 - \mathbf{k}^2 + i0)^2} \right], \quad (10.20)$$

where, by the same token as before, in performing the integral over dk_0, only the residue at $k_0 = |\mathbf{k}| - i0$ should be taken into account. It is not difficult to see that the latter vanishes at $\xi = 3$ (*Yennie gauge*). Infrared divergences in the multiloop corrections to $G(p)$ also vanish in Yennie gauge due to the factorization (10.2) and exponentiation as in Eq. (10.9).

10.2   Non-Abelian Complications. Coherent States.

The structure of infrared-divergent graphs in QCD is much more involved. First of all, the factorization property (10.2) does not hold anymore. Consider the graphs in Fig. 10.4 describing the emission of two soft gluons from an external quark line. The first two graphs have different color structure, $t^a t^b \neq t^b t^a$, and their sum does not give the factorized expression $\propto 1/(p \cdot k_1)(p \cdot k_2)$ anymore. The graph in Fig. 10.4c makes the things even worse. In QCD soft gluons are not emitted independently from each other, but in an intricate correlated manner.

As a result, the Bloch–Nordsieck theorem, which would claim that infrared divergences cancel out of the "physical cross sections" involving emission of extra soft gluons as defined in Eq. (10.17), is simply wrong in QCD. Its violation is already seen at the 1-loop level. Consider the process of quark scattering in an external non-Abelian gauge field. The Born amplitude has the color structure $M_{ji}^{(0)} \propto (t^a)_{ji}$, where i and j are initial and final quark colors and a is the color of the static gluon field. The Born differential cross section is

$$d\sigma^{(0)} \propto | M^{(0)}|^2 \propto \sum_j (t^a)_{ji}(t^a)_{ij} = (t^a t^a)_{ii}, \quad (10.21)$$

where we have performed the sum over the final quark colors j while keeping a and the initial quark color i fixed.

Consider the virtual correction associated with a soft gluon loop attached to an outgoing quark line as in Eq. 10.2a, paying attention to its color structure only. We have

$$d\sigma^{\text{virt}}(out) \propto 2\text{Re}\, M^{\text{virt}}(out) M^{(0)*} \propto \sum_{jb}(t^b t^b t^a)_{ji}(t^a)_{ij}$$
$$= c_F(t^a t^a)_{ii} \, . \qquad (10.22)$$

The color structure of the differential cross section describing the emission of a real soft gluon from the outgoing quark line is the same:

$$d\sigma^{\text{real}}(out) \propto \sum_{jb}(t^b t^a)_{ji}(t^a t^b)_{ij} = c_F(t^a t^a)_{ii} \, , \qquad (10.23)$$

and the cancellation takes place by the same token as in QED. Consider, however, the graphs describing the soft gluon emission from an *initial* line. The color structure is now

$$d\sigma^{\text{real}}(in) \propto \sum_{jb}(t^a t^b)_{ji}(t^b t^a)_{ij} = \sum_b (t^b t^a t^a t^b)_{ii} \, , \qquad (10.24)$$

which *is* not cancelled with the corresponding virtual contribution $\propto c_F(t^a t^a)_{ii}$. Thus, the "physical cross section" $\sum_n d\sigma_n$ is not so physical in QCD, since it is infrared-divergent.

How do we resolve this apparent paradox? The first idea which comes to mind is to say that quarks or gluons do not exist anyway (as asymptotic states), due to confinement, and the infrared divergences are absent because the external lines in Feynman graphs never go onto the mass shell. This answer is not satisfactory, however. As we discussed before, confinement is related to the asymptotic freedom, and one can well imagine a non-Abelian gauge field theory without confinement (it is enough to have a number of quark flavors sufficiently large). But the infrared structure of such a theory would be exactly the same.

A somewhat better try is to blame the observed absence of the cancellation on the fact that we fixed the color of the initial quark. Confinement or not, the notion of a quark or gluon asymptotic state with a fixed color is not physical: an emission or absorption of a very soft gluon can change it. Indeed, if we define a physical cross section such that the averaging over initial colors is performed, then $\sum_b \text{Tr}\{t^b t^a t^a t^b\} = c_F \text{Tr}\{t^a t^a\}$ and the mismatch between Eq. (10.24) and Eq. (10.22) disappears. One can show, however, that even though the BN cancellation occurs in QCD at the 1-loop level in this case, the mismatch strikes back at the 2-loop level [22].

The fact that summing over initial colors still makes the infrared behavior of the cross section more benign is not accidental, however. The true resolution of the paradox is achieved if one observes not only that asymptotic state color is not physical, but that asymptotic states as such are not a completely physical notion. And this refers not only to QCD, but also to QED!

An asymptotic electron state involves not only the electron, but also its Coulomb electromagnetic "coat". The Coulomb field extends at infinite distance. The point is that, to form such a state, an infinite time is required, which is never available. Consider a process when the electron is scattered and changes the direction of its motion. At first this change concerns only the electron itself and not its coat. The electromagnetic field is neutral and continues to move in the original direction (this is nothing but bremsstrahlung). As for the scattered electron, it eventually grows a new coat, but before this happens, we are dealing not with a true asymptotic state or an eigenfunction of the QED Hamiltonian, but with a wave packet, a coherent superposition of many states involving, besides the asymptotic electron state, also an indefinite number of very soft photons. The physical cross section (10.17) can be interpreted as a cross section to create such a coherent state, with the parameter ω_m characterizing the width of the wave packet in functional space.

In QED the cross section of creating a set of such coherent states in the collision of asymptotic particles happens to be finite. In QCD it is not. What is always finite, however, is the probability to create coherent states *when one also has coherent states initially*. Applied to the quark scattering process, one can claim that the infrared divergences drop out in the probability if we add to the "physical cross section" a properly normalized probability of processes involving the *absorption* of soft gluons in the initial state. The corresponding diagrams are depicted in Figs. 10.5, 10.6.

Consider first the graphs of Fig. 10.5. When the gluon is soft, the Lorentz structure of M^{abs} coincides, up to a sign, with that of the emission amplitude, M^{emis}. Integrating $|M^{\text{abs}}|^2$ over the phase space of the initial soft gluon with the restriction $\omega \leq \omega_m$, we obtain roughly the same result as for the emission cross section. An important subtlety is, however, that the color structures of M^{abs} and M^{emis} are different. We have seen that the absence of the BN cancellation for this process was associated with the

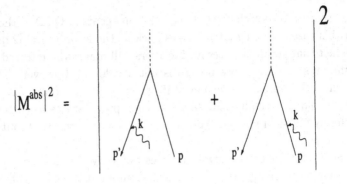

Fig. 10.5   Absorption of initial gluons: conventional graphs.

initial quark line of fixed color. Consider now the part of M^{abs} describing the absoption of a soft gluon by the initial quark. The color structure of the corresponding contribution in the probability [¶] is

$$d\sigma^{\mathrm{abs}}(in) \propto \sum_{jkb} (t^a)_{ji}(t^b)_{ik}(t^b)_{ki}(t^a)_{ij} = c_F(t^a t^a)_{ii}. \qquad (10.25)$$

When writing Eq. (10.25), we summed over the colors k, b of the initial quark and gluon, requiring that together they form a fundamental color state with fixed color index i.

The infrared-divergent part in $d\sigma^{\mathrm{abs}}(in)$ cancels the divergence in virtual corrections, $d\sigma^{\mathrm{virt}}(in)$, which $d\sigma^{\mathrm{real}}(in) \equiv d\sigma^{\mathrm{emis}}(in)$ failed to do. Speaking of the divergences in $d\sigma^{\mathrm{emis}}(in)$, they are also cancelled by the divergences in the contribution of the *disconnected* graphs depicted in Eq. 10.6. Disconnected graphs are, of course, rather exotic animals. Usually, they do not come in from the cold because, in the standard scattering problem involving only two particles in the initial state, such disconnected graphs only affect the real part of the forward elastic scattering amplitude, not a very physical quantity. But in our case, disconnected graphs are absolutely crucial. Without them, the cancellation would not occur.

We leave it as a simple exercise for the reader to show that the dis-

---

[¶] We emphasize again that $d\sigma^{\mathrm{abs}}(in)$ is, in fact, not a cross section, but a probability of the process involving three particles in the initial state. But we assume it to be normalized such that it has the dimension of differential cross section, and we are going to add it to $d\sigma^{\mathrm{BN}}$ to observe the cancellation of the infrared divergences at the end of the day. This explains the notation $d\sigma^{\mathrm{abs}}(in)$.

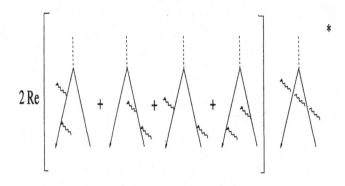

Fig. 10.6 Absorption of initial gluons: disconnected graphs.

connected contribution of Eq. 10.6 involving the absorption and emission of a soft gluon from the initial quark line has the same color structure as $d\sigma^{\text{emis}}(in)$ and the divergences cancel in the sum of these two contributions just as they do in the sum $d\sigma^{\text{virt}}(in) + d\sigma^{\text{abs}}(in)$. Another exercise is to show that the infrared divergences also cancel in the contributions of the cross graphs where the soft gluon interacts with both the initial and the final quark.

This cancellation observed is a particular case of the so-called *Kinoshita–Lee–Nauenberg* theorem. The content of this theorem is, however, broader. The papers of Kinoshita and of Lee and Nauenberg appeared in the first half of the sixties, well before the advent of QCD, and the primary subject of their studies was the structure of the so-called *collinear singularities* in QED. Collinear divergences in QED and in QCD will be the subject of our next lecture.

Lecture 11

Collinear Singularities: Theory and Phenomenology

11.1 Double Logarithmic Asymptotics.

Let us look again at Eqs. (10.12) and (10.19) and ask what happens in the limit $Q^2 \gg m^2$. We see that $A(Q^2 \gg m^2) \approx 2\ln(Q^2/m^2)$, so that

$$d\sigma^{\text{el}} \sim d\sigma_0 \exp\left\{ -\frac{2\alpha}{\pi} \ln \frac{\sqrt{Q^2}}{\lambda} \ln \frac{Q^2}{m^2} \right\} ,$$

$$d\sigma^{\text{BN}} \sim d\sigma_0 \exp\left\{ -\frac{2\alpha}{\pi} \ln \frac{\sqrt{Q^2}}{\omega_m} \ln \frac{Q^2}{m^2} \right\} . \tag{11.1}$$

This is what is called *double logarithmic asymptotics* (there are two powers of large logarithm for each power of α). When deriving (11.1), we assumed that $\mu \sim E_{\text{char}} \sim \sqrt{Q^2}$. In this case, $d\sigma^{\text{hard}}(\mu = \sqrt{Q^2}) = A(\alpha)d\sigma_0$, and one can show that $A(\alpha)$ does not involve double logarithms. The asymptotics (11.1) are interesting both from theoretical and phenomenological viewpoint: at the energies achieved at present accelerators, $\sqrt{Q^2} \sim 200$ GeV, the exponent for $d\sigma^{\text{BN}}$ in Eq. (11.1) is not small and *all* orders in α should be taken into account.

What is the origin of the second logarithm $\sim \ln(Q^2/m^2)$ in Eq. (11.1)? To understand it, look at the integrals (10.6) determining the coefficient A. In particular, let us consider the third integral in Eq. (10.6), describing the contribution of the graph in Fig. 10.2c to the amplitude, and let us estimate it in the *Breit frame* $\mathbf{p} = -\mathbf{p}'$, assuming $Q^2 \gg m^2$. The electrons are then ultrarelativistic $p_0 = p_0' \equiv E \sim |\mathbf{p}| = |\mathbf{p}'| \sim \sqrt{Q^2}/2$. We see that, on top of the infrared divergence $\sim \int d|\mathbf{k}|/|\mathbf{k}|$, the angular integral

157

diverges logarithmically in the limit $Q^2 \to \infty$, being saturated by the regions $\cos\theta_{pk} \sim \pm 1$, i.e. when the virtual photon goes in the same direction as either the outgoing or the incoming electron. We are dealing here with a collinear kinematic singularity. An elementary calculation gives

$$
\begin{aligned}
M^{(1)}(p,p') &\approx -\frac{\alpha}{\pi}\ln\frac{E}{\lambda}\int_{-1}^{1}\frac{dz}{1-v^2z^2}M^{(0)}(p,p') \\
&\approx -\frac{\alpha}{\pi}\ln\frac{E}{\lambda}\ln\frac{Q^2}{m^2}M^{(0)}(p,p') ,
\end{aligned}
\tag{11.2}
$$

where $v = \sqrt{1 - m^2/E^2}$ is the velocity of the incoming and outgoing electron in the Breit frame, to be distinguished from the relative velocity β entering Eq. (10.7). Exponentiating the correction $1 - \frac{\alpha}{\pi}\ln\frac{E}{\lambda}\ln\frac{Q^2}{m^2}$ and going over to the cross sections, we reproduce Eq. (11.1).

Up to now, we have considered only the amplitudes on mass shell, which vanish when the infrared cutoff is lifted $\lambda \to 0$. It often makes a lot of sense to also consider off-mass-shell amplitudes which are regular in this limit. Consider the electron formfactor $S(Q^2, M^2)$ in the so called *Sudakov kinematic region* $Q^2 \gg -p^2 = -p'^2 = M^2 \gg m^2$. The novelty here is that a single parameter M^2 cuts off both infrared and collinear singularities. Repeating all the steps of the previous calculation, we obtain at the 1-loop level

$$
S^{(1)}(Q^2, M^2) \sim -\frac{\alpha}{\pi}\int_0^1\int_0^1\frac{d\rho\, d\theta^2_{pk}}{\rho\theta^2_{pk} + 4M^2/Q^2} \approx -\frac{\alpha}{2\pi}\ln^2\frac{Q^2}{M^2}
\tag{11.3}
$$

($\rho = |\boldsymbol{k}|/E$). The leading logarithmic terms $\sim \alpha^n L^{2n}$ in the Sudakov formfactor exponentiate by the same token as before. This follows from the fact that the factorization property (10.2) is easily generalized to the case when the external lines are virtual. In particular, for the amplitude of emission of two soft photons with momenta k and q from the external line p, we have

$$
M_{\mu\nu} = e^2 p_\mu p_\nu \times
$$

$$
\left[\frac{1}{\left(p\cdot k \pm \frac{M^2}{2}\right)\left(p\cdot k + p\cdot q \pm \frac{M^2}{2}\right)} + \frac{1}{\left(p\cdot q \pm \frac{M^2}{2}\right)\left(p\cdot k + p\cdot q \pm \frac{M^2}{2}\right)}\right]
$$

$$
= \frac{e^2 p_\mu p_\nu}{\left(p\cdot k \pm \frac{M^2}{2}\right)\left(p\cdot q \pm \frac{M^2}{2}\right)}\left[1 \pm \frac{M^2}{2p\cdot k + 2p\cdot q \pm M^2}\right] ,
\tag{11.4}
$$

where the sign of M^2 depends on whether the line is incoming or outgoing.

The leading logarithmic contributions come from the region $Q^2 \gg p \cdot k, p \cdot q \gg M^2$, so that the second term in the square brackets is suppressed and the amplitude is factorized. We obtain

$$S(Q^2, M^2) = \exp\left\{-\frac{\alpha}{2\pi} \ln^2 \frac{Q^2}{M^2}\right\} + O(\alpha^n L^{2n-1}). \qquad (11.5)$$

We just mention here without proof that the classical QED result (11.5) can be generalized to QCD. In particular, the Sudakov electromagnetic formfactor of a quark has the same form as Eq. (11.5), with a trivial substitution $\alpha \to \alpha_s c_F$ [23]. Moreover, also nonleading (in the first place, next-to-leading $\sim \alpha^n L^{2n-1}$) logarithms can be summed over both in QED [24] and in QCD [25]. The formula for the Sudakov formfactor that takes into account the terms $\sim \alpha_s^n L^{2n}$ and $\sim \alpha_s^n L^{2n-1}$ has the form

$$S(Q^2, M^2) = \exp\left\{-\frac{\alpha_s(M^2) c_F}{2\pi} \ln^2 \frac{Q^2}{M^2} + \frac{3\alpha_s c_F}{4\pi} \ln \frac{Q^2}{M^2}\right\}$$
$$+ \quad O(\alpha_s^n L^{2n-2}), \qquad (11.6)$$

the exponential of the 1-loop result with the coupling constant being normalized at the low virtuality M^2 (the normalization point for α_s in the second term cannot be fixed at the leading logarithmic level as the dependence $\alpha_s(M^2)$ involves only one power of L for each power of α_s).

11.2 Jet Cross Sections.

Let us look again at Eq. (11.1). We see that, in *massless electrodynamics*, infrared and collinear divergences appear on exactly the same footing. Both have kinematic nature and are not *immediately* related to the ultraviolet divergences reflecting the behavior of the theory at ultrahigh energies.*
It is noteworthy that the physical cross sections (10.19) are singular in the massless QED which resembles the infrared situation in non-Abelian theories discussed in the previous lecture.

---

*After a certain analysis, a relationship between collinear and ultraviolet divergences can be unravelled; see the discussion below.

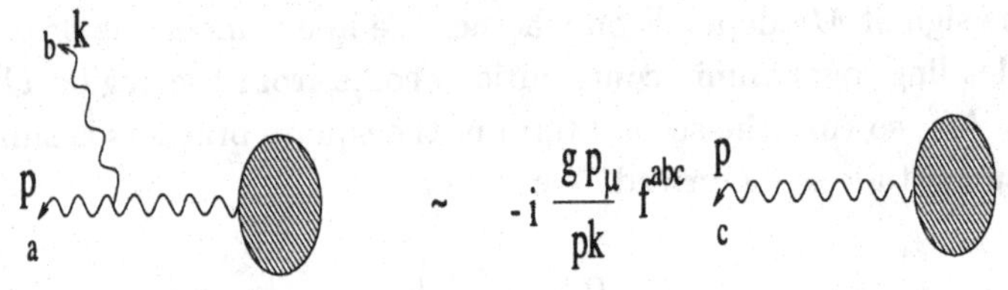

Fig. 11.1   Emission of soft gluon from an external gluon line.

Collinear singularities appear also in QCD. First of all, there are light quarks and the logarithms $\sim \ln(Q^2/m_q^2)$ that arise by the same token as in QED. Second, QCD also involves gluons, which carry color charge and are *inherently* massless. Consider the amplitude of the emission of a soft gluon from an external gluon line. Its Lorentz and color structure is displayed in Fig. 11.1. The factor $\sim p_\mu/(p \cdot k)$ is the same as in QED. The amplitude becomes singular both in the infrared limit $k \to 0$ and in the collinear limit $\cos\theta_{pk} \to 1$. Collinear singularities in QCD are in many respects similar to those in massless QED. In particular, one could consider a "gluon Sudakov formfactor" describing scattering of a gluon with virtuality M^2 in an external colorless (say, gravitational) field such that the relation $Q^2 \gg M^2$ holds. In the leading logarithmic order, it is given by Eq. (11.5) with $\alpha \to \alpha_s c_V$.

In the previous lecture we learned how to treat infrared singularities. One can (and should) wonder now: whether collinear singularities are tractable as well? In other words, whether it is possible to define generalized cross sections and/or probabilities with a clear physical meaning which would not be sensitive to fictitious infrared and collinear cutoffs? With some reservations, the answer to this question is positive.

As we have learned in the previous lecture, the presence of colored particles in the initial state brings about extra complications. We will discuss it later, and for the time being consider the processes with production of colored particles in collisions of color neutral objects. The simplest example is the process of e^+e^- annihilation into a quark–antiquark pair. The elastic cross section $d\sigma_{e^+e^- \to \bar{q}q}$ is given by the expression in Eq. (11.1), with the change $\alpha \to \alpha_s c_F$. It involves both infrared and collinear divergences. The Bloch–Nordsieck cross section (10.19), taking into account the emission of extra soft gluons, is finite in the infrared, but the collinear divergences are still there. The statement we want to make now is that these divergences cancel if we add to the Bloch–Nordsieck cross section the cross section of the processes involving emission of some number of real hard gluons going in the same direction as the outgoing quark or outgoing antiquark.

Consider the case when we have only one such gluon. Our first observation is that the cross section of its production involves the collinear divergence. Indeed, the amplitude of emission of a real hard collinear gluon with momentum k from the outgoing quark line with momentum p involves the factor $1/(p+k)^2 \sim 1/p \cdot k \sim 1/\theta_{pk}^2$. The vertex involves, however, an extra factor $\sim \theta_{pk}$ due to suppression by chirality.[†] Thus, the amplitude behaves as $\sim 1/\theta_{pk}$, and the corresponding cross section is

$$\int |M|^2 dV_{\mathrm{ph}} \sim \int \left(\frac{1}{\theta_{pk}}\right)^2 \theta_{pk} d\theta_{pk} \sim \int \frac{d\theta_{pk}}{\theta_{pk}}, \qquad (11.7)$$

which diverges logarithmically at the lower limit. Also, the amplitude of the emission of a hard collinear gluon from the outgoing antiquark line is singular $\sim 1/\theta_{p'k}$ and its contribution to the cross section diverges logarithmically. The cross term $\sim \int d\Omega_k/\theta_{pk}\theta_{p'k}$ is regular.

It is worthwhile to note that the chirality suppression argument works straightforwardly and the last statement is exactly correct if only physical polarizations of the real hard gluon are taken into account and the gluon density matrix is chosen in the spatially transverse form

$$\rho_{ij} = \delta_{ij} - \frac{k_i k_j}{k^2}. \qquad (11.8)$$

If we choose $\rho_{\mu\nu} = -\eta_{\mu\nu}$, and allow the emission of longitudinal and scalar gluons, the final answer would, of course, be the same, but the same singularities would distribute over Feynman graphs in a different manner. In particular, the collinear singularity in the cross section of the hard gluon production would come exclusively from the cross term.

The choice (11.8) is, however, more physical and more convenient. It is also convenient to choose the *temporal gauge*, $A_0 = 0$,[‡] where the gluon

---

[†] To emit a real gluon with the helicity ± 1 in the forward direction, a quark should change its own helicity. But this is not allowed in the massless limit due to the chiral invariance of the lagrangian [see Eq. (12.19) below].

[‡] The temporal gauge is not Lorentz invariant and calculations are usually somewhat simpler in Feynman gauge. Usually, but not in this case. As far as the analysis of the collinear divergences in QCD and in massless QED is concerned, the gauge (11.9) (and some close relatives thereof) are very strongly preferred. Note in passing that the temporal gauge is ghost-free.

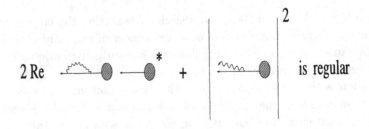

Fig. 11.2   Cancellation of collinear divergences in the temporal gauge.

propagator has the form

$$D_{00}(k) = D_{i0}(k) = 0 , \qquad D_{ij}(k) = i\frac{\delta_{ij} - \text{v.p.}\frac{k_i k_j}{k_0^2}}{k^2 + i0} . \qquad (11.9)$$

After this, the cancellation of the real and virtual collinear divergences holds graph by graph. In particular, the real collinear divergences brought about by the emission of a hard collinear gluon from the outgoing quark line cancel the virtual collinear divergences associated with the external line correction as is shown in Fig. 11.2 (as usual, the dashed blob emphasizes that such a cancellation holds not only for the process $e^+e^- \to \bar{q}q$ we are discussing now, but in a general case). The same cancellation holds for the outgoing antiquark, and all other graphs do not involve collinear divergences. If the imaginary part of any QCD graph is calculated with the Cutkosky rules, cutting the gluon line (11.9) gives the transverse gluon density matrix (11.8), and this is a basic reason why the cancellation in Fig. 11.2 takes place. A formal proof of this is given at the end of the lecture as the solution to the **Problem**.

Thus, we see that the *true* physical cross section for e^+e^--annihilation should involve the emission of hard collinear gluons. To understand it, let us ask ourselves how to physically distinguish a hard quark and a state where the energy and momentum are split between a quark and a hard gluon moving almost in the same direction. After a little meditation, we find out that it is very difficult, and if the directions of the outgoing particles coincide — just impossible. Indeed, quarks and gluons are not observed directly but only via a set of hadrons in the final state they give birth to. An energetic quark creates a *jet* of hadrons moving roughly in the direction of the quark as shown in Fig. 0.1, and exactly the same jet is produced by the system of a quark and a gluon moving in parallel directions with the same total energy.§

One can suggest the following definition of the *jet cross section* which is *infrared stable* (i.e. free from both infrared and collinear singularities). We define the process $e^+e^- \to$ (2 *jets*) as a process where almost all particles carrying the fraction $1 - \epsilon$ or more of the total energy go out within two opposite cones with half-angle δ. The parameters $\epsilon \ll 1$ and $\delta \ll 1$ are the energetic and angular resolutions.

Sterman and Weinberg calculated the fraction of the two-jet events thus defined in the total number of events $e^+e^- \to$ *hadrons* to first order in α_s. To this end, one should add to the elastic cross section $\sigma^{\text{el}}_{e^+e^- \to \bar{q}q}$ the cross section of the process $e^+e^- \to \bar{q}qg$ integrated over the kinematical region where either the energy of the extra gluon does not exceed $\omega_m = \epsilon\sqrt{s}$, or one of the angles θ_{qg}, $\theta_{\bar{q}g}$ is less than 2δ, or both. The result is

$$f(\epsilon, \delta) = 1 - \frac{\alpha_s C_F}{\pi} \left[4\ln(2\epsilon)\ln\delta + 3\ln\delta + \frac{\pi^2}{3} - \frac{7}{4} + O(\epsilon, \delta) \right] + o(\alpha_s) .$$

(11.10)

Naturally, the better the resolution and the smaller ϵ and/or δ are, the less fraction $f(\epsilon, \delta)$ of the events which fit out jet definition is. Practically, ϵ and δ cannot be made too small. One of the reasons is that, if $\epsilon, \delta \lesssim \mu_{\text{hadr}}/\sqrt{s}$, nonperturbative effects of fragmentation of quarks and gluons into hadrons become relevant and, even knowing the structure of colored particles jets, one cannot make conclusions on the structure of hadron jets.

We have discussed so far only the collinear divergences associated with external quark lines, but the same reasoning also applies to gluons. A gluon can split into a pair of parallel gluons or into a pair of parallel quark and antiquark. These states are physically indistinguishable from the parent one-gluon state. Collinear divergences cancel in properly defined cross sections by the same mechanism as in Fig. 11.2.

The Sterman–Weinberg formula (11.10) can be generalized to more complicated processes with production of several jets generated by quarks or gluons [26]. Consider, for example, the process $e^+e^- \to \bar{q}qg$ where the final quark, antiquark, and gluon are hard and not parallel to each other, and allow for the emission of an additional gluon which is either soft (with

§Also in the theories without confinement like massless QED such collinear degenerate states cannot be physically distinguished. We will explain why a little bit later.

energy $\leq \omega_m$), or, if hard, forms a small angle $\leq 2\delta_q$ with the quark, or a small angle $\leq 2\delta_{\bar{q}}$ with the antiquark, or a small angle $\leq 2\delta_g$ with the gluon. The result for the jet cross section is

$$
f(\omega_m, \delta_i) =
$$
$$
1 - \frac{\alpha_s}{\pi} \left\{ c_F(R_q + R_{\bar{q}}) + c_V R_g + (c_V - 2c_F) \ln \epsilon \ln \frac{1 - \cos \theta_{\bar{q}q}}{2} \right.
$$
$$
\left. - \frac{c_V}{2} \ln \epsilon \left[\ln \frac{1 - \cos \theta_{qg}}{2} + \ln \frac{1 - \cos \theta_{\bar{q}g}}{2} \right] + O(1) \right\} + o(\alpha_s) ,
$$

$$\tag{11.11}$$

where

$$
R_{q(\bar{q})} = 2 \ln \delta_{q(\bar{q})} \ln \epsilon_{q(\bar{q})} + \frac{3}{2} \ln \delta_{q(\bar{q})} ,
$$
$$
R_g = 2 \ln \delta_g \ln \epsilon_g + \left(\frac{11}{6} - \frac{N_f}{3N_c} \right) \ln \delta_g , \tag{11.12}
$$

ϵ_q is the ratio of ω_m and the energy of the hard quark E_q, δ_q is the half-angle of the cone associated with the quark jet, and $\epsilon_{\bar{q}}$, $\epsilon_g, \delta_{\bar{q}}$, and δ_g are defined analogously.[¶]

The collinear logarithms brought about by the different outgoing jets are factorized in accordance with the discussion above. As $c_V = N_c > c_F$, the gluon jets are broader than the quark ones. As for the purely infrared divergences, they appear in the "cross" contributions also in the physical gauge (11.9). As a result, the logarithms $\sim \ln \epsilon$ in the jet cross section depend on geometry.

The fractions $f(\omega_m, \delta_i)$ are infrared stable to all orders: that is a particular case of the KLN theorem to be discussed in its full form in the last section. An explicit two-loop calculation [27] revealed that the leading logarithmic terms in $f(\omega_m, \delta_i)$ exponentiate. For the process $e^+ e^- \to \bar{q}q$, we

---

[¶]The reader has noticed, of course, that the coefficient of $\ln \delta_g$ in R_g is proportional to the coefficient b_0 in the β-function [see Eq. (9.27)]. We only mention here that it is not an accidental coincidence: in the temporal gauge, the charge renormalization to leading order happens to be determined by the graphs in Fig. 9.1a,d. The same graphs determine R_g and there is actually a close relationship between the ultraviolet and collinear divergences ... We will say few more words about it when discussing the evolution of the quark and gluon fragmentation functions in the next section.

have

$$f(\omega_m, \delta_i) = \exp\left\{-\frac{4\alpha_s c_F}{\pi} \ln \epsilon \ln \delta\right\} + O(\alpha_s^n L^{2n-1}) .\tag{11.13}$$

There is little doubt that the result (11.13) holds to all orders in α_s, though I do not know how to prove it.

The Sterman–Weinberg definition of jet cross section is not the only possible one. One can alternatively define the process $e^+e^- \to$ (2 jets) as the process where all final particles can be separated into two groups, with invariant mass of each group not exceeding $M^2 \ll s \equiv Q^2$. In that case, the fraction of the two-jet events (to all orders in α_s with account of the leading logarithms $\sim \alpha_s^n L^{2n}$ and next-to-leading ones $\sim \alpha_s^n L^{2n-1}$) just coincides with the square of the expression (11.6) for the quark Sudakov formfactor [25]! This formula works in the region $\mu_{\text{hadr}} \sqrt{Q^2} \ll M^2 \ll Q^2$. The lower restriction for M^2 has the same nature as the restriction $\epsilon, \delta \gg \mu_{\text{hadr}}/\sqrt{Q^2}$ for the Sterman–Weinberg cross section. First, for $M^2 \lesssim \mu_{\text{hadr}} \sqrt{Q^2}$, the condition $\alpha_s(M^2) \ln(Q^2/M^2) \ll 1$ is not fullfilled, which means that the neglection of higher-order terms in Eq. (11.6) is not justified. Second, hadronization effects change M^2 exactly by the amount $\sim \mu_{\text{hadr}} \sqrt{Q^2}$ (imagine adding a soft pion with energy $\sim \mu_{\text{hadr}}$ in the final state). Finally, there are some nonperturbative (and not calculable) QCD corrections estimated to be of order $\sim M^2/(\mu_{\text{hadr}} \sqrt{Q^2})$.

11.3 DIS and KLN.

Let us discuss now the processes with hadrons (at the perturbative level — quarks and gluons) in the final state. The simplest example is the process of deep inelastic eN scattering. To lowest order in α_s, the DIS cross section can be symbolically written as

$$\sigma^{\text{DIS}}(x) = \sum_j q_j(x) \sigma_{\text{quark}}(x) ,\tag{11.14}$$

where

$$x = \frac{Q^2}{(2m_N)(E_e - E'_e)}$$

is the Bjorken scaling variable, $q_j(x)$ is the quark *fragmentation function* — the probability to find in the nucleon a quark of type j carrying the fraction

x of the total momentum of the target in the Breit frame (with $p_q = -p_q'$), and $\sigma_{\text{quark}}(x)$ is the cross section of elastic electron–quark scattering.[||]

In higher orders in α_s, the graphs with gluon loops and also the graphs describing emission of real gluons appear. Both types of graphs involve infrared and collinear divergences. Infrared divergences happen to cancel in the sum of all contributions, but the collinear ones survive in this case.

It is not that the cross section becomes really singular. Quarks inside the nucleon are not free and always carry a certain virtuality which cuts off the angular integrals. But the effect of these collinear "not quite" singularities on σ^{DIS} is rather essential. The quark and gluon fragmentation functions get effectively renormalized, and this renormalization is Q^2-dependent. To first order in α_s, the renormalized quark fragmentation function is

$$q^{(1)}(x, Q^2) = q^{(0)}(x) \left[1 - \frac{\alpha_s C_F}{2\pi} \ln \frac{Q^2}{\mu_{\text{hadr}}^2} \int_0^1 dy \frac{1 + y^2}{1 - y} \right]$$
$$+ \frac{\alpha_s C_F}{2\pi} \ln \frac{Q^2}{\mu_{\text{hadr}}^2} \int_x^1 \frac{dy}{y} \frac{1 + y^2}{1 - y} q^{(0)} \left(\frac{x}{y} \right) , \qquad (11.15)$$

where $q^{(0)}(x)$ is a "primordial" fragmentation function. The first term comes from the renormalization of the external line [see Eq. (11.29) below] and the second term describes the process when a "primordial" quark carrying the fraction x/y of the nucleon momentum splits in a collinear quark + gluon pair, so that the quark with momentum xP^{target}, which is ready to be scattered by the electron, carries the fraction y of the momentum $(x/y)P^{\text{target}}$ of the primordial quark.

The integrals in Eq. (11.15) diverge logarithmically at $y = 1$; this is the good old infrared divergence studied extensively in the previous lecture. For sure, this divergence cancels out of the sum, which is best seen if rewriting Eq. (11.15) as

$$q^{(1)}(x, Q^2) - q^{(0)}(x) \equiv \Delta_q^{(1)} q(x) = \frac{\alpha_s C_F}{2\pi} \ln \frac{Q^2}{\mu_{\text{hadr}}^2} \int_0^1 dy \frac{1 + y^2}{1 - y} B_q(x, y),$$
$$B_q(x, y) \equiv \left[\frac{1}{y} q^{(0)} \left(\frac{x}{y} \right) \theta(y - x) - q^{(0)}(x) \right] . \qquad (11.16)$$

Eqs. (11.15) and (11.16) have a clear probabilistic interpretation. The probability for a quark to split into a collinear pair of quark and gluon carrying

---

[||] We do not go into details of the DIS kinematics here referring the reader to the textbooks [5; 6] and the monography [28].

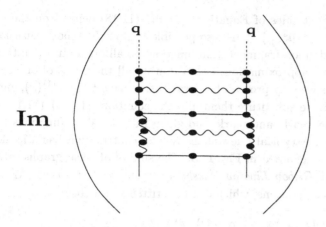

Fig. 11.3 A typical leading-order contribution in the deep inelastic scattering cross section. The "blobbed" vertices and the propagators involve corrections due to renormalization.

the fractions y and $1 - y$ of the initial momentum, respectively, is proportional to $(1 + y^2)/(1 - y)$, $(y \neq 1)$. Note that $\int_0^1 q^{(1)}(x)dx = \int_0^1 q^{(0)}(x)dx$: the total probability to find a quark in the nucleon is the same before and after splitting.

Actually, Eq. (11.15) is not complete. Besides quarks, there are also gluons in the nucleon. A primordial gluon can split into a collinear quark–antiquark pair, after which one of the quarks would be scattered on the external electromagnetic source. It brings about an additional term

$$
\begin{aligned}
\Delta_g q(x) &= \Delta_g \bar{q}(x) \\
&= \frac{\alpha_s}{4\pi} \ln \frac{Q^2}{\mu_{\text{hadr}}^2} \int_x^1 \frac{dy}{y} \left[y^2 + (1 - y)^2 \right] g^{(0)} \left(\frac{x}{y} \right) , \quad (11.17)
\end{aligned}
$$

where the kernel is interpreted as the probability for a gluon to split into a $\bar{q}q$ pair. The gluon fragmentation functions $g(x)$ are also modified due to the splitting processes $g \to gg$ and $q(\bar{q}) \to q(\bar{q})g$. In particular,

$$
\Delta_g g(x) = \frac{\alpha_s c_V}{\pi} \ln \frac{Q^2}{\mu_{\text{hadr}}^2} \int_0^1 dy \left[\frac{y}{1 - y} + \frac{1 - y}{y} + y(1 - y) \right] B_g(x, y) ,
$$

$$
(11.18)
$$

with $B_g(x, y)$ being defined in the same way as in Eq. (11.16).

The right sides of Equations (11.16)–(11.18) depend on the combination $\alpha_s \ln(Q^2/\mu_{\text{hadr}}^2)$. It is also possible to find the renormalization of the quark and gluon fragmentation functions to all orders in α_s in the leading logarithmic approximation, i.e. to sum up all the terms of order $\sim \alpha_s^n L^n$. To do this we may proceed by iterations. Once $q^{(1)}(x)$, $\bar{q}^{(1)}(x)$, and $g^{(1)}(x)$ are found, we substitute them into the equations (11.16)–(11.18) again to obtain the quark, antiquark, and gluon fragmentation functions to second order, etc. A typical relevant many-loop contribution has a ladder structure and is depicted in Fig. 11.3. The sum of all such graphs satisfies the system of *Gribov–Lipatov–Dokshitser–Altarelli–Parisi* (DGLAP) integro-differential equations, which can be written in the form

$$Q^2 \frac{dq(x, Q^2)}{dQ^2} = \frac{\alpha_s(Q^2)}{2\pi} \int_0^1 dy \left[P_{qq}(y) B_q(x, y) \right.$$

$$\left. + P_{qg}(y) \frac{1}{y} g^{(0)} \left(\frac{x}{y}\right) \theta(y - x) \right],$$

$$Q^2 \frac{d\bar{q}(x, Q^2)}{dQ^2} = \frac{\alpha_s(Q^2)}{2\pi} \int_0^1 dy \left[P_{qq}(y) B_{\bar{q}}(x, y) \right.$$

$$\left. + P_{qg}(y) \frac{1}{y} g^{(0)} \left(\frac{x}{y}\right) \theta(y - x) \right],$$

$$Q^2 \frac{dg(x, Q^2)}{dQ^2} = \frac{\alpha_s(Q^2)}{2\pi} \int_0^1 dy \left[P_{gg}(y) B_g(x, y) \right.$$

$$\left. + P_{gq}(y) \frac{1}{y} (q^{(0)} + \bar{q}^{(0)}) \left(\frac{x}{y}\right) \theta(y - x) \right], \quad (11.19)$$

where $P_{qq}(y)$ etc. are the splitting probabilities:

$$P_{qq}(y) = P_{gq}(1 - y) = c_F \frac{1 + y^2}{1 - y}, \quad P_{qg}(y) = \frac{1}{2}[y^2 + (1 - y)^2],$$

$$P_{gg}(y) = 2c_V \left[\frac{y}{1 - y} + \frac{1 - y}{y} + y(1 - y) \right]. \quad (11.20)$$

Eq. (11.19) reminds us a lot of Eq. (9.60), which we derived earlier when reproducing by the resummation of relevant graphs the result (9.56), obtained before in a simple way with the renormalization group technique. This is not a coincidence, and the problem of resummation of collinear logarithms in $q(x, Q^2)$ can be reformulated as the problem of resummation of ultraviolet logarithms in the properly defined Green's functions related to the *moments* of the fragmentation function

$$M_n(Q^2) = \int_0^1 q(x, Q^2) x^{n-1} dx , \qquad (11.21)$$

with Q^2 playing the role of the normalization point [cf. the footnote after Eq. (11.12)]. Actually, this last method is more powerful than the purely diagrammatic one. It allows one, for example, to deduce without many tears (which otherwise must be shed when analyzing the "blobs" in Fig. 11.3) that the coupling constant multiplying the integrals on the right side of Eq. (11.19) should run, indeed. Also, generalizations to nonleading orders are much simpler to derive with the renormalization group methods. This is all described in detail in the books [5; 6], where we refer the reader, but the main subject of this particular lecture are *collinear* rather than ultraviolet divergences, and let us stay within this framework.

We see that the DIS cross section involves uncancelled collinear divergences. Being prepared by the discussion at the end of the previous lecture, the reader can guess that the divergences cancel in the total properly defined probabilities involving also absorption of collinear gluons in the initial state. This guess is correct. At the 1-loop level, the singularities in the virtual corrections to the initial external line cancel with those in the absorption probability described by the "conventional" graph like in Fig. 10.5. (We have now only *one* such graph where the gluon is attached to the initial quark line. The integral $\int |M|^2 dV_{\text{ph.}}^{\text{in}}$ should be done with the kinematic restriction that either gluon is soft, or it forms a small angle $\leq 2\delta$ with the initial quark.) The logarithm in the second term in Eq. (11.15) is cancelled if the contribution of the disconnected graph associated with the initial line (the second term in Fig. 10.6) is added.

This cancellation is guaranteed by the general KLN theorem which can be formulated as follows:

Theorem. Consider, together with a given process of (massless) QED or (massless) QCD, the processes where one or several final or initial masless particles are replaced by sets of particles with the same net charge (color), moving nearly in parallels to each other, and whose total momentum is roughly the same as the momentum of the "parent" particle. Also, both the final and initial state may include an arbitrary number of extra soft photons (gluons) or soft e^+e^- ($\bar{q}q$) pairs if the fermions are massless. Infrared and collinear divergences *disappear* in the sum of the probabilities of all such processes integrated over the phase spaces of initial and final particles with a proper kinematical restriction.

A mathematician would not like, of course, words like "roughly", "nearly", or "proper" in the formulation, but we are, however, physicists and the meaning of what has to be done was explained earlier in examples in sufficient details and it is clear, I hope.

Anyway, a theorem should have a proof. The original arguments of Lee and Nauenberg were somewhat sloppy. Weinberg transformed them into a more or less rigorous proof, which is presented in his book [3], where we refer the reader. Weinberg's proof uses actually the philosophy (if not directly the language) of coherent states discussed in the previous lecture. The corresponding wave packets now involve *all* physically almost indistingushable nearly degenerate in energy states as is spelled out in the previous paragraph.

The last comments which we are going to make refer not to QCD, but to the theories involving unconfined massless charged particles, like massless QED. Let us first explain why a massless electron is physically indistinguishable from a collinear pair $e + \gamma$ with the same net momentum. Basically, to *detect* an electron means to scatter it on some external field. It is obvious, however, that an electron can emit a hard collinear bremsstrahlung photon before being scattered, so that the energy of the scattered electron would be much less than that of the initial one, and one can never say whether the hard collinear photon was emitted during the scattering, or it preexisted as a component of a collinear pair while the other component scattered without bremsstrahlung. Also, if the energy of the scattered electron is large, it is never possible to tell whether it was the primordial energetic electron which scattered without bremsstrahlung or we had originally a less energetic electron which absorbed a hard collinear photon before being scattered. If you think more about it, you will find out that the very notion of asymptotic states is ill-defined in massless QED, and only coherent superpositions of nearly degenerate states specified above have physical meaning.

The cross section of deep inelastic scattering in QCD involves collinear logarithms $\ln(Q^2/\mu_{\text{hadr}}^2)$. This is quite correct; the KLN probability, where these logarithms do not appear, does not have a particular physical meaning here. But what happens in massless QED where the cross section of the scattering of an asymptotic on-mass-shell electron state seems to be singular? The answer is that the electron *never* goes on the mass shell. As was discussed in the previous section, an electron just produced in a hard process is "young", it has not yet grown its electromagnetic coat and it does not interact with other particles in the same way as its elder comrades. The characteristic electron virtuality (or, in other language, the width of the wave packet) decreases with time as $(p^2)_{\text{eff}} \sim E/t$. This means that in massless QED we should substitute $\ln(E^2/m^2) \to \ln(Et)$ in all formulae.**

Interesting and nontrivial physics (not studied yet with proper care!) starts in the region $(\alpha/\pi)\ln(Et) \gtrsim 1$. The corresponding characteristic times and distances rapidly grow when α is decreased. For $\alpha = 1/137$ they would be well beyond the scale of the Universe.

Problem. By an explicit calculation check that the collinear divergences cancel in the sum of the contributions in Fig. 11.2.

Solution. Let us analyze the real contribution first. The amplitude of the emission of an extra gluon is $M_\mu^a(\to qg) = -g\bar{u}(p)t^a\gamma_\mu(\not{p}+\not{k})X/(2pk)$, and we have

$$|M_{\to qg}|^2 = \frac{g^2 c_F}{4(pk)^2}\text{Tr}\left|\{(\not{p}+\not{k})\gamma_\nu\not{p}\gamma_\mu(\not{p}+\not{k})X\bar{X}\}\rho_{\mu\nu}(k)\right., \quad (11.22)$$

where X is a nonsingular part of the amplitude and $\rho_{\mu\nu}(k)$ is the transverse density matrix (11.8), which is convenient to write in the form

$$\rho_{\mu\nu} = -\eta_{\mu\nu} + \frac{k_\mu\delta_{\nu 0} + k_\nu\delta_{\mu 0}}{|k|} - \frac{k_\mu k_\nu}{k^2}. \quad (11.23)$$

We are only interested in the terms which are singular in the collinear limit $\theta_{pk} \to 0$. Then the third term in Eq. (11.23) does not contribute at all, and the contributions of the first and the second terms are

$$|M_{\to qg}|_{\text{I}}^2 \sim \frac{g^2 c_F}{p\cdot k}\text{Tr}\{\not{k}X\bar{X}\},$$

$$|M_{\to qg}|_{\text{II}}^2 \sim \frac{2g^2 c_F p_0}{|k|p\cdot k}\text{Tr}\{(\not{p}+\not{k})X\bar{X}\}. \quad (11.24)$$

**We should do it also in the conventional QED whenever a characteristic time between subsequent collisions is less than the *formation time* $t_{\text{form}} \sim E/m^2$. This is the famous *Landau–Pomeranchuk–Migdal* effect.

We thus obtain

$$|M_{\to qg}|^2 \sim \left[\beta + \frac{2(1-\beta)}{\beta}\right] \frac{g^2 c_F}{p \cdot k} |M^{(0)}|^2 , \qquad (11.25)$$

where $\beta = |k|/|p + k|$ is the fraction of momentum carried away by the gluon. Next, we have to integrate it over phase space involving the factor

$$dV_{\text{ph}}^{qg} = \frac{dp}{2|p|(2\pi)^3} \frac{dk}{2|k|(2\pi)^3} . \qquad (11.26)$$

Introducing

$$k = \beta P + p_\perp, \qquad k_0 \approx \beta|P| + \frac{p_\perp^2}{2\beta|P|} ,$$

$$p = (1-\beta)P - p_\perp, \qquad p_0 \approx (1-\beta)|P| + \frac{p_\perp^2}{2(1-\beta)|P|} , \qquad (11.27)$$

with $p_\perp P = 0$, we find

$$dpdk = dP|P|d\beta dp_\perp , \quad \text{and} \quad pk \approx \frac{p_\perp^2}{2\beta(1-\beta)} . \qquad (11.28)$$

Then

$$\int |M_{\to qg}|^2 dV_{\text{ph}}^{qg} \sim \frac{g^2 c_F}{8\pi^2} \int \frac{dp_\perp^2}{p_\perp^2} \int_0^1 d\beta \left[\beta + \frac{2(1-\beta)}{\beta}\right] \int |M^{(0)}|^2 dV_{\text{ph}}^q$$

$$= -\frac{\alpha_s c_F}{\pi} \ln\eta \int_0^1 dy \frac{1+y^2}{1-y} d\sigma^{\text{el}} = \frac{2\alpha_s c_F}{\pi} \left[\ln\eta \ln\lambda + \frac{3}{4}\ln\eta\right] d\sigma^{\text{el}} , \qquad (11.29)$$

where $y = 1 - \beta$, η and λ are the collinear and infrared cutoffs, and $dV_{\text{ph}}^q = dP/(16\pi^3|P|)$.

Now, consider the virtual contribution. When k is not soft, no particularly simple formula for $M^{(0)}(p; k, -k)$ is known to us, and the representation (10.4) is useless. The virtual external line correction involves an uncertainty $0/0$. To resolve it, we should first assume the external line to have a nonzero virtuality p^2 and take the limit $p^2 \to 0$. The extra factor $1/2$ coming from the fact that we are dealing here not with the quark propagator, but with the quark external line should not be forgotten.[††] We

---

[††]According to the reduction formula (6.2), each external line brings about the factor

$$\sqrt{Z} = \sqrt{1 + C\alpha_s + \dots} = 1 + \frac{1}{2}C\alpha_s + \dots$$

obtain

$$d\sigma^{\text{virt}} = \lim_{p^2 \to 0} \frac{ig^2 c_F}{p^2} \times$$

$$\int \frac{d^4 k \, \rho_{\mu\nu}(k)}{(2\pi)^4 (k^2 + i0)[(k-p)^2 + i0]} \text{Tr} \left\{ p\!\!\!/ \gamma_\mu (p\!\!\!/ - k\!\!\!/) \gamma_\nu p\!\!\!/ X \bar{X} \right\} , \quad (11.30)$$

where $\rho_{\mu\nu}(k)$ coincides with Eq. (11.8) [again, the form (11.23) is more convenient] up to an irrelevant change $|k| \to k_0$ in the denominators.

We perform the integral over dk_0 closing the contour in the lower half-plane and picking up the contribution of *two* poles: $k_0 = |k| - i0$ and $k_0 = p_0 + |p - k| - i0$. The result is

$$d\sigma^{\text{virt}} = \lim_{p^2 \to 0} \frac{g^2 c_F}{16\pi^3 p^2} \int dk \, \text{Tr}\{p\!\!\!/ \gamma_\mu (p\!\!\!/ - k\!\!\!/) \gamma_\nu p\!\!\!/ X \bar{X}\} \rho_{\mu\nu}(k) \times$$

$$\left[\frac{1}{|k|(p^2 - 2p \cdot k)} \Big|_{k^2 = 0} - \frac{1}{|p - k|[p^2 + 2p \cdot (k - p)]} \Big|_{(k-p)^2 = 0} \right]. \quad (11.31)$$

The contribution of each pole is singular $\sim 1/p^2$ when we go on the mass shell and each integral diverges linearly at large momenta. But the sum of two contributions involves no such singularity (a massless quark stays massless after renormalization), neither the divergence. We can therefore expand

$$\frac{1}{-2p \cdot k + p^2} = -\frac{1}{2p \cdot k} - \frac{p^2}{4(p \cdot k)^2} + \cdots ,$$

$$\frac{1}{2p \cdot (k - p) + p^2} = \frac{1}{2p \cdot (k - p)} - \frac{p^2}{4[p \cdot (k - p)]^2} + \cdots \quad (11.32)$$

and drop the leading terms. Picking out only collinear divergent pieces, which come from the kinematic region $k = \beta p$, with $\beta \in (0, \infty)$ in the first integral and $\beta \in (1, \infty)$ in the second integral, we arrive at the expression (11.29), but with the negative sign.

Remark. When we sum up $d\sigma^{\text{virt}} + d\sigma^{\text{real}}$, the collinear cutoff η is substituted by the device angular resolution δ and infrared cutoff λ by

in the exact amplitude, where Z is the residue of the corresponding propagator at the pole. It is instructive also to look at Eq. (11.4): when $k = -q$, the second term in the square brackets contribute as much as the first. To reproduce $M_{\mu\nu}(k, -k)$ for small k correctly, we should take the limit $M^2 \to 0$ of Eq. (11.4) and divide the result by 2.

$\omega_m/E_q = 2\epsilon$. Adding the contributions from the external quark and external antiquark lines, we are reproducing thereby the terms $\sim \ln \delta$ and $\sim \ln \delta \ln(2\epsilon)$ in the Sterman–Weinberg formula (11.10).

PART 3

NONPERTURBATIVE QCD

Lecture 12
Symmetries: Anomalous and Not

Symmetry is probably the most fundamental, rich, and important notion in physics (as well as in mathematics). We start by briefly recalling some generalities.

Suppose the action of a classical field theory is invariant with respect to some global transformation. This means that, under an infinitesimal transformation characterized by a set of small x-independent parameters α^i, the Lagrangian density $\mathcal{L}$ is either invariant or changes by a total divergence $\delta_\alpha \mathcal{L} = \alpha^i \partial_\mu f^{i\mu}$. Renowned Noether's theorem then implies the existence of a set of conserved currents $J^{i\mu}$, which are extracted from the variation of the action under the same transformation, but with slowly varying parameters:

$$ J^{i\mu} = \frac{\delta \mathcal{L}}{\delta(\partial_\mu \alpha^i)} - f^{i\mu} . \tag{12.1} $$

We have $\partial_\mu J^{i\mu} = 0$ on any classical trajectory, and the corresponding charges $Q^i = \int J^{i0}(x) d^3 x$ are integrals of motion.

If the Q^i do not depend explicitly on time, the conditions $\dot{Q}^i = 0$ imply that, in the Hamiltonian formalism, the Poisson brackets $\{Q^i, H\}_{P.B.}$ of the charges with the Hamiltonian vanish. For quantum theory, this implies that the commutators of the corresponding operators $[\hat{Q}^i, \hat{H}]$ vanish, too.* And that means that many energy levels of the Hamiltonian are degenerate (the degeneracy occurs for the levels whose wave functions are not invariant under the symmetry transformations and on which the generators of such

---

*Strictly speaking, a direct classical counterpart of the quantum commutator is not the Poisson bracket, but rather the so called *Moyal bracket* [29]. This distinction has no relevance for QCD and will not bother us here.

transformations $\hat{Q}^i$ act nontrivially).

A simple illustrative example is provided by the theory (4.6) (the *unconstrained* version thereof) discussed earlier. The Lagrangian (4.10) (with $\lambda \equiv 0$) is invariant under the $O(2)$ rotations $x_i' = O_{ij}(\chi)x_j$ with constant χ. The corresponding conserved charge is $p_\phi = x\dot{y} - y\dot{x}$. In the quantum theory, it is upgraded to the angular momentum operator $\hat{p}_\phi = x\hat{p}_y - y\hat{p}_x$, which commutes with the Hamiltonian. The energy levels of the Hamiltonian with positive and negative eigenvalues m of $\hat{p}_\phi$ (and a given radial quantum number n_r) are degenerate: $E_m = E_{-m}$.

We hasten to comment that not all symmetries relevant to field theory applications are as simple as that. A well-known example for a somewhat less trivial symmetry is Lorentz symmetry. Conserved charges corresponding to Lorentz boosts are $M_{0i} = tP_i - x_iH$. They explicitly depend on time, which results in a nonzero commutator $[M_{0i}, H] = -iP_i$. In practice, this means that a Lorentz boost relates states with different energies (like an electron at rest to a moving electron). Lorentz symmetry belongs to the class of so-called *dynamical* symmetries.

Let us first say a few words about gauge symmetry discussed at length in the previous lectures. We want to emphasize here that it is actually *not* a symmetry in the same sense as rotational or Lorentz symmetry. Namely, in the case of gauge symmetry, we are not allowed to consider states which are not invariant under symmetry transformations. The constraint $\hat{G}^a\Psi_n = 0$ dictates that all physical states Ψ_n are gauge singlets. Gauge symmetry simply does not act on the Hilbert space of physical states, it exhibits itself only in the Lagrangian formulation, involving some extra unphysical variables which can be disposed of, in principle. In other words: *gauge symmetry is not a symmetry, but rather a convenient way to describe constrained systems.*

What *are* the true symmetries of QCD? QCD is a relativistic field theory and its Lagrangian is invariant with respect to the Poincare group (involving rotations, Lorentz boosts, spatial and time translations). There is nothing new, compared to, say, $\lambda\phi^4$ theory or QED, and we will not discuss it here. Relevant and interesting symmetries, which constitute the subject of this lecture and which are specific to QCD and to some its "relatives", are *conformal symmetry* and *chiral symmetries*.

12.1 Conformal Symmetry and its Breaking.

As was already mentioned before, the bare Lagrangian (1.17) of pure Yang–Mills theory and also the bare Lagrangian (1.19) of QCD with strictly massless quarks involve no scale. That means that the classical action is invariant under the transformations

$$\left\{ \begin{array}{l} x^\mu \to \lambda x^\mu \\ A_\mu \to \lambda^{-1} A_\mu \\ \psi \to \lambda^{-3/2} \psi \end{array} \right. . \qquad (12.2)$$

Here we took into account the fact that the gauge field A_μ has the canonical dimension of mass and the canonical dimension of the quark field is $[\psi] = m^{3/2}$ (the canonical field dimensions are found by requiring that the dimension of the Lagrangian is equal to 4, so that the action is dimensionless).

For pure gauge theory, a somewhat stronger form of the scaling symmetry (12.2) holds. Consider the action of Yang–Mills theory on a curved four-dimensional manifold

$$S^{\rm YM}_{\rm curved} = -\frac{1}{2g_0^2} {\rm Tr} \int d^4 x \sqrt{-\det\|g\|}\ F_{\mu\nu} F_{\alpha\beta} g^{\alpha\mu} g^{\beta\nu}\ , \qquad (12.3)$$

where g_0^2 is the coupling constant, not to be confused with the metric tensor $g_{\mu\nu}(x)$.[†] Note now that the action (12.3) is invariant under the local conformal transformations

$$\left\{ \begin{array}{l} g_{\mu\nu}(x) \to \lambda^{-2}(x) g_{\mu\nu}(x) \\ g^{\mu\nu}(x) \to \lambda^{2}(x) g^{\mu\nu}(x) \end{array} \right. , \qquad (12.4)$$

while the fields and coordinates are not transformed. For a flat metric $g_{\mu\nu}(x) = \eta_{\mu\nu}$ and $\lambda(x) = \lambda$, the transformation (12.4) amounts to a homogeneous scale dilatation and is equivalent to (12.2). Local conformal invariance is specific to gauge theories. For example, $\lambda\phi^4$ theory with the Lagrangian (6.1) is invariant under global scale transformations, but no locally conformal invariant generalization to curved manifolds exists in that case.

[†]We assume that the reader is familiar with the basic notions of Riemannian geometry. Eq. (12.3) is an obvious non-Abelian generalization for the action of an electromagnetic field on a curved background (see e.g. Ref. [10]).

As far as Yang–Mills theory is concerned, the symmetry (12.4) has an important consequence:

$$\frac{\delta S^{\text{YM}}_{\text{curved}}}{\delta \lambda(x)} = -2g_{\mu\nu}(x)\frac{\delta S^{\text{YM}}_{\text{curved}}}{\delta g_{\mu\nu}(x)} = -\Theta^{\mu}_{\mu}(x)\sqrt{-\det\|g\|} = 0 \ , \qquad (12.5)$$

where $\Theta^{\mu\nu}(x)$ is (the symmetric version of) the energy–momentum tensor:

$$\Theta^{\mu\nu} = \frac{2}{g_0^2}\text{Tr}\left[-F^{\mu\rho}F^{\nu}_{\rho} + \frac{g^{\mu\nu}}{4}F_{\rho\sigma}F^{\rho\sigma}\right] \ . \qquad (12.6)$$

Having obtained Eq. (12.5), we can now forget about its derivation and set $g_{\mu\nu} \equiv \eta_{\mu\nu}$: the identity (12.5) will also hold for the theory on the flat background we are primarily interested in. Eq. (12.5) can be interpreted as a local conservation law of the dilatation current

$$J^{\mu}_D = x_{\nu}\Theta^{\mu\nu} \qquad (12.7)$$

($\partial_{\mu}J^{\mu}_D = \Theta^{\mu}_{\mu} + x_{\nu}\partial_{\mu}\Theta^{\mu\nu} = 0$). The expression (12.7) can also be derived without going to curved space, just by a direct application of Noether's theorem. Let us derive this for pure photodynamics allowing for the reader to work it out himself in the theory with matter as well as in the non-Abelian case, where the calculations are more cumbersome.

Let us first recall that the *canonical* energy–momentum tensor of the electromagnetic field, which is obtained by Noether's method exploiting the symmetry of the Lagrangian with respect to time and spatial translations, does not coincide with the Abelian version of Eq. (12.6) and has the form

$$(\Theta^{\mu\nu})^{\text{can}} = \frac{\delta\mathcal{L}}{\delta(\partial_{\mu}A_{\rho})}\partial^{\nu}A_{\rho} - \eta^{\mu\nu}\mathcal{L} = -F^{\mu\rho}\partial^{\nu}A_{\rho} + \frac{\eta^{\mu\nu}}{4}F_{\rho\sigma}F^{\rho\sigma}. \qquad (12.8)$$

In contrast to Eq. (12.6), it is not symmetric with respect to the indices μ, ν (*à propos*, it not gauge invariant either). For fields satisfying the classical equations of motion, it differs from (12.6) by a term $\partial_{\rho}B^{\mu\rho\nu}$ where $B^{\mu\rho\nu} = -A^{\nu}F^{\mu\rho}$ is antisymmetric in the indices μ, ρ. As $\partial_{\mu}\partial_{\rho}B^{\mu\rho\nu} \equiv 0$, both $(\Theta^{\mu\nu})^{\text{can}}$ and $(\Theta^{\mu\nu})^{\text{sym}}$ are conserved.

Let us next derive the canonical expression for the dilatation current for a general theory with Lagrangian $\mathcal{L}(\phi_i, \partial_{\mu}\phi_i)$, depending on some set of bosonic fields $\phi_i(x)$ with canonical dimension 1, and their derivatives. To this end, consider an infinitesimal scale transformation

$$\begin{cases} \delta x^\mu = \alpha x^\mu \\ \delta \phi_i = -\alpha \phi_i \end{cases} \tag{12.9}$$

($\alpha \ll 1$) and, by a change of variables, rewrite this in a form where only the fields, but not the coordinates, are transformed. We have

$$\delta \phi_i = \alpha(-x^\nu \partial_\nu \phi_i - \phi_i) \quad \overset{\alpha\,=\,\mathrm{const}}{\Longrightarrow}$$
$$\delta(\partial_\mu \phi_i) = \alpha(-x^\nu \partial_\nu \partial_\mu \phi_i - 2\partial_\mu \phi_i) \,, \tag{12.10}$$

where the first term in $\delta\phi_i(x)$, $\delta[\partial_\mu\phi_i(x)]$ comes from the shift of argument due to the change of variables $x_\mu^{\mathrm{old}} = x_\mu^{\mathrm{new}}(1-\alpha)$ [the sign of shift is opposite compared to that in Eq. (12.9)], and the second term reflects the canonical dimension 1 of the fields ϕ_i and the canonical dimension 2 of the fields $\partial_\mu\phi_i$. The variation of the Lagrangian under the global transformation (12.10) is

$$\delta\mathcal{L} = -\alpha\left\{ x^\nu \partial_\nu \mathcal{L} + \sum_i \left[\frac{\delta\mathcal{L}}{\delta\phi_i}\phi_i + 2\frac{\delta\mathcal{L}}{\delta(\partial_\mu\phi_i)}\partial_\mu\phi_i \right] \right\} \,. \tag{12.11}$$

For a scale invariant Lagrangian, the second term is just $-4\alpha\mathcal{L}$ (4 is the canonical dimension of $\mathcal{L}$) and the variation gives a total derivative $\delta\mathcal{L}_{\mathrm{sc.inv.}}$ $= -\alpha\partial_\nu(x^\nu \mathcal{L}_{\mathrm{sc.inv.}})$. Assuming now that the parameters $\alpha(x)$ are not constant and calculating the canonical Noether current (12.1), we obtain, up to an irrelevant sign

$$(J_D^\mu)^{\mathrm{can}} = \sum_i \phi_i \frac{\delta\mathcal{L}}{\delta(\partial_\mu\phi_i)} + x^\nu(\Theta_\nu^\mu)^{\mathrm{can}} \,, \tag{12.12}$$

where

$$(\Theta_\nu^\mu)^{\mathrm{can}} = \frac{\delta\mathcal{L}}{\delta(\partial_\mu\phi_i)}\partial_\nu\phi_i - \delta_\nu^\mu \mathcal{L}$$

is the canonical energy–momentum tensor. For photodynamics the expression (12.12) differs from $x^\nu(\Theta_\nu^\mu)^{\mathrm{sym}}$ by the term

$$(J_D^\mu)^{\mathrm{can}} - x^\nu(\Theta_\nu^\mu)^{\mathrm{sym}} = -F^{\mu\rho}A_\rho - x^\nu F^{\mu\rho}\partial_\rho A_\nu \,. \tag{12.13}$$

Taking into account the equations of motion $\partial_\rho F^{\mu\rho} = 0$, we obtain $\Delta J_D^\mu = -\partial_\rho(x^\nu F^{\mu\rho}A_\nu)$, i.e. the canonical dilatation current (12.12) coincides with the current (12.7) up to a term whose divergence is zero. Thus, one can use the current (12.7) instead of (12.12) in all calculations. The deeper reason for this equivalence is again the local conformal invariance of the action (12.3) in a curved background.

For a theory involving also massless quarks, an extra term

$$(\Theta_\mu^\mu)^q = -3i \sum_f \bar\psi_f \slashed{D} \psi_f \tag{12.14}$$

appears in the trace.[‡] It is not zero off-shell [because a theory with quarks does not obey the local conformal invariance (12.4)], but it is still zero for fields satisfying the classical equations of motion $\slashed{D}\psi_f = 0$. For massive quarks, scale invariance is lost and $\Theta_\mu^\mu \sim \sum_f m_f \bar\psi_f \psi_f \neq 0$.

Up to now, we were only discussing the classical theory. What happens in the quantum case? One may ask whether conformal symmetry is still present and, in particular, whether the trace of the quantum *operator* corresponding to the classical expression (12.6) for the energy–momentum tensor still vanishes?

First of all, note that the neglection of terms which vanish due to the classical equations of motion is quite justified. The counterpart of the latter in quantum theory are the Heisenberg operator equations of motion which tell us that all the matrix elements of operators like $(i\slashed{D} - m)\psi$ between the physical states vanish. Still, the answer to the question above is *negative*. Actually, we have already seen this when discussing the phenomenon of dimensional transmutation in Lecture 9. Quantum Yang–Mills theory (and QCD) involves an intrinsic mass scale — the scale where the running coupling constant is of order 1. This scale determines the masses of all physical hadron states, etc. At the formal level: even though the action is invariant under scale transformations, the *regularized* path integral (4.34) no longer is, because a finite lattice introduces a scale $\Lambda_0 = a^{-1}$. Combined with the bare coupling constant according to Eq. (9.40), this leads to the physical mass scale $\Lambda_{\rm QCD}$.

The absence of scale invariance at the quantum level results in that the trace $\hat\Theta_\mu^\mu$ of the Heisenberg operator describing the energy-momentum tensor is not zero anymore. Accurately defining what the operator $\hat\Theta^{\mu\nu}$

[‡]When deriving this by canonical methods, do not forget that $\partial_\rho F^{\mu\rho}$ ($D_\rho F^{\mu\rho}$ in the non-Abelian case) is no longer zero and the expression (12.8) is no longer equivalent to (12.6), but involves an extra term with the fermion current.

actually means, one can derive [§]

$$\hat{\Theta}^{\mu}_{\mu} = \frac{\beta(g^2)}{2g^4} \text{Tr}\{F_{\mu\nu}F^{\mu\nu}\} . \tag{12.15}$$

This is the renowned *conformal anomaly*. It means that the classically conserved dilatation current (12.7) is no longer conserved in the full quantum theory. Conformal symmetry is broken explicitly by quantum effects. Eq. (12.15) is an operator equality, i.e. all matrix elements on the left and right hand sides, taken between some physical states, coincide.

We will derive Eq. (12.15) with path integral methods. Consider a regularized Euclidean Yang–Mills path integral depending on an ultraviolet regulator Λ_0 and a bare coupling constant g_0^2. The scale transformation affects only the regulator: $\delta\Lambda_0 = -\alpha\Lambda_0$. Due to Eq. (9.40), this results in the same change of all physical mass scales $\sim \Lambda_{\text{QCD}}$ as the shift of the bare coupling constant $\delta g_0^2 = \alpha\beta(g_0^2)$, with Λ_0 kept fixed. The corresponding modification of the integrand

$$\exp\{-S^E\} = \exp\left\{-(1/2g_0^2)\int \text{Tr}\{F_{\mu\nu}F_{\mu\nu}\}d^4x\right\}$$

in the path integral is then described by the shift

$$\delta S^E = -\frac{\alpha\beta(g_0^2)}{g_0^2}S^E . \tag{12.16}$$

On the other hand, $\delta S^E/\delta\alpha = -\int \Theta^{\mu}_{\mu}d^4x$. Comparing this with Eq. (12.16), we derive [¶].

$$\int \hat{\Theta}^{\mu}_{\mu}d^4x = \frac{\beta(g_0^2)}{2g_0^4}\int \text{Tr}\{F_{\mu\nu}F^{\mu\nu}\}d^4x . \tag{12.17}$$

[§]The expression more often quoted in the literature involves g^2 rather than g^4 in the denominator. This is just due to a different normalization of the fields A_μ chosen.

[¶]A note for pundits: in supersymmetric gauge theories, the most natural definition of the Wilsonian β-function entering Eq. (12.17) is such that it involves only the leading term and the higher-order corrections vanish (see the footnote on p. 135). This makes the expression for the conformal anomaly (12.15) similar to that of the *chiral* anomaly, Eq. (12.21), which we will discuss below . (The coefficient in Eq. (12.21) has a geometric interpretation and does not involve a series in g^2.) This is not accidental. In supersymmetric theories the chiral current and the dilatation current belong to the same supermultiplet and their anomalies are intimately related to each other.

Strictly speaking, this alone does not guarantee that the corresponding *integrands* also coincide. To show this, one has to find the variation of the path integral under a *local* scale transformation (12.4). We should imagine a lattice whose spacing (in physical units) depends on x: $a(x) = a_0[1 + \alpha(x)]$. Somewhat heuristically,[||] we might say that a lattice with small x-dependent spacing and a constant coupling g_0^2 describes the same physics as a lattice with constant spacing and x-dependent coupling constant $g^2(x) = g_0^2 + \alpha(x)\beta(g_0^2)$, on distance scales which are much larger than $a(x)$. Thereby, the variation of the action is

$$\delta S^E = \frac{\beta(g_0^2)}{2g_0^4} \int \alpha(x) \, \mathrm{Tr}\{F_{\mu\nu} F^{\mu\nu}\} d^4 x \; . \tag{12.18}$$

Varying it over $\alpha(x)$, we derive Eq. (12.15).

This could also be derived by operator methods, which we will tackle here, but illustrate in the following section devoted to

12.2 Anomalous Chiral Symmetry.

Consider Yang–Mills theory with just one massless quark. The term $i\bar{\psi}\slashed{D}\psi$ in the Lagrangian is invariant under global chiral transformations[**]

$$\delta\psi = -i\alpha\gamma^5\psi, \quad \delta\bar{\psi} = -i\alpha\bar{\psi}\gamma^5 \; . \tag{12.19}$$

The corresponding canonical Noether current,

$$j^{\mu 5} = \bar{\psi}\gamma^\mu\gamma^5\psi \; , \tag{12.20}$$

is conserved upon applying the equations of motion.

An important fact is that the symmetry (12.19) exists only in the classical case. The full quantum path integral is *not* invariant under the transformations (12.19). Like it was also the case with the conformal symmetry, this symmetry breaking due to quantum effects can be expressed as an

---

[||] We do not know whether an accurate derivation exists. One of the problems here is that the space with an x-dependent scale is no longer flat, since the scale transformation (12.4) of the flat metric brings about nonzero curvature.

[**] It is also invariant under the symmetry $\delta\psi = i\beta\psi$, but it is just a special case of gauge symmetry which, as was mentioned, is not a symmetry and is not a subject of this lecture.

operator identity involving an anomalous divergence:

$$\partial_\mu j^{\mu 5} = -\frac{1}{16\pi^2}\epsilon^{\alpha\beta\mu\nu}\text{Tr}\{F_{\alpha\beta}F_{\mu\nu}\} . \tag{12.21}$$

There are many ways to derive and understand this relation. Historically, it was first derived by purely diagrammatic methods. We will explain how this is done in Lecture 14 when we discuss 't Hooft's self-consistency conditions. Here we will concentrate on two other ways which are more modern and more general. First, we derive (12.21) as an operator equality (we skipped an analogous derivation when discussing the conformal anomaly) and, second, we will show that the path integral measure is actually not invariant under chiral transformations, but is modified in such a way that the relation (12.21) is satisfied.

Let us start with the operator derivation and, for simplicity, first consider the Abelian case. To begin with, we need to define a quantum operator corresponding to the classical axial current (12.20). The problem is that one cannot harmlessly multiply field operators $\psi(x)$ at coinciding points. Indeed, e.g. the vacuum average $\langle\psi(x)\bar\psi(0)\rangle_0$ behaves as $i\not{x}/(2\pi^2 x^4)$ at small x, and the limit $x \to 0$ is singular.

Following Schwinger, we *define* the axial current operator of QED as

$$j^{\mu 5} = \lim_{\epsilon \to 0} \bar\psi(x + \epsilon)\gamma^\mu\gamma^5 \exp\left\{i \int_x^{x+\epsilon} A_\alpha(y)dy^\alpha\right\} \psi(x) . \tag{12.22}$$

The factor $\exp\left\{i \int_x^{x+\epsilon} A_\alpha(y)dy^\alpha\right\}$ is very important and makes the expression gauge invariant, in spite of the different arguments of $\bar\psi(x + \epsilon)$ and $\psi(x)$ [see Eq. (1.5)]. The limit $\epsilon \to 0$ is taken, assuming averaging over the directions of ϵ^μ (otherwise the current (12.22) would not be a Lorentz vector). Expanding the exponential up to terms $O(\epsilon)$, differentiating the whole expression (12.22) with respect to x^μ, and making use of the operator equations of motion $\not{D}\psi = \gamma^\mu(\partial_\mu - iA_\mu)\psi = 0$, we obtain

$$\partial_\mu j^{\mu 5} = \lim_{\epsilon \to 0} \bar\psi(x + \epsilon)\left[-i\gamma^\mu A_\mu(x + \epsilon) + i\gamma^\mu A_\mu(x)\right.$$
$$\left. +i\epsilon^\nu\gamma^\mu\partial_\mu A_\nu(x)\right]\gamma^5\psi(x) = \lim_{\epsilon \to 0} i\epsilon^\nu F_{\mu\nu}(x)\bar\psi(x + \epsilon)\gamma^\mu\gamma^5\psi(x) . \tag{12.23}$$

Superficially, it seems to be zero in the limit $\epsilon \to 0$. This is not the case, however. Let us average Eq. (12.23) over a state involving a classical background field $A_\mu(x)$. The fermion Green's function is the free Green's function plus the term describing one insertion of the external field plus terms

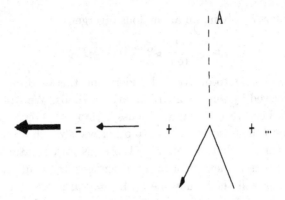

Fig. 12.1   Fermion propagator in an external field.

with multiple insertions (see Fig. 12.1). The free Green's function $\sim \not\!\epsilon/\epsilon^4$ does not contribute in our case because the corresponding spinor trace $\mathrm{Tr}\{\gamma^\alpha\gamma^\mu\gamma^5\}$ is zero in four dimensions. To calculate the graph with one field insertion, it is convenient to choose the fixed point (or Fock–Schwinger) gauge $(y-x)^\alpha A_\alpha(y) = 0$, so that

$$A_\alpha(y) = -\frac{1}{2}(y-x)^\beta F_{\alpha\beta} + o(y-x) \ . \tag{12.24}$$

An explicit calculation (see **Problem 1**) gives

$$\langle\psi(x)\bar\psi(x+\epsilon)\rangle_A = -\frac{i\not\!\epsilon}{2\pi^2\epsilon^4} - \frac{1}{32\pi^2\epsilon^2}F_{\alpha\beta}(\not\!\epsilon\gamma^\alpha\gamma^\beta + \gamma^\alpha\gamma^\beta\not\!\epsilon) + \dots \tag{12.25}$$

Multiplying this by $-i\epsilon^\nu\gamma^\mu\gamma^5 F_{\mu\nu}$ (the extra minus sign comes from the permutation of anticommuting field operators), taking the trace with (N.9), and averaging over the directions $\epsilon^\nu\epsilon_\rho/\epsilon^2 \to \delta^\nu_\rho/4$, we arrive at the result

$$\partial_\mu j^{\mu5} = -\frac{1}{16\pi^2}\epsilon^{\alpha\beta\mu\nu}F_{\alpha\beta}F_{\mu\nu} \ . \tag{12.26}$$

Note that the terms describing multiple insertions in the Green's function are less singular in ϵ, and their contribution to $\partial_\mu j^{\mu5}$ vanishes in the limit $\epsilon \to 0$.

Let us briefly discuss the situation in other even dimensions. In two dimensions, the term with one insertion of the external field in the propagator is $O(\epsilon)$ and does not contribute to the anomaly. The anomaly is still there, however. It appears due to the leading term $i\rlap{/}{\partial}/(2\pi\epsilon^2)$ in the expansion of the fermion Green's function (in two dimension, $\mathrm{Tr}\{\gamma^\mu\gamma^\nu\gamma^5\} \neq 0$!).

A simple calculation with the conventions (N.17) gives

$$\partial_\mu j^{\mu 5} = -\frac{1}{2\pi}\epsilon^{\mu\nu}F_{\mu\nu} \ . \tag{12.27}$$

In six dimensions, both the leading term and the term linear in the external field do not contribute because the corresponding spinor traces vanish. The anomaly is there due to the term with 2 field insertions. For $d = 8$, the term with 3 insertions works, etc. In general, one can derive

$$\partial_\mu j^{\mu 5} = -\frac{2}{n!(4\pi)^n}\epsilon^{\mu_1\cdots\mu_{2n}}F_{\mu_1\mu_2}\cdots F_{\mu_{2n-1}\mu_{2n}} \ , \tag{12.28}$$

where $d = 2n$. For odd dimensions, there is no γ^5 matrix, no axial current and an anomaly of this kind is absent: *no man — no problem* [30].

Let us return to four-dimensional QCD, however. We again define the axial current as in Eq. (12.22) and the whole derivation can be repeated. The only difference is that, for the terms in $\partial_\mu j^{\mu 5}$ coming from differentiating the fermion fields, we have also to take into account the term $\sim O(\epsilon)$ in the expansion of the exponential. That will give

$$\Delta\left(\partial_\mu j^{\mu 5}\right)^{\text{non-ab}} = \lim_{\epsilon\to 0}\epsilon^\nu\bar\psi(x + \epsilon)\gamma^\mu\gamma^5[A_\mu(x), A_\nu(x)]\psi(x) \ ,$$

which, being combined with the other terms, just gives the non-Abelian field strength tensor. Also the relation (12.25) still holds with the full non-Abelian $F_{\alpha\beta}$, which is part of the magics of the fixed point gauge method. The only distinction is that the quark Green's function is now not only a nontrivial matrix in spinor, but also in color indices. Thereby, the result (12.21) is reproduced.

Let us now derive the anomaly relation (12.21) with path integral methods. As we have seen, the anomaly appears due to the necessity to regularize theory in the ultraviolet. The politically most correct approach would be to study a path integral regularized by a lattice. We mentioned, however, that an accurate definition of the fermionic path integral on the lattice is not so easy. We will address this last issue in the next lecture and for the time being instead use the more standard and habitual *finite mode regularization*.

Consider an Euclidean path integral for the partition function for QCD with one massless quark flavor. The fermionic part of the integral is

$$\int \prod_x d\bar{\psi}(x)d\psi(x) \exp\left\{i\int d^4x\, \bar{\psi}\not{D}^E\psi\right\} ,$$

which formally coincides with the determinant of the Dirac operator $-i\not{D}^E$. Let us assume that the theory is somehow regularized in the infrared so that the spectrum of the operator $\not{D}^E$ is discrete. Let us expand $\psi(x)$, $\bar{\psi}(x)$ as in Eq. (4.30). Then

$$\prod_x d\bar{\psi}(x)d\psi(x) \equiv \prod_k d\bar{c}_k dc_k . \tag{12.29}$$

Suppose the field variables are transformed by an infinitesimal global chiral transformation (12.19).[tt]

Now, $\psi' = \psi + \delta\psi$ and $\bar{\psi}' = \bar{\psi} + \delta\bar{\psi}$ can again be expanded in the series (4.30). The new expansion coefficients are related to the old ones:

$$c'_k = c_k - i\alpha\sum_m c_m \int d^4x\, u_k^\dagger(x)\gamma^5 u_m(x) \equiv \sum_m (\delta_{km} - i\alpha A_{km})c_m$$

$$\bar{c}'_k = \bar{c}_k - i\alpha\sum_m \bar{c}_m \int d^4x\, u_m^\dagger(x)\gamma^5 u_k(x) \equiv \sum_m \bar{c}_m(\delta_{km} - i\alpha A_{mk}) .$$

$$\tag{12.30}$$

The point is that the transformation (12.30) has a nonzero Jacobian. We have

$$\prod_k d\bar{c}_k{}' dc'_k = J^{-2}\prod_k d\bar{c}_k dc_k , \tag{12.31}$$

where

$$J = \det(1 - i\alpha A) \approx \exp\left\{-i\alpha\sum_k A_{kk}\right\} \tag{12.32}$$

[tt] As was discussed in Lecture 4, we keep the Euclidean $\bar{\psi}$ and ψ independent and are not worried by the fact that $\delta\bar{\psi} \neq (\delta\psi)^\dagger$. Actually, in continuum theory we could manage somehow to keep the relation $\bar{\psi} \equiv \psi^\dagger$ if we choosed the chiral transformation in the form $\delta\psi = -\alpha\gamma^5\psi$ (without the factor i). But, as was already mentioned in Lecture 4, in a viable lattice version of the fermionic Euclidean path integral, to be discussed in the next lecture, $\bar{\psi}$ and ψ *should* be considered independent.

[the Jacobian appears in the denominator due to the rule (3.41)]. Or in other words

$$\ln J = -i\alpha \int d^4x \sum_k u_k^\dagger(x)\gamma^5 u_k(x) + o(\alpha) . \qquad (12.33)$$

One's first (wrong!) impression might be that $\int d^4x \sum_k u_k^\dagger(x)\gamma^5 u_k(x)$ is just zero. Indeed, as was already mentioned earlier [see the discussion preceding Eq. (4.32)], the function $u_k' = \gamma^5 u_k$ is also an eigenfunction of the Dirac operator with eigenvalue $-\lambda_k$. If $\lambda_k \neq 0$, $u_k(x)$ and $\gamma^5 u_k(x)$ thereby represent *different* eigenfunctions and the integral $\int d^4x\, u_k^\dagger(x)\gamma^5 u_k(x)$ vanishes.

A nonzero value of (12.33) is due to the fact that, for intricate enough *topologically nontrivial* gauge fields, the spectrum of the Dirac operator involves some number of exact zero modes for which $\gamma^5 u_0(x) = \pm u_0(x)$ (depending on whether the modes are right-handed or left-handed) and their contribution in the sum (12.33) is responsible for the whole effect. A famous theorem of Atiyah and Singer dictates

$$\int d^4x \sum_k u_k^\dagger(x)\gamma^5 u_k(x) = n_R^{(0)} - n_L^{(0)} = q , \qquad (12.34)$$

where $n_{R,L}^{(0)}$ is the number of the right-handed (left-handed) zero modes and q is the Pontryagin index (2.6) of the gauge field configuration.

Let us derive it now. As all nonzero modes contribute zero to the sum, we are allowed to consider instead of (12.34) the sum

$$\int d^4x \sum_k u_k^\dagger(x)\gamma^5 u_k(x) e^{-\lambda_k^2/M^2} . \qquad (12.35)$$

This is exactly the finite mode regularization mentioned above: the contribution of modes with large λ_k^2 is suppressed. To calculate the sum, consider a quantum mechanical problem with the Hamiltonian

$$H = (\not{P})^2 = -[\gamma_\mu^E(\hat{p}_\mu - A_\mu)]^2 , \qquad \hat{p}_\mu = -i\partial_\mu$$

(it is a matrix in both spinor and color indices). The sum (12.35) is nothing but $\int d^4x \mathrm{Tr}\{\gamma^5 \mathcal{K}(x, x; 1/M^2)\}$, where $\mathcal{K}$ is the (matrix) evolution operator of our quantum mechanical system with imaginary time $\beta = 1/M^2$. [Cf. Eq. (3.12). Note that the phase space of our system is 8-dimensional, involving the coordinates x_μ, p_μ, and the evolution occurs in some unphysical "fifth" time.] The trace is done over the spinor and color indices. The evolution operator can be expressed as a path integral. Assume now that M is very large, so that β is very small. We are then dealing with an infinitesimal Euclidean evolution operator which can be represented as a finite-dimensional phace space integral

$$\int d^4x \mathrm{Tr}\{\gamma^5 \mathcal{K}(x, x; 1/M^2)\} = \int \frac{d^4x\, d^4p}{(2\pi)^4} \mathrm{Tr}\left\{\gamma^5 e^{-(\mathcal{P})^2/M^2}\right\} . \quad (12.36)$$

Bearing in mind that $(\mathcal{P})^2 = -\mathcal{D}^2 - (i/2)\gamma_\mu^E \gamma_\nu^E F_{\mu\nu}$ and that

$$\mathrm{Tr}\{\gamma^5 \gamma_\mu^E \gamma_\nu^E \gamma_\alpha^E \gamma_\beta^E\} = -4\epsilon_{\mu\nu\alpha\beta}$$

[see the conventions (N.13), (N.15)], and shifting the variable of the integration $p_\mu - A_\mu \to p_\mu$, the integral can be easily calculated to leading order in $1/M^2$ [recall that we are allowed to trade a functional integral for an ordinary integral (12.36) only in the limit $M^2 \to \infty$]. The result exactly coincides with q. Thus, the index theorem (12.34) is proven.[‡‡]

Substituting (12.34) in Eqs. (12.33, 12.31), we now see that the change of the measure under a chiral transformation can be represented as a shift of the effective action

$$\delta S^E = -\frac{i\alpha}{16\pi^2}\epsilon_{\alpha\beta\mu\nu}\int d^4x \mathrm{Tr}\{F_{\alpha\beta}F_{\mu\nu}\} . \quad (12.37)$$

In other words, a global chiral transfomation is equivalent to leaving the fermionic fields intact, but shifting instead the parameter θ in the original theory defined in Eq. (5.30): $\theta \to \theta - 2\alpha$.

Just like in the case of the conformal anomaly, we have studied up to now only the change of the measure under global symmetry transformations. In the chiral symmetry case, it is not too difficult to find out how the measure changes under local transformations with x-dependent parameters $\alpha(x)$. We have

$$\ln J[\alpha(x)] = -i\int d^4x\, \alpha(x) \sum_k u_k^\dagger(x)\gamma^5 u_k(x) + o(\alpha) . \quad (12.38)$$

[‡‡]By the way, the Atiyah–Singer index for the Dirac operator is identical to the Witten index (3.38) of a certain supersymmetric quantum-mechanical system. Our method for deriving Eq. (12.34) coincides with the known derivation of an integral representation for the Witten index due to Cecotti and Girardello. See e.g. [31] for more details.

Repeating all the steps of the above derivation, we find[†]

$$\lim_{M \to \infty} \sum_k u_k^\dagger(x)\gamma^5 u_k(x)e^{-\lambda_k^2/M^2} = \frac{1}{32\pi^2}\epsilon_{\alpha\beta\mu\nu}\text{Tr}\{F_{\alpha\beta}F_{\mu\nu}\}(x) \ ,$$

which gives

$$\delta S^E = -\frac{i}{16\pi^2}\epsilon_{\alpha\beta\mu\nu}\int d^4x \ \alpha(x)\text{Tr}\{F_{\alpha\beta}F_{\mu\nu}\}(x) \ . \qquad (12.39)$$

The shift of the Minkowski action is given by an analogous expression without the prefactor i [cf. Eqs. (5.30) and (5.31)]. Varying it with respect to $\alpha(x)$, we obtain the anomalous divergence of the axial current in accordance with (12.21).

The last comment is that, in real QCD with several light quarks, each flavor contributes on equal footing to the anomaly of the singlet axial current $j^{\mu 5(\text{singl})} = \sum_f \bar\psi_f \gamma^\mu \gamma^5 \psi_f$, and the result (12.21) is just multiplied by N_f.

Problem 1. Using the fixed point gauge (12.24), derive the expression (12.25) for the fermion Green's function.

Solution. With the gauge choice (12.24), the Green's function depends only on the difference of coordinates $x - (x + \epsilon) = -\epsilon$. We can set $x = 0$ and write

$$\langle \psi(0)\bar\psi(\epsilon) \rangle_A = G_0(-\epsilon) + \int d^4u G_0(-u)i\gamma^\alpha A_\alpha(u)G_0(u - \epsilon) \ , \quad (12.40)$$

where $G_0(\epsilon)$ is the free fermion Green's function with the convention (N.6). Substituting here $A_\alpha(u) = -\frac{1}{2}u^\beta F_{\alpha\beta}$ and going over into momentum space, we obtain

$$G_A(p) = \frac{i\slashed{p}}{p^2} - \frac{F_{\alpha\beta}}{2}\left(\frac{\partial}{\partial p_\beta}\frac{\slashed{p}}{p^2}\right)\gamma^\alpha \frac{\slashed{p}}{p^2} = \frac{i\slashed{p}}{p^2} + \frac{F_{\alpha\beta}}{4p^4}(\slashed{p}\gamma^\alpha\gamma^\beta + \gamma^\alpha\gamma^\beta\slashed{p}) \ .$$

$$(12.41)$$

Its Fourier image is Eq. (12.25).

[†]Note that this is now a *local* quantity not directly related to the global properties of a gauge field configuration like a nonzero net topological charge q and the presence of fermion zero modes.

Problem 2. For the instanton field configuration (2.18), show that the function

$$u_0(x) = \frac{\rho}{\pi(x^2 + \rho^2)^{3/2}} \begin{pmatrix} \epsilon_{i\alpha} \\ 0 \end{pmatrix} \qquad (12.42)$$

($i = 1, 2$ is the color and $\alpha = 1, 2$ is the spinor index) represents an exact normalized right-handed zero mode of the Dirac equation ('t Hooft).

Solution. It is convenient to express the Euclidean γ matrices (N.13) as

$$\gamma_\mu^E = \begin{pmatrix} 0 & -\sigma_\mu^\dagger \\ \sigma_\mu & 0 \end{pmatrix} \qquad (12.43)$$

with $\sigma_\mu = (i, \boldsymbol{\sigma})$. The matrices σ_μ satisfy the relations

$$\begin{aligned}
\sigma_\mu^\dagger \sigma_\nu + \sigma_\nu^\dagger \sigma_\mu &= \sigma_\mu \sigma_\nu^\dagger + \sigma_\nu \sigma_\mu^\dagger = 2\delta_{\mu\nu} , \\
\sigma_\mu^\dagger \sigma_\nu - \sigma_\nu^\dagger \sigma_\mu &= 2i\eta_{\mu\nu}^a \sigma^a , \\
\sigma_\mu \sigma_\nu^\dagger - \sigma_\nu \sigma_\mu^\dagger &= 2i\bar\eta_{\mu\nu}^a \sigma^a .
\end{aligned} \qquad (12.44)$$

Note also the corollaries

$$\sigma_\mu^\dagger \sigma_\nu \sigma_\mu^\dagger = -2\sigma_\nu^\dagger , \qquad \sigma_\mu \sigma_\nu^\dagger \sigma_\mu = -2\sigma_\nu .$$

Let us first convince ourselves that only the right-handed solutions to the equation $\not{D}^E u = 0$ are admissible. Indeed, a left-handed solution should satisfy $\sigma_\mu^\dagger D_\mu u_L = 0$. Act upon this with the operator $\sigma_\nu D_\nu$. We obtain

$$\begin{aligned}
\sigma_\nu D_\nu \sigma_\mu^\dagger D_\mu &= (D_\mu)^2 + \frac{1}{4}[\sigma_\nu \sigma_\mu^\dagger - \sigma_\mu \sigma_\nu^\dagger][D_\nu, D_\mu] \\
&= (D_\mu)^2 - (\bar\eta_{\nu\mu}^a \sigma^a)(\eta_{\nu\mu}^b \sigma^b)\frac{\rho^2}{(x^2 + \rho^2)^2} = (D_\mu)^2 , \quad (12.45)
\end{aligned}$$

where we have substituted the instanton field strength $[D_\nu, D_\mu] = -iF_{\nu\mu}$ from Eq. (2.19) and used the property $\bar\eta_{\nu\mu}^a \eta_{\nu\mu}^b = 0$. But the operator $-(D_\mu)^2 = (iD_\mu)^2$ is a sum of squares of Hermitian operators and is thus positive definite. The equation $D^2 u_L = 0$ and hence the equation $\sigma_\mu^\dagger D_\mu u_L = 0$ have no solutions.

For the right-handed spinors, the zero mode equation has the form

$$(\sigma_\mu)_{\alpha\beta} \left[\partial_\mu - \frac{i\eta_{\mu\nu}^a x_\nu \sigma^a}{x^2 + \rho^2}\right]_{ij} u_{j\beta}^R(x) = 0 . \qquad (12.46)$$

Using the relations (12.44) and the property $\sigma_\mu^T = -\sigma_2 \sigma_\mu^\dagger \sigma_2$, we can rewrite this equation as

$$\left[\partial_\mu - \frac{x_\nu}{2(x^2+\rho^2)}(\sigma_\mu^\dagger \sigma_\nu - \sigma_\nu^\dagger \sigma_\mu)\right] u\sigma_2\sigma_\mu^\dagger = 0 \ . \tag{12.47}$$

The natural ansatz $u_{i\alpha} = (\sigma_2)_{i\alpha} F(x^2) = -i\epsilon_{i\alpha} F(x^2)$ satisfies the equation provided

$$2F'(x^2) + \frac{3F(x^2)}{x^2+\rho^2} = 0 \ \text{ and hence } \ F(x^2) = \frac{C}{(x^2+\rho^2)^{3/2}} \ .$$

The particular value of C in (12.42) normalizes the solution to one:

$$\int d^4x \ u_{i\alpha}^\dagger u_{i\alpha} = 1 \ .$$

Due to the index theorem (12.34), and since left-handed solutions are absent, the instanton field does not admit other zero mode solutions.

12.3 Nonsinglet Chiral Symmetry and its Spontaneous Breaking.

A theory with several flavors of massless quarks enjoys, besides the singlet axial symmetry $\delta\psi_f = -i\alpha\gamma^5\psi_f$ (which, as we have seen, it *does* not actually enjoy), a set of flavor-nonsinglet symmetries:

$$\delta\psi_f = -i\alpha_a[t^a\psi]_f \tag{12.48}$$

and

$$\delta\psi_f = -i\beta_a\gamma^5[t^a\psi]_f \ , \tag{12.49}$$

where t^a are the generators of the flavor $SU(N_f)$ group (N_f is the number of flavors). The symmetry (12.48) is the ordinary isotopic symmetry. It is still present even if the quarks are endowed with a mass (of the same magnitude for all flavors). The symmetry (12.49) holds only in massless theory. The corresponding Noether currents are

$$(j^\mu)^a = \bar\psi t^a \gamma^\mu \psi, \quad (j^{\mu 5})^a = \bar\psi t^a \gamma^\mu \gamma^5 \psi \ . \tag{12.50}$$

They are not anomalous, i.e. they are duly conserved not only at the classical level, but also in full quantum theory. To describe a finite element of the symmetry group, it is convenient to define left-handed and right-handed quark fields as in Eq. (N.12).

One can easily see that the Lagrangian of massless QCD is invariant under the transformations

$$\psi_L \to V_L \psi_L, \quad \psi_R \to V_R \psi_R, \qquad (12.51)$$

where V_L and V_R are two different $U(N_f)$ matrices. The singlet axial transformations with $V_L = V_R^* = e^{i\phi}$ are anomalous by the same token as in the theory with only one quark flavor. Therefore, the true fermionic symmetry group of massless QCD is

$$\mathcal{G} = SU_L(N_f) \times SU_R(N_f) \times U_V(1) . \qquad (12.52)$$

A fundamental *experimental fact* is that the symmetry (12.52) is actually *spontaneously broken*, which means that the vacuum state is not invariant under the action of the group $\mathcal{G}$. The symmetry $\mathcal{G}$ is, however, not broken completely. The vacuum is still invariant under the transformations with $V_L = V_R$, generated by the vector isotopic currents.[‡] Thereby, the pattern of breaking is

$$SU_L(N_f) \times SU_R(N_f) \to SU_V(N_f) . \qquad (12.53)$$

Spontaneous breaking of the axial symmetry shows up in the appearance of nonzero vacuum expectation values,

$$\Sigma^{fg} = \langle \psi_L^f \bar{\psi}_R^g \rangle_0 \qquad (12.54)$$

(the *quark condensate* matrix). Nonbreaking of the vector symmetry implies that the matrix order parameter (12.54) can be brought into the form

$$\Sigma^{fg} = \frac{1}{2} \Sigma \delta^{fg} \qquad (12.55)$$

by the group transformations (12.51). This means that a generic condensate matrix Σ^{fg} is a unitary $SU(N_f)$ matrix multiplied by Σ.

In general, Σ could be any complex number. It can be made real by a global $U_A(1)$ rotation which, according to Eq. (12.37) (with the factor N_f), amounts to a shift of the vacuum angle θ. In other words, in the theory with quarks the physics does not depend on the parameter θ and

---

[‡]An exact *theorem* that the vector symmetry cannot be broken spontaneously in QCD will be proven in Lecture 14.

the phase of Σ separately, but only on their combination $\theta - N_f \arg(\Sigma)$. The earlier mentioned fact that the experimental value of θ is very small actually refers to *this* particular combination. It is convenient then to choose Σ real and positive and $\theta = 0$. From experiment, we know that $\Sigma \approx (240 \text{ MeV})^3/2$ with about 30% uncertainty (this value refers to some particular normalization point $\mu \sim .5$ GeV on which the operator $\bar{\psi}\psi$ and its vacuum expectation value depend).

By Goldstone's theorem, if a global continuous symmetry is broken spontaneously, purely massless particles called *Goldstone bosons* appear in the spectrum. The simplest example for this is the theory of a complex scalar field $\phi(x)$ with a "Mexican hat" potential $\lambda(\bar{\phi}\phi - \mu^2)^2$. The Lagrangian is invariant with respect to global phase rotations $\phi(x) \to e^{i\alpha}\phi(x)$, but the vacuum state (say, the state with $\langle\phi\rangle_{\text{vac}} = \mu$) is characterized by a particular value of the phase playing the role of an order parameter. The point is that a field configuration involving small fluctuations of the phase $\phi(x) = \mu e^{i\alpha(x)}$, $\alpha(x) \ll 1$, still has zero potential energy while the kinetic energy is proportional to $\int d^4x (\partial_\mu \alpha)^2$. Quantizing this effective Lagrangian gives massless particles.

In general, the number of Goldstone particles is equal to the dimension of the vector space formed by the generators of the gauge group which are "broken", i.e. act nontrivially on the vacuum state. By construction, for any such generator J, a Goldstone branch in the spectrum $|\alpha_J\rangle_p$ exists, so that the matrix elements $\langle 0|J|\alpha_J\rangle_p$ are not zero. In our case, the breaking (12.53) is associated with the appearance of $N_f^2 - 1$ Goldstone particles [$N_f^2 - 1$ is the dimension of the original group $\mathcal{G}$ minus the dimension of the residual group $U_V(N_f)$]. As it is the axial symmetry which is broken, the Goldstone particles are pseudoscalars.

An analogy with an ordinary iron bar is very instructive here. The Hamiltonian of a ferromagnet is rotationally invariant. Spontaneous magnetization signals the spontaneous breaking of the rotational invariance $SO(3)$ down to $SO(2)$. The direction n of the magnetization $\langle M \rangle = M_0 n$, with $n^2 = 1$, is arbitrary. Rotating the reference frame, we can choose, say, $n = (0, 0, 1)$ [cf. Eq. (12.55)]. Fluctuations of the vector n in space and time are described by $3 - 1 = 2$ parameters and correspond to magnon massless excitations.

12.4   Effective Chiral Langrangian.

It is a fundamental and important fact that spontaneous breaking of a continuous symmetry not only leads to massless Goldstone particles, but also fixes the *interactions* of the latter at low energies. To see this, let us first recall that the Goldstone field describes fluctuations of the order parameter: $\Sigma^{fg} \to \Sigma^{fg}(x) = \frac{1}{2}\Sigma U(x)$ with $U(x) \in SU(N_f)$. It is convenient to express $U(x)$ as an exponential

$$U(x) = \exp\left\{\frac{2i\phi^a(x)t^a}{F_\pi}\right\}, \qquad (12.56)$$

where $\phi^a(x)$ are the physical meson fields [so that $\phi^a(x) = 0$ corresponds to the vacuum (12.55)] and F_π is a constant of dimension of mass.

Goldstone particles are massless whereas all other states in the physical spectrum have nonzero mass. Therefore, we are exactly in the Born–Oppenheimer situation: there are two distinct energy scales and one can write down an effective Lagrangian depending only on *slow* Goldstone fields, with the *fast* degrees of freedom corresponding to all other particles being integrated out.

To fix the exact form of this Lagrangian, note that the transformations (12.51) are realized at the level of the effective Lagrangian as $U \to V_L U V_R^\dagger$. Any scalar function depending on U and invariant under this symmetry is also a function of $U^\dagger U = 1$, i.e. it is just a constant. There is only one invariant structure involving two derivatives:

$$\mathcal{L}_{\text{eff}}^{(2)} = \frac{F_\pi^2}{4}\text{Tr}\left\{\partial_\mu U \partial^\mu U^\dagger\right\}. \qquad (12.57)$$

Take $N_f = 2$. The perturbative expansion of Eq. (12.57) in powers of ϕ reads

$$\mathcal{L}_{\text{eff}}^{(2)} = \frac{1}{2}(\partial_\mu \phi^a)^2 + \frac{1}{6F_\pi^2}\left[(\phi^a \partial_\mu \phi^a)^2 - (\phi^a \phi^a)(\partial_\mu \phi^b)^2\right] + \dots. \qquad (12.58)$$

Also for $N_f \geq 3$, we have, on top of the standard kinetic term, a quartic term involving two derivatives, with somewhat more complicated group structure. We see that the symmetry dictates rather specific interactions between the Goldstone bosons. They do not interact at the S wave level, which means that the amplitude vanishes at zero momenta, but the strength of interaction grows rapidly with energy.

Equation (12.57) describes the effective chiral Lagrangian to leading order. The first corrections involve 4 derivatives and there are 3 different invariant functions

$$\mathcal{L}_{\text{eff}}^{(4)} = L_1 \text{Tr} \left\{ \partial_\mu U \partial^\mu U^\dagger \right\}^2 + L_2 \text{Tr} \left\{ \partial_\mu U \partial_\nu U^\dagger \right\} \text{Tr} \left\{ \partial^\mu U \partial^\nu U^\dagger \right\}$$
$$+ L_3 \text{Tr} \left\{ \partial_\mu U \partial^\mu U^\dagger \partial_\nu U \partial^\nu U^\dagger \right\} \quad (12.59)$$

(only 2 linearly independent structures are left for $N_f = 2$).

The relevant Born–Oppenheimer expansion parameter is

$$\kappa_{\text{chir}} \sim p^{\text{char}} / F_\pi \;.$$

When $\kappa_{\text{chir}} \sim 1$ (in practice, one should rather take $\kappa_{\text{chir}} \sim 2\pi$), the Born–Oppenheimer approximation, as well as the whole effective Lagrangian approach, breaks down, and non-Goldstone degrees of freedom become important. The physical meaning of F_π is thus clarified. It characterizes the gap in the spectrum and sets a scale below which massive degrees of freedom can be disregarded.

Let us discuss the real world now. The Lagrangian of real QCD (1.19) is not invariant under the axial symmetry transformations just because quarks have nonzero masses. The symmetry (12.52) is still very much relevant for QCD because *some* of the quarks happen to be very light. This is especially so for u and d quarks whose masses given in Eq. (0.2) are much smaller than the characteristic hadron scale $\mu_{\text{hadr}} \approx .5$ GeV: the symmetry (12.52) is almost there!

Spontaneous breaking of an exact $SU_L(2) \times SU_R(2)$ symmetry would lead to the existence of three strictly massless Goldstone bosons. As the symmetry is not quite exact, the Goldstone particles have a small mass. However, their mass M goes to zero in the chiral limit $m_{u,d} \to 0$. Indeed, trading the mass term

$$-m_u \bar{u} u - m_d \bar{d} d = m_u (u_L \bar{u}_R + u_R \bar{u}_L) + m_d (d_L \bar{d}_R + d_R \bar{d}_L) \quad (12.60)$$

in the QCD Lagrangian for the contribution [§]

$$\mathcal{L}_{\text{eff}}^{(m)} = \Sigma \, \text{Re} \left[\text{Tr} \{ \mathcal{M} U \} \right] \quad (12.61)$$

[§]Equation (12.61) is the leading chiral-noninvariant contribution in $\mathcal{L}_{\text{eff}}$. Also, terms of higher order in $\mathcal{M}$, as well as terms of first order in $\mathcal{M}$ but involving derivatives of U, are allowed.

[$\mathcal{M}$ is the quark mass matrix which is chosen here in the form $\mathcal{M} = \text{diag}(m_u, m_d)$ with real m_u, m_d] in the effective chiral Lagrangian and expanding Eq. (12.61) in ϕ^a, we obtain the *Gell-Mann–Oakes–Renner relation*,

$$F_\pi^2 M^2 = (m_u + m_d)\Sigma + O(m_q^2) . \qquad (12.62)$$

These light pseudo-Goldstone particles are well know to experimentalists. They are nothing but the pions. Experimentally, $F_\pi \approx 93$ MeV. The gap $\approx 8F_\pi$ is of order of the mass of ρ meson. The constant F_π appears also in the matrix element $\langle \text{vac}|A_\mu^+|\pi\rangle = i\sqrt{2}F_\pi p_\mu^\pi$ of the axial current $A_\mu^+ = \bar{d}\gamma_\mu\gamma^5 u$ and determines the charged pion decay rate.

In the real world, there is also a third relatively light quark — the strange quark. Its mass, $m_s \approx 150$ MeV, is still small enough for the symmetry (12.52) to make sense. Thus, QCD enjoys an approximate $SU_L(3) \times SU_R(3)$ symmetry which is broken spontaneously with the appearance of the quark condensate (12.55), and also explicitly due to nonzero quark masses. As the latter are relatively small, one can build up a Born-Oppenheimer expansion (alias, *chiral perturbation theory* [32]) over the small parameters $p^{\text{char}}/\mu_{\text{hadr}}$ and m_q/μ_{hadr}. The spectrum of QCD includes 8 light pseudo-Goldstone mesons, the well-known pseudoscalar octet (π, K, η).

Exploiting the symmetry (12.52) allows one to obtain many nontrivial predictions for the properties of these mesons, but this, as well as many other wonderful achievements in describing the physics of hadrons through QCD, is beyond the scope of this book.

Lecture 13

Quarks on Euclidean Lattice

We now have enough funds to pay our old debt and to define what QCD *is*. In other words, we are in a position now to give a correct definition of the path integral in QCD involving, besides the gauge fields, fermion fields as well.

13.1 Nielsen–Ninomiya's No-go Theorem.

We have started to discuss this issue in Lecture 4, where the naïve fermion action (4.37) was written with a remark that it gives rise to at least 16 degenerate light fermion species in the continuum limit and does not reproduce QCD.

To understand it, consider first tree massless fermions. Let

$$(\partial_\mu^+ \psi)_n \;=\; \frac{1}{a}[\psi_{n+e_\mu} - \psi_n], \quad (\partial_\mu^- \psi)_n \;=\; \frac{1}{a}[\psi_n - \psi_{n-e_\mu}] \qquad (13.1)$$

be the forward and backward lattice derivative operators. The naïve free massless Dirac operator is*

$$\mathfrak{D}^0_{\text{free}} \;=\; -\frac{i}{2}\gamma_\mu(\partial_\mu^+ + \partial_\mu^-) \,. \qquad (13.2)$$

The eigenfunctions of $\mathfrak{D}^0_{\text{free}}$ are characterized by the Euclidean 4-momentum p_μ,

$$u_p(n) \;=\; C_p e^{iap_\mu n_\mu} \,,$$

*$\gamma_\mu \equiv \gamma_\mu^E$ throughout this lecture.

where C_p is a constant Grassmann bi-spinor. The eigenvalue equation $\mathfrak{D}^0_{\text{free}} u_p(n) = -i\lambda_p u_p(n)$ implies

$$\left[\frac{1}{a}\gamma_\mu \sin(ap_\mu)\right] C_p = -i\lambda_p C_p \tag{13.3}$$

with

$$\lambda_p = \pm\frac{1}{a}\sqrt{\sum_\mu \sin^2(ap_\mu)} \ . \tag{13.4}$$

[The operator (13.2) is anti-Hermitian and its eigenvalues are purely imaginary.]

When $ap_\mu \ll 1$, we reproduce the continuum massless fermions with the spectrum $\lambda_p = \pm\sqrt{p_\mu^2}$. Each eigenvalue (13.4) is doubly degenerate due to 2 possible polarizations. The eigenfunctions with negative λ_p are obtained from the ones with positive λ_p by multiplication by γ^5 (see the discussion at the end of Section 4.1).

Note, however, that the lattice Dirac equation (13.3) has an *additional* discrete symmetry $(Z_2)^4$: for any eigenfunction u_p, the function $\hat{Q}_\mu u_p \equiv \gamma_\mu\gamma^5 u_{p+(\pi/a)e_\mu}$ (no summation over μ) is also the eigenfunction of $\mathfrak{D}^0_{\text{free}}$ with the same eigenvalue λ_p. The operators $\hat{Q}_\mu$ commute with $\mathfrak{D}^0_{\text{free}}$ and anticommute with each other:

$$\hat{Q}_\mu\hat{Q}_\nu + \hat{Q}_\nu\hat{Q}_\mu = 2\delta_{\mu\nu} \ .$$

The functions

$$u_p, \ \hat{Q}_\mu u_p, \ \hat{Q}_{[\mu}\hat{Q}_{\nu]}u_p, \ \hat{Q}_{[\mu}\hat{Q}_\nu\hat{Q}_{\lambda]}u_p, \ \hat{Q}_{[\mu}\hat{Q}_\nu\hat{Q}_\lambda\hat{Q}_{\rho]}u_p \tag{13.5}$$

form a degenerate 16-plet.

In the free case, each eigenstate of the naïve Dirac operator is not just 16-fold, but 32-fold degenerate due to polarizations. In the interacting case (on a generic gauge field background), polarization is not a good quantum number, but the 16-fold degeneracy (13.5) still holds. The naïve lattice Dirac operator in Eq. (4.37), which can be written in the form $\mathfrak{D}^0 = -\frac{i}{2}\gamma_\mu(\mathcal{D}^+_\mu + \mathcal{D}^-_\mu)$, where

$$(\mathcal{D}^+_\mu\psi)_n = \frac{1}{a}\left(\psi_{n+e_\mu}U_{n,n+e_\mu} - \psi_n\right) \ ,$$

$$(\mathcal{D}^-_\mu\psi)_n = \frac{1}{a}\left(\psi_n - \psi_{n-e_\mu}U_{n,n-e_\mu}\right) \tag{13.6}$$

are the covariant lattice forward and backward derivatives, still enjoys the symmetries

$$\hat{Q}_\mu : \psi_n \longrightarrow (-1)^{n_\mu} \gamma_\mu \gamma^5 \psi_n \tag{13.7}$$

(no summation over μ). We see that if ψ_n changes smoothly from node to node, its 15 doublers (13.5) wildly oscillate on the microscopic lattice spacing scale. We might call these modes "unphysical", but they would not listen to us and contaminate with a vengeance any numerical lattice calculation we might wish to do. Some way to get rid of them should be suggested, otherwise QCD, the theory involving only 6 quarks with different masses, would not be operationally defined.

The problem is that it is not so simple. Let us look for some other lattice Dirac operator $\mathfrak{D} \neq \mathfrak{D}^0$ satisfying the following 4 natural conditions:

(1) At distances much larger than the lattice spacing a, $\mathfrak{D} \to -i\slashed{p}$, giving rise to a massless fermion in the continuum limit.[†]

(2) All the modes of $\mathfrak{D}$ not associated with the latter are of order $1/a$ (no doublers!).

(3) $\mathfrak{D}$ is local. In other words, the matrix elements $\mathfrak{D}_{nn'}$ decay exponentially fast at large distances $|n - n'| \gg 1$.

(4) Chiral symmetries (12.19), (12.49) of the massless fermionic action are not broken by the regularization explicitly. [The singlet axial symmetry (12.19) is eventually going to be broken due to noninvariance of the fermionic measure, but we require the absence of explicit breaking in the regularized Lagrangian.] This seems to imply the condition $\mathfrak{D}\gamma^5 + \gamma^5\mathfrak{D} = 0$.

The no-go theorem due to Nielsen and Ninomiya tells us, however, that such a $\mathfrak{D}$ *does not exist*. To understand it, consider first the free fermion case. The momentum p_μ is then a good quantum number, and the Dirac operator in the momentum representation has the form

$$\mathfrak{D}(p) = \gamma_\mu F_\mu(p) + G(p) . \tag{13.8}$$

Condition 4 tells us that $G(p) = 0$. Condition 1 implies that $F_\mu(p) = p_\mu + O(ap^2)$ for $ap_\mu \ll 1$. Now, $F_\mu(p)$ is a periodic function of its four arguments p_μ with period $2\pi/a$. It thus realizes a smooth map $T^4 \to R^4$, where R^4 is the tangent space. A look at Fig. 13.1 can convince the reader

[†] Adding a finite mass term to $\mathfrak{D}$ presents no difficulties [see Eq. (4.37)].

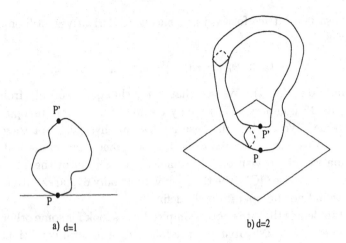

P' $\qquad$ P'

P

a) d=1   b) d=2

Fig. 13.1   Nielsen–Ninomiya theorem. P and P' are the different zeros of the lattice Dirac operator.

that the point on the tangent space where it touches our torus always has at least one more pre-image. His/her intuition would not betray them: this statement can be proven in a rigorous mathematical manner. Basically, it follows from the fact that the degree of a map $T^d \to R^d$ is zero, which means that [33]

$$\sum_{\substack{\text{pre-images} \\ \text{of } P}} \text{sign} \left[\det \|\partial_\nu F_\mu(p)\|\right] = 0 . \tag{13.9}$$

As the Jacobian of the mapping $p_\mu \to F_\mu(p)$ is equal to 1 at the point $p_\mu = 0$, Eq. (13.9) implies that some other pre-images of zero, i.e. some other solutions of the equation system $F_\mu(p) = 0$ should be present (one can have just one extra solution as in Fig. 13.1 or more: for the "round upright torus" $F_\mu(p) = \sin(ap_\mu)/a$, there are $2^d - 1$ extra solutions). And that means the presence of doublers in contradiction with the condition 2.

The only remaining possibility is that $F_\mu(p)$ are not continuous. Besides being ugly, it also contradicts condition 3: the matrix elements $\mathfrak{D}_{nn'} = \mathfrak{D}(n-n') = \gamma_\mu F_\mu(n-n')$ are actually the Fourier coefficients of the periodic function $\gamma_\mu F_\mu(p)$. If the latter is discontinuous, the Fourier coefficients cannot decay faster than $1/|n - n'|$ (otherwise, the Fourier series would converge uniformly on the torus $p_\mu \in \left[0, \frac{2\pi}{a}\right)$ and the sum of such a series would be continuous).

We have proven that a lattice Dirac operator satisfying the above four conditions cannot be found for free fermions, but this also means that it cannot be found for QCD: any $\mathfrak{D}$ having this property should satisfy our list for any smooth set of link variables [‡] and, in particular, for the set $U_{\text{each link}} = 1$ corresponding to free theory.

13.2 Ways to Go. The Ginsparg–Wilson Way.

If we still want to build up a lattice version of QCD, we have to relax at least one of our four conditions. Conditions 1 and 2 are, however, indispensible: a lattice theory where they do not hold just has nothing to do with QCD. Therefore, either locality or chiral invariance of the lattice action should be abandoned.

One of the possible procedures is that only *one mode* of each degenerate 16-plet of $\mathfrak{D}^0$ is taken into account in the fermionic determinant and in the spectral decomposition of fermion Green's functions,

$$\langle \psi_n \bar{\psi}_{n'} \rangle = \sum_k \frac{u_k(n) u_k^\dagger(n')}{m - i\lambda_k}, \tag{13.10}$$

etc., where $u_k(n)$ describe the k-th eigenmode of $\mathfrak{D}^0$ as a function of the node. This amounts to choosing the lattice Dirac operator in the form $(\mathfrak{D}^0)^{1/16}$, which is not local. A similar method is sometimes used in practical lattice calculations, but besides purely technical inconveniences it is unsatisfactory from a philosophical viewpoint: we *would* like to have a local lattice approximation for a local field theory.

But then the chiral invariance (12.19) is necessarily lost. Though re-

---

[‡] We *believe* that, in the continuum limit $a \to 0$, the characteristic fields $\{U\}$ contributing to the path integral can be *gauge-transformed* into the form where $U = 1 + O(a)$ for all links. Note that the statement that the characteristic fields *are* always smooth is simply wrong: there are instantons which, in the singular gauge, involve singularities of the gauge potential $\sim x_\mu/x^2$ at the instanton center. Note also that a gauge transformation removing the singularity at $x = 0$ moves it to some other point or to infinity. For Euclidean torus that means that we cannot *simultaneously* require $U_{\text{all links}} = 1 + O(a)$ and the periodicity of U. To the best of our knowledge, a rigorous proof of this crucial assumption is absent, but it can be justified by arguing that the action of field configurations which are "essentially singular" (so that the singularity cannot be removed by a gauge transformation) would be infinite in the continuum limit.

nouncing chiral invariance is also not desirable — when regularizing the
theory, we should try to preserve as much of its symmetries as possible —
it is still considered the least of evils.

Two ways of chiral noninvariant lattice regularization have been known
for some time and used in practical calculations: (i) Wilson fermions and
(ii) Kogut–Susskind or staggered fermions. We will describe here the first
method which consists in adding to $\mathfrak{D}^0$ the term $\sim a\mathcal{D}_\mu^- \mathcal{D}_\mu^+ = a\mathcal{D}_\mu^+ \mathcal{D}_\mu^-$,
where the covariant lattice derivatives $\mathcal{D}_\mu^\pm$ are defined in Eq. (13.6). Thus,
the Wilson–Dirac operator is defined as

$$
\mathfrak{D}^W = -\frac{i}{2}\gamma_\mu \left(\mathcal{D}_\mu^+ + \mathcal{D}_\mu^-\right) - \frac{ra}{2}\mathcal{D}_\mu^+ \mathcal{D}_\mu^- \, ,
\tag{13.11}
$$

where r is an arbitrary nonzero real constant.

The second term in Eq. (13.11) (call it A) is Hermitian while the first
term $\equiv iB$ is anti-Hermitian. A and B do not commute for a generic field
configuration and cannot be simultaneously diagonalized. A generic matrix
$\mathfrak{D}^W = A + iB$ still can be diagonalized, but the corresponding transformation matrix is not unitary and the eigenvectors of $\mathfrak{D}^W$ are not orthogonal
to each other. This might be not so nice, but it does not prevent one to
determine the spectrum of $\mathfrak{D}^W$ (the roots of the characteristic equation)
and compute the determinant of $-i\mathfrak{D}^W + m$ which enters the lattice approximation of the partition function of QCD.

The essential properties of $\mathfrak{D}^W$ can be understood by studying the case
of free fermions. The situation is much simpler here than in the interacting
case. The Hermitian and anti-Hermitian parts of $\mathfrak{D}_{\text{free}}^W$ can now be simultaneously diagonalized, and this can be done explicitly. We have $\mathcal{D}_\mu^\pm \to \partial_\mu^\pm$
and

$$
(\partial_\mu^+ \partial_\mu^- \psi)_n = \frac{1}{a^2} \sum_\mu \left(\psi_{n+e_\mu} + \psi_{n-e_\mu} - 2\psi_n\right) \, ,
$$

which is the lattice laplacian. Passing to the momentum representation, we
obtain

$$
\mathfrak{D}_{\text{free}}^W(p) = \frac{1}{a}\gamma_\mu \sin(ap_\mu) + \frac{2r}{a}\sum_\mu \sin^2\left(\frac{ap_\mu}{2}\right) \, .
\tag{13.12}
$$

The second term has the form of momentum-dependent mass. For small
$p_\mu \ll 1/a$, it can be neglected and the continuum massless Dirac operator is
reproduced. In contrast to $\mathfrak{D}^0$, the operator (13.12) is not anti-Hermitian,

and its eigenvalues are complex. What is important is that, for p_μ that are not small, the absolute values of the eigenvalues of $\mathfrak{D}^W$,

$$-i\lambda_p^W = \pm \frac{i}{a}\sqrt{\sum_\mu \sin^2(ap_\mu) + \frac{2r}{a}\sum_\mu \sin^2\left(\frac{ap_\mu}{2}\right)}, \qquad (13.13)$$

are of order $1/a$. The doublers disappear. At $p_\mu = \left(\frac{\pi}{a}, 0, 0, 0\right)$, the eigenvalue (13.13) is $\frac{2r}{a}$; at $p_\mu = \left(0, \frac{\pi}{a}, 0, \frac{\pi}{a}\right)$, it is $\frac{4r}{a}$, etc.

The chiral symmetry is broken, however, and it is messy. In principle, when the continuum limit $a \to 0$ is taken, the effects due to the breaking of γ^5 invariance must be suppressed, but in this particular problem the continuum limit with restoration of chiral symmetry is rather slow to reach, and, though the most experts think that it is eventually reached, this has not been shown quite rigorously. In particular, it is difficult to make the pions light. In practical calculations this is achieved by introducing a large bare quark mass of order $g^2(a)/a$ and fine-tuning it so that the effects due to two chiral noninvariant terms — the Wilson term and the bare quark term — would cancel each other. Needless to say, this is a rather artificial and unaesthetic procedure.

As we see, this Nielsen–Ninomiya puzzle defies attempts to solve it. A recent remarkable observation [34] is that the best strategy here is to follow the example of the Alexandre the Great and just cut through it! The adequate sword was forged back in 1982 by Ginsparg and Wilson. They suggested to consider the lattice Dirac operators satisfying the relation

$$\gamma^5 \mathfrak{D} + \mathfrak{D}\gamma^5 = a\mathfrak{D}\gamma^5\mathfrak{D}. \qquad (13.14)$$

The anticommutator $\{\mathfrak{D}, \gamma^5\}$ does not vanish which means that the lattice Lagrangian is not invariant with respect to the chiral transformations

$$\delta\psi_n = -i\alpha\gamma^5\psi_n, \qquad \delta\bar{\psi}_n = -i\alpha\bar{\psi}_n\gamma^5, \qquad (13.15)$$

a lattice Euclidean counterpart of Eq. (12.19).

It took 16 years to realize that the lattice fermion action

$$S_F = a^4 \sum_{nn'} \bar{\psi}_n \mathfrak{D}_{nn'} \psi_{n'} \qquad (13.16)$$

(color and spinor indices being suppressed), with $\mathfrak{D}$ satisfying the relation

(13.14), is invariant with respect to the following transformations:

$$\delta\psi = -i\alpha\gamma^5 \left[1 - \frac{1}{2}a\mathfrak{D}\right]\psi ,$$

$$\delta\bar{\psi} = -i\alpha\bar{\psi}\left[1 - \frac{1}{2}a\mathfrak{D}\right]\gamma^5 . \tag{13.17}$$

($\delta\bar{\psi} \neq (\delta\psi)^\dagger$ *quite* definitely here.) If $\mathfrak{D}$ is local (in the sense of condition 3 in the Nielsen–Ninomiya list), Eq. (13.17) is as good a lattice approximation of the continuous chiral symmetry (12.19) as the trivial (13.15). In particular, the pions would automatically be light (massless in the chiral limit), and no fine tuning is required. But condition 4 above is no longer satisfied and one can now hope to find a local $\mathfrak{D}$ not involving doublers. The problem is still not trivial: as we will see a bit later, many solutions of the Ginsparg–Wilson relation (13.14) can be found and most of them *are* not what we are looking for. The simplest *good* solution was suggested by Neuberger. It has the form

$$\mathfrak{D} = \frac{1}{a}\left[1 - A\frac{1}{\sqrt{A^\dagger A}}\right] , \tag{13.18}$$

where $A = 1 - a\mathfrak{D}^W$ and $\mathfrak{D}^W$ is the Wilson–Dirac operator (13.11) with $r > 1/2$. As A and $A^\dagger$ do not necessarily commute, the order of the factors in Eq. (13.18) is essential. The operator (13.18) satisfies the Ginsparg–Wilson relation and has a nice property $[\mathfrak{D}, \mathfrak{D}^\dagger] = 0$, allowing for a simultaneous diagonalization of its Hermitian and anti-Hermitian part.

In particular, for $r = 1$ and for the free fermions, we have

$$a\mathfrak{D}(p) = 1 - \frac{1 - 2\sum_\mu \sin^2\left(\frac{ap_\mu}{2}\right) - \sum_\mu \gamma_\mu \sin(ap_\mu)}{\left[1 + 8\sum_{\mu<\nu}\sin^2\left(\frac{ap_\mu}{2}\right)\sin^2\left(\frac{ap_\nu}{2}\right)\right]^{1/2}} . \tag{13.19}$$

The eigenvalues of (13.19) are different from zero provided $p_\mu \neq 0$. In particular, for

$$p_\mu = \left(\frac{\pi}{a},0,0,0\right), \quad p_\mu = \left(\frac{\pi}{a},\frac{\pi}{a},0,0\right), \quad p_\mu = \left(\frac{\pi}{a},\frac{\pi}{a},\frac{\pi}{a},0\right) ,$$

and

$$p_\mu = \left(\frac{\pi}{a},\frac{\pi}{a},\frac{\pi}{a},\frac{\pi}{a}\right) ,$$

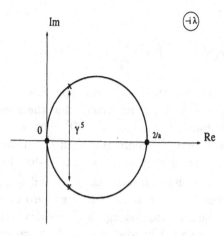

Fig. 13.2 The circle of eigenvalues for Neuberger's operator. The eigenmodes with the eigenvalues marked by crosses are related by a γ^5 transformation.

the eigenvalues $-i\lambda$ of $\mathfrak{D}$ are all equal to $2/a$. The doublers are absent. A second look at Eq. (13.19) reveals a beautiful feature displayed in Fig. 13.2: the eigenvalues of $\mathfrak{D}$ lie on the circle

$$(\text{Re }\lambda)^2 + \left(\text{Im }\lambda - \frac{1}{a}\right)^2 = \frac{1}{a^2} . \tag{13.20}$$

This property holds also in the interacting case. Using the property $\mathfrak{D}^\dagger = \gamma^5\mathfrak{D}\gamma^5$ and the Ginsparg–Wilson relation (13.14), it is not difficult to see that the operator $V = 1 - a\mathfrak{D}$ is unitary, i.e. its eigenvalues lie on the circle $\{e^{i\phi}\}$. And the eigenvalues of $\mathfrak{D} = (1 - V)/a$ lie on the circle in Fig. 13.2.

The function (13.19) has singularities associated with the square root, but they all occur at complex values of p_μ. It is analytic on the torus $p_\mu \in [0, \frac{2\pi}{a})$, which means that its Fourier image decay exponentially at large distances. The Dirac operator thus constructed is local. In the interacting case, it stays local if the gauge field is smooth enough, i.e. the link variables $U_{n,n+e_\mu}$ are sufficiently close to 1.

As was mentioned, the singlet axial symmetry (12.19) is anomalous, which shows up in the noninvariance of the fermionic measure. The measure $\prod_n d\bar{\psi}_n d\psi_n$ is obviously invariant, however, with respect to the ultralocal transformations (13.15). This follows from the fact that $\text{Tr}\{\gamma^5\} = 0$. On the other hand, Eq. (12.33) relates the modification of the measure to the operator trace of γ^5,

$$\mathbf{Tr}\{\gamma^5\} = \int d^4x \sum_k u_k^\dagger(x)\gamma^5 u_k(x) , \qquad (13.21)$$

which is thus also zero. For the naïve Dirac operator, only the zero modes contribute to the sum (13.21), which means that the number of the right-handed and the left-handed zero modes of $\mathfrak{D}^0$ should be equal. This also follows from the fact that $\mathfrak{D}^0$ commutes with $\hat{Q}_\mu$ defined in Eq. (13.7). It is not difficult to see that if u_k is, say, right-handed, $\hat{Q}_{[\mu}\hat{Q}_{\nu]}u_k$ and $\hat{Q}_{[\mu}\hat{Q}_\nu\hat{Q}_\lambda\hat{Q}_{\rho]}u_k$ are right-handed too, but $\hat{Q}_\mu u_k$ and $\hat{Q}_{[\mu}\hat{Q}_\nu\hat{Q}_{\lambda]}u_k$ are left-handed. Thus, instead of a single right-handed zero mode in (the lattice approximation for) the instanton background, we have a degenerate 16-plet with 8 right-handed and 8 left-handed modes which are not necessarily zero modes anymore. The vanishing of the index of $\mathfrak{D}^0$ is closely related to the identity (13.9). Indeed, the sign of the Jacobian $\det\|\partial_\nu F_\mu(p)\|$ describes the orientation of the neighbourhood of a pre-image $P_i \in T^4$ with respect to the orientation of the tangent space R^4 into which it is mapped. This orientation is obviously related to chirality.

The absence of anomaly is one of the deseases of the naïve lattice Dirac operator. For the operators of Ginsparg–Wilson type, the situation is different. The chiral symmetry is now implemented as in Eq. (13.17) and, generically, the measure is *not* invariant with respect to these transformations. We have, instead of Eq. (12.33),

$$\ln J = -i\alpha\mathbf{Tr}\left\{\gamma^5\left(1 - \frac{1}{2}a\mathfrak{D}\right)\right\} . \qquad (13.22)$$

Even though $\mathbf{Tr}\{\gamma^5\} = 0$ (we derived this in the basis involving the eigenvalues of $\mathfrak{D}^0$, but it is true in any basis), $\mathbf{Tr}\{\gamma^5\mathfrak{D}\}$ need not vanish and the anomaly is there.

It is instructive to see how a nonzero operator trace on the right side of Eq. (13.22) is obtained in the basis involving the eigenvalues of $\mathfrak{D}$. First, it is still true that, for any eigenfunction u_k of $\mathfrak{D}$, $\gamma^5 u_k$ is also an eigenfunction with the eigenvalue $\lambda_k' = -\lambda_k^*$ (complex conjugation appears and this is what distinguishes the Ginsparg–Wilson case from the continuum or naïve lattice Dirac operators). Thus, for almost all eigenstates on the circle in Fig. 13.2, u_k and $\gamma^5 u_k$ have *different* eigenvalues and are orthogonal to each other. Their contribution to the trace vanishes. The only exception are two points on the circle, $\lambda = 0$ and $\lambda = 2i/a$, where $\lambda_k = \lambda'_k$ and the eigenstates can have a definite chirality. But the doublers $-i\lambda = \frac{2}{a}$ obviously give zero contribution in Eq. (13.22). Only zero modes of $\mathcal{D}$ are relevant. We have derived the *lattice index theorem*

$$n_R^0(\mathcal{D}) - n_L^0(\mathcal{D}) = -\frac{1}{2}\mathrm{Tr}\{\gamma^5 a\mathcal{D}\} . \tag{13.23}$$

The right side of Eq. (13.23) is a functional depending of the link variables $\{U\}$. By definition, it is given by a sum over all lattice nodes of some local expression. After some work, one can be convinced that it goes over to the topological charge (2.6) in the continuum limit.

Problem. Consider the Wilson–Dirac operator (13.12) with arbitrary r and analyze the corresponding Neuberger's operator (13.18). Show that only the values $r > 1/2$ are admissible.

Solution. As was mentioned above, the eigenvalues of the free Wilson–Dirac operator (13.11) which correspond to the naïve doublers are $-i\lambda^W = 2rl/a$, where l is the number of nonzero components π/a of the momenta p_μ. The corresponding eigenvalues of the operator (13.18) are

$$-i\lambda = \frac{1}{a}\left[1 - \frac{1 - 2rl}{|1 - 2rl|}\right] . \tag{13.24}$$

We see that, if $r < 1/2$, four doublers with $l = 1$ become massless again, and that is not what we want.

Massless doublers may appear also for r slightly exceeding $1/2$ if gauge fields are present. Note that the operator (13.18) satisfies the Ginsparg–Wilson relation for any r, which means that its eigenvalues still lie on the circle in Fig. 13.2. But only a part of the circle is covered.

Lecture 14

Aspects of Chiral Symmetry

We now abandon the lattice and will discuss in this section various aspects of chiral symmetry in the continuum limit. Sometimes, we will think in terms of quarks, of the symmetries (12.49) and (12.19), and of the order parameter (12.54) associated with the spontaneous breaking of the flavor-nonsinglet symmetry. Sometimes, we will describe the system in terms of the pseudo-Goldstone degrees of freedom and the effective Lagrangians(12.57), (12.59), and (12.61). Sometimes, we will confront the two languages using the philosophy of the *quark–hadron duality* — that is how most of the results discussed in this section, a bunch of beautiful *exact theorems of QCD*, will be obtained.

14.1 QCD Inequalities. Vafa–Witten Theorem.

As was discussed above, the octet of pseudoscalar mesons (π, K, η) can be interpreted as that of the pseudo-Goldstone particles appearing due to the spontaneous chiral symmetry breaking according to the pattern (12.53) in the massless limit. This is the reason why the pseudoscalar mesons are lighter than those with other quantum numbers. It is interesting that the latter statement can be formulated as an exact theorem of QCD without any reference to the (experimental!) fact that the chiral symmetry is broken.

Consider a QCD-like theory with at least two quark flavors and assume that these quarks (denote them by u and d) have equal masses $m_u = m_d = m$. Consider a set of Euclidean correlators,

$$C_\Gamma(x, y) = \langle J^{\bar{u}d}(x) J^{\bar{d}u}(y) \rangle_{\text{vac}}, \qquad (14.1)$$

where $J^{\bar{u}d}(x)$ are flavor-changing bilinear quark currents $J^{\bar{u}d} = \bar{u}\Gamma d$ with Hermitian

$$\Gamma = 1,\ \gamma^5,\ i\gamma_\mu,\ \gamma_\mu\gamma^5,\ \text{and}\ i\sigma_{\mu\nu}.$$

At large distances, the correlators (14.1) decay exponentially

$$C_\Gamma(x,y) \propto \exp\{-M_\Gamma|x-y|\}, \qquad (14.2)$$

where M_Γ is the mass of the lowest meson state in the corresponding channel.*

On the other hand, the correlators (14.1) of the quark currents can be expressed as

$$C_\Gamma(x,y) = -Z^{-1}\int d\mu_A \text{Tr}\{\Gamma\mathcal{G}_A(x,y)\Gamma\mathcal{G}_A(y,x)\}, \qquad (14.3)$$

where

$$d\mu_A = \prod_{x\mu a} dA_\mu^a(x)\prod_f \det\|m_f - i\slashed{D}\|\exp\left\{-\frac{1}{4g^2}\int F_{\mu\nu}^a F_{\mu\nu}^a d^4x\right\} \quad (14.4)$$

is the standard QCD measure and $\mathcal{G}_A(x,y)$ is the Euclidean Green function of the u and d quarks in a given gauge-field background. Note that, when writing down Eq. (14.3), we used the fact that $J^{\bar{u}d}$ is not a singlet in flavor [otherwise, the disconnected contribution

$$\propto \text{Tr}\{\Gamma\mathcal{G}_A(x,x)\}\,\text{Tr}\{\Gamma\mathcal{G}_A(y,y)\}$$

would appear on the right side]. In addition, the assumption $m_u = m_d$ was made [otherwise, we would have two different Green's functions $\mathcal{G}_A^u(x,y) \neq \mathcal{G}_A^d(x,y)$].

An important nontrivial relation

$$\gamma^5\mathcal{G}_A(x,y)\gamma^5 = \mathcal{G}_A^\dagger(y,x) \qquad (14.5)$$

holds. To understand it, write the spectral decomposition for $\mathcal{G}_A(x,y)$,

$$\mathcal{G}_A(x,y) = \langle\psi(x)\bar{\psi}(y)\rangle^A = \sum_k \frac{u_k(x)u_k^\dagger(y)}{m - i\lambda_k} \qquad (14.6)$$

*We assume here that the quarks are confined, otherwise the whole discussion is pointless.

(cf. Eq. (13.10)). Using the symmetry $u_k \to \gamma^5 u_k$ and $\lambda_k \to -\lambda_k$, we obtain

$$
\gamma^5 \mathcal{G}_A(x,y)\gamma^5 = \sum_k \frac{[\gamma^5 u_k(x)][\gamma^5 u_k(y)]^\dagger}{m - i\lambda_k} = \sum_p \frac{u_p(x)u_p^\dagger(y)}{m + i\lambda_p}
$$

$$
= \left[\sum_p \frac{u_p(y)u_p^\dagger(x)}{m - i\lambda_p}\right]^\dagger = \mathcal{G}_A^\dagger(y,x), \tag{14.7}
$$

as annonced. We see that the pseudoscalar correlator

$$
\sim \langle \mathrm{Tr}\{\gamma^5 \mathcal{G}_A(x,y)\gamma^5 \mathcal{G}_A(y,x)\}\rangle = \langle \mathrm{Tr}\{|\mathcal{G}_A(x,y)|^2\}\rangle
$$

plays a distinguished role — it represents an absolute upper bound for any other such correlator. The fastest way to show this is to expand the 4×4 matrix $\mathcal{G}_A(x,y)$ over the full basis

$$
\mathcal{G}_A(x,y) = s(x,y) + \gamma^5 p(x,y) + i\gamma_\mu v_\mu(x,y) + \gamma_\mu \gamma^5 a_\mu(x,y)
$$

$$
+ \frac{1}{2}i\sigma_{\mu\nu} t_{\mu\nu}(x,y) . \tag{14.8}
$$

Then

$$
-\frac{1}{4}C_{\gamma^5}^A(x,y) = \frac{1}{4}\mathrm{Tr}\{\gamma^5 \mathcal{G}_A(x,y)\gamma^5 \mathcal{G}_A(y,x)\} = \frac{1}{4}\mathrm{Tr}\{|\mathcal{G}_A(x,y)|^2\}
$$

$$
= |s|^2 + |p|^2 + |v_\mu|^2 + |a_\mu|^2 + \frac{1}{2}|t_{\mu\nu}|^2 , \tag{14.9}
$$

but, say,

$$
-\frac{1}{4}C_1^A(x,y) = \frac{1}{4}\mathrm{Tr}\{\mathcal{G}_A(x,y)\mathcal{G}_A(y,x)\} = \frac{1}{4}\mathrm{Tr}\{\mathcal{G}_A(x,y)\gamma^5 \mathcal{G}_A^\dagger(x,y)\gamma^5\}
$$

$$
= |s|^2 + |p|^2 - |v_\mu|^2 - |a_\mu|^2 + \frac{1}{2}|t_{\mu\nu}|^2 . \tag{14.10}
$$

The inequalities

$$
|C_{\gamma^5}^A(x,y)| \geq |C_\Gamma^A(x,y)| \tag{14.11}
$$

in any given gauge background, the positivity of the measure (14.4) in Eq. (14.3), and the asymptotics (14.2) imply that the mass M_{PS} of the (lightest) pseudoscalar meson in the $\bar{u}d$ channel is less or may be equal to the masses M_S, M_V, M_A, M_T of the lightest scalar, vector, axial, and tensor states.[†]

Let us emphasize again that this statement can be justified only in the theory with the positive measure (14.4) (e.g. in the theory with nonzero vacuum angle $\theta \neq 0$, the measure *is* not positive and pseudoscalar states *need* not to be the lightest), with equal quark masses, and only for those states that are not flavor-singlet. For flavor-singlet states it need not be true. Consider, for example, the theory with just one quark of a large mass. Then the lowest meson states would be made of gluons and would know nothing about quarks. The lowest glueball state is believed to be scalar rather than pseudoscalar.

In Lecture 12 we have mentioned already the Vafa–Witten theorem saying that the vector isotopic symmetry is not broken spontaneously in QCD. Now we are ready to prove it. Indeed, if such a breaking occured, the massless Goldstone scalar particles would appear in the spectrum. The inequality $M_{PS} \leq M_S$ implies that a massless *pseudoscalar* particle would also exist. But the theory with $m_u = m_d \neq 0$ (which duly enjoys the exact isovector symmetry, a possible spontaneous breaking of which is under discussion now) has no exact axial isotopic symmetry, and there are *no reasons* for the massless pseudoscalar state to exist. So, it does not exist, hence the massless scalar does not exist either, and the isovector symmetry is not broken.

Many more inequalities of this kind (e.g. $M_N \geq M_\pi$ or $M_{\pi^+} \geq M_{\pi^0}$) can be formulated, but their proof relies on some extra assumptions and even though the assumptions are very natural, the status of these results is a little bit less solid. We address the reader to Ref. [36] for a nice recent review.

14.2   Euclidean Dirac Spectral Density.

Consider the Euclidean Dirac operator $\not{D}$ in a given gauge field background $A_\mu(x)$. We assume that the system is placed in a finite 4-volume so that

---

[†]We were a little bit sloppy here. The inequalities (14.11) hold strictly speaking only for unrenormalized correlators, not for the renormalized ones. However, renormalization only brings about multiplicative factors, which do not depend on distance. Thus, taking $|x - y|$ to infinity *before* the limit $\Lambda_{UV} \to \infty$ is done, we can ensure that the inequalities (14.11) for renormalized correlators at large distances are fulfilled.

the spectrum of $\not{P}$ is discrete. Let $\{\lambda_k\}$ be the background-dependent set of eigenvalues of $\not{P}$. The *spectral density*[‡] is defined as follows:

$$\rho(\lambda) = \left\langle \frac{1}{V} \sum_k \delta\left(\lambda - \lambda_k[A_\mu(x)]\right) \right\rangle , \qquad (14.12)$$

where the average is done with the weight function (14.4). The γ^5 symmetry of the spectrum implies that $\rho(\lambda)$ is an even function of λ.

In contrast to solids or nuclei, the spectral density (14.12) is defined in Euclidean space and seems to have no direct physical meaning. There are, however, a set of remarkable identities which relate the spectral density of the Euclidean Dirac operator to physical observables. The simplest such identity relates the spectral density at "zero virtuality" $\lambda = 0$ to the quark condensate.

To derive it, set $x = y$ in the spectral decomposition (14.6), integrate it over $\frac{1}{V}d^4x$, and perform the averaging over the gauge fields with weight (14.4). In view of the definitions (12.54), (12.55), and (14.12), using the symmetry $\rho(-\lambda) = \rho(\lambda)$, and assuming the reality of Σ, we obtain

$$\Sigma = \left\langle \sum_k \frac{1}{m - i\lambda_k} \right\rangle = \int_{-\infty}^{\infty} \frac{\rho(\lambda)d\lambda}{m - i\lambda} = 2m \int_0^\infty \frac{\rho(\lambda)d\lambda}{\lambda^2 + m^2} . \quad (14.13)$$

To better understand this formula, let us first look at what happens for free fermions. As there is no physical dimensionfull scale in this case [remember that λ_k in Eq. (14.12) are eigenvalues of the *massless* Dirac operator], $\rho(\lambda) = C|\lambda|^3$ on dimensional grounds. By counting the eigenvalues of the free Dirac operator,

$$\lambda(n_\mu) = \pm\frac{2\pi}{L} \sqrt{\sum_\mu \left(n_\mu + \frac{1}{2}\right)^2} , \qquad (14.14)$$

[‡]The notion of spectral density and the definition (14.12) are also widely used in condensed matter physics and nuclear physics. It is especially useful if a system is disordered or involves elements of disorder like it is the case for electron spectra in most solids or for energy levels in complicated nuclei. It makes sense also for ordered systems (such as metals). In this case, rather than averaging over stochastic external field, one averages over some interval of eigenvalues $\Delta\lambda$ much larger than the characteristic level spacing, but much less than the characteristic scale of λ on which $\rho(\lambda)$ is essentially changed.

[antiperiodic boundary conditions for the fermions in all four directions are chosen, n_μ are integer, and each level (14.14) involves an extra $2N_c$-fold degeneracy] in the 4D ball $1/L \ll |\lambda| < \Lambda$, it is not difficult to determine $C = N_c/(4\pi^2)$. Thus,

$$\rho^{\text{free}}(\lambda) \;=\; \frac{N_c}{4\pi^2}|\lambda|^3 \,. \tag{14.15}$$

In the interacting theory the spectral density behaves as $\rho(\lambda) \propto \lambda^3$ for λ much greater than the characteristic hadron scale μ_{hadr} so that interaction is weak. To be more precise, the power λ^3 is multiplied by the anomalous dimension factor

$$\sim \left[\frac{\alpha_s(\lambda)}{\alpha_s(\mu)}\right]^{\gamma/b_0} \,, \tag{14.16}$$

where $\mu \sim \mu_{\text{hadr}}$ is the normalization point and $b_0 = 11 - 2N_f/3$ [cf. Eq. (9.56)]. A recent one-loop calculation [35] implies that $\gamma = 32$.

We see that the integral in Eq. (14.13) diverges quadratically in the ultraviolet. The same result can be obtained directly by calculating the fermion bubble graph in the momentum representation

$$\langle \psi(0)\bar{\psi}(0)\rangle \;=\; \int \frac{d^4 p_E}{(2\pi)^4}\,\text{Tr}\left\{\frac{\not{p}_E + m}{p_E^2 + m^2}\right\} \propto m\Lambda_{UV}^2 \,. \tag{14.17}$$

Thus, strictly speaking, the formula (14.13) does not make much sense as it stands. Note, however, that even though the (purely perturbative) contribution (14.17) diverges in the ultraviolet, it vanishes in the chiral limit $m \to 0$. The whole point is that in QCD the integral (14.13) acquires an additional *nonperturbative* contribution coming from the region of small λ which survives in the "continuum chiral thermodynamic limit" (*first* $V \to \infty$, *then* $m \to 0$, and only then the ultraviolet cutoff is lifted $\Lambda_{UV} \to \infty$). The fact that chiral symmetry is broken spontaneously *means* that the vacuum expectation value $\langle \psi(0)\bar{\psi}(0)\rangle$ is nonzero in this particular limit.

Obviously, the necessary condition for the condensate to develop is $\rho(0) \neq 0$. Neglecting all terms which vanish in the continuum chiral thermodynamic limit defined above, we finally obtain the famous *Banks–Casher relation*

$$\langle \psi(0)\bar{\psi}(0)\rangle_{\text{vac}} \;\equiv\; \Sigma \;=\; \pi\rho(0) \,. \tag{14.18}$$

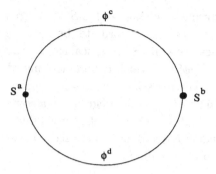

Fig. 14.1 Pseudogoldstone loop in scalar correlator.

Note that the result does not depend on flavor, which tells us again that the flavor vector symmetry is not broken.

Not only $\rho(0)$, but also the form of $\rho(\lambda)$ at small $\lambda \ll \mu_{\text{hadr}}$ can be determined [37]. Consider the theory with $N_f \geq 2$ light quarks of common mass m. Let us study the integrated correlator

$$\int d^4x \langle S^a(x)S^b(0)\rangle = \frac{1}{V}\int d^4x d^4y \langle S^a(x)S^b(y)\rangle , \qquad (14.19)$$

where $S^a(x) = \bar{\psi}(x)t^a\psi(x)$ and t^a is the generator of the $SU(N_f)$ flavor group. Fix a particular gluon background and define

$$C^{ab}\Big|_A = -\frac{1}{V}\int d^4x d^4y \, \text{Tr}\left\{t^a \mathcal{G}_A(x,y)t^b \mathcal{G}_A(y,x)\right\} . \qquad (14.20)$$

Substitute here the spectral decomposition (14.6) for $\mathcal{G}_A(x,y)$, do the integration, and perform averaging over the gluon fields trading the sum over eigenvalues for the integral over the spectral density (14.12). We obtain

$$C^{ab} = -\frac{\delta^{ab}}{2V}\left\langle \sum_k \frac{1}{(m-i\lambda_k)^2}\right\rangle = -\frac{\delta^{ab}}{2}\int_{-\infty}^{\infty}\frac{\rho(\lambda)d\lambda}{(m-i\lambda)^2}$$

$$= -\delta^{ab}\int_0^{\infty}\frac{\rho(\lambda)(m^2-\lambda^2)}{(m^2+\lambda^2)^2}d\lambda , \qquad (14.21)$$

where the property $\rho(-\lambda) = \rho(\lambda)$ was used.

On the other hand, the same correlator can be saturated by physical states, among which the pseudo-Goldstone states play a distinguished role. Consider the 1-loop graph in Fig. 14.1 describing the contribution of the 2-Goldstone intermediate state $\sim \int \langle 0|S^a|\phi^c\phi^d\rangle\langle\phi^c\phi^d|S^b|0\rangle$ (obviously, the one-particle state does not contribute, because the pseudo-Goldstone mesons are pseudoscalars while $S^a(x)$ is scalar). To calculate it, we need to know the vertex $\langle 0|S^a|\phi^c\phi^d\rangle$, which can be determined via the generating functional of QCD involving scalar sources u^a coupled to the current S^a. Adding the source term $u^a S^a$ to the Lagrangian amounts to adding $u^a t^a$ to the quark mass matrix $\mathcal{M}$. The latter also enters the mass term (12.61) in the effective Lagrangian. Expanding U up to second order in ϕ^a and varying $\mathcal{L}_{\text{eff}}^{(m)}$ with respect to u^a, we obtain

$$\langle 0|S^a|\phi^c\phi^d\rangle = -\frac{\Sigma}{F_\pi^2}d^{abc}, \tag{14.22}$$

with d^{abc} defined in Eq. (A.27). The vertex is nonzero only for three or more flavors.

Now we can calculate the graph in Fig. 14.1. Actually, we cannot because the integral diverges logarithmically in the ultraviolet, but anyway the effective theory is not valid at high momenta (technically, the divergence is absorbed into local counterterms of higher order in p^{char} and m). Only the infrared-sensitive part of the integral is relevant. A simple calculation using Eq. (A.28) gives

$$\left(C^{ab}\right)^{\text{infrared}} = -\frac{N_f^2-4}{32\pi^2 N_f}\left(\frac{\Sigma}{F_\pi^2}\right)^2 \delta^{ab}\ln\frac{M_\phi^2}{\mu_{\text{hadr}}^2}. \tag{14.23}$$

Now compare it with Eq. (14.21). Note first of all that the constant part $\rho(0)$ does not contribute here,

$$\int_0^\infty \frac{m^2-\lambda^2}{(m^2+\lambda^2)^2}d\lambda = 0.$$

Thus, only the difference $\rho(\lambda)-\rho(0)$ is relevant. It is easy to see that, in order to reproduce the singularity $\sim \ln M_\phi^2 \sim \ln m$, we should have $\rho(\lambda)-\rho(0) = C|\lambda|$ at small $|\lambda|$. Substituting it in Eq. (14.21) and comparing the coefficient of $\ln m$ with the coefficient of $\ln M_\phi^2$ in Eq. (14.23), we finally obtain [37]

$$\rho(\lambda) = \frac{\Sigma}{\pi} + \frac{N_f^2-4}{32\pi^2 N_f}\left(\frac{\Sigma}{F_\pi^2}\right)^2 |\lambda| + o(\lambda). \tag{14.24}$$

Thus, for $N_f \geq 3$, the spectral density has a nonanalytic "dip" at $\lambda = 0$. The behavior is smooth in the theory with two light flavors. Physically, it is

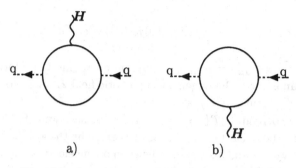

Fig. 14.2   Anomalous triangle in diangle kinematics.

rather natural that the greater the number of flavors, the stronger the suppression of $\rho(0)$ is. The determinant factor in the measure (14.4) punishes small eigenvalues, and the larger N_f, the more important this factor is. By "analytic continuation" of this argument, one should expect a nonanalytic bump rather than a nonanalytic dip at $\lambda = 0$ in the case $N_f = 1$. Indeed, Eq. (14.24) displays such a bump. One should not forget, of course, that the whole derivation was based on the effective chiral Lagrangianapproach and does not directly apply to the case $N_f = 1$. Some additional, more elaborate reasoning shows, however, that a bump at $N_f = 1$ as predicted by Eq. (14.24) is there [38]. The existence of the bump was also confirmed by a numerical calculation in the instanton liquid model [39].

14.3   Infrared Face of Anomaly.

As was explained in Lecture 12, the symmetry (12.19) of the classical theory involving massless fermions is broken down by quantum effects. The $U_A(1)$ breaking is introduced by an ultraviolet regularization [so that the measure in the functional integral is not $U_A(1)$ invariant] and the effects due to this breaking do not go away in the limit $\Lambda_{UV} \to \infty$. This is the ultraviolet face of the anomaly. The latter also has, however, a different, infrared face: one can understand its origin exclusively in terms of the low energy dynamics of the theory. We have already seen how the anomaly is related to the presence of fermion zero modes in an Euclidean topologically nontrivial gauge background. Let us now find out what happens in Minkowski space.

Let us discuss the massless QED first. Consider the correlator

$$T_{\mu\nu}^H(q) = -i \int \langle T\{j_{\mu5}(x)j_\nu(0)\}\rangle_H \, e^{iq\cdot x} d^4x \,, \tag{14.25}$$

where $j_\nu = \bar{\psi}\gamma_\nu\psi$, $j_{\mu5} = \bar{\psi}\gamma_\mu\gamma_5\psi$, and the averaging is performed in the presence of an external homogeneous magnetic field H. The correlator can be calculated perturbatively as a series in e^2 and H. If the electron is massless, the correlator $\langle T\{j_{\mu5}(x)j_\nu(0)\}\rangle_{\text{vac}}$ vanishes and the expansion in H starts with the linear term. The latter is given by the graphs in Fig. 14.2, where the wavy line signals the modification of the electron propagator due to the external field to leading order,

$$\Delta G(p) = e\epsilon_{ijk}H_k \frac{\gamma_i(\not{p}-m)\gamma_j}{2(p^2-m^2)^2} \,. \tag{14.26}$$

[in the massless limit it coincides with Eq. (12.41)]. Actually, the graphs in Fig. 14.2 describe the 3-point correlator

$$T_{\mu\nu\alpha}(q,k) = -i \int \langle T\{j_{\mu5}(x)j_\nu(0)j_\alpha(y)\}\rangle \, e^{iq\cdot x - ik\cdot y} d^4x d^4y \tag{14.27}$$

in a special kinematics: α is spacelike and $k = (0, \boldsymbol{k} \to 0)$. This kinematics is somewhat simpler and, theoretically, more instructive than the standard symmetric kinematics $k^2 = (q-k)^2 = 0$ (which is, on the other hand, well adapted to describe the phenomenology of the decay $\pi^0 \to \gamma\gamma$). To make things simpler still, assume that the vectors q and H are parallel and direct them along the $3^{\underline{d}}$ axis. Calculating the integral with, say, the Pauli–Villars regularization method,[§] we obtain

$$T_{\mu\nu}^H(q) = \frac{eH}{2\pi^2} \frac{q_\mu \tilde{\epsilon}_{\nu\alpha}q^\alpha}{q^2} \,, \tag{14.28}$$

with $H = |H|$ and $\tilde{\epsilon}_{\nu\alpha}$ living in the two-dimensional (03)-space so that $\tilde{\epsilon}_{03} = -\tilde{\epsilon}_{30} = -1$, and $\tilde{\epsilon}_{\perp\alpha} = 0$.

The amplitude (14.28) satisfies the property $q^\nu T_{\mu\nu}^H = 0$ which reflects the conservation of the vector current. However,

$$q^\mu T_{\mu\nu}^H = \frac{eH}{2\pi^2}\tilde{\epsilon}_{\nu\alpha}q^\alpha \neq 0 \,, \tag{14.29}$$

[§]Calculational details will be given at the end of this section in the two-dimensional case, which is somewhat simpler.

and this is a manifestation of the anomaly (12.26). Indeed, Eq. (14.29) means that the average of the operator $\partial^\mu j_{\mu5}$ in the presence of a magnetic *and* an electric field $E_i = \partial_0 A_i - \partial_i A_0$ is equal to ¶

$$\langle \partial^\mu j_{\mu5} \rangle_{E,H} = \frac{e^2}{2\pi^2} \boldsymbol{E} \cdot \boldsymbol{H} . \tag{14.30}$$

This agrees with Eq. (12.26) (where the electric charge is absorbed into field normalization).

The important observation is that the amplitude (14.25) is singular at $q^2 = 0$, and this singularity can only be explained by the presence of *massless particles* in the spectrum. It is very instructive to see what happens if electrons are endowed with a small mass, which explicitly breaks the $U_A(1)$ invariance and also smears out the singularity in Eq. (14.28). The direct calculation of Im $T_{\mu\nu}^{H,m}$ by the graphs in Fig. 14.2 with nonzero mass gives

$$\text{Im } T_{\mu\nu}^{H,m} = -\frac{eH}{\pi} \frac{m^2 \theta(q^2 - 4m^2)}{\sqrt{q^2(q^2 - 4m^2)}} \frac{q_\mu \tilde{\epsilon}_{\nu\alpha} q^\alpha}{q^2} \xrightarrow{m \to 0} -\frac{eH}{2\pi} \delta(q^2) q_\mu \tilde{\epsilon}_{\nu\alpha} q^\alpha .$$

$$\tag{14.31}$$

This means that anomalous nonconservation of the axial charge in QED is associated with the creation of massless e^+e^- pairs of zero energy in the presence of electric and magnetic fields with $\boldsymbol{E} \cdot \boldsymbol{H} \neq 0$. These pairs carry nonzero axial charge. If $\boldsymbol{E}$ and $\boldsymbol{H}$ are constant and homogeneous, the pairs are created all the time and everywhere. If the fields die out fast enough at spatial infinity and also in the limits $t \to \pm\infty$, the number of created pairs is finite and coincides with the change of the axial charge which, according to Eq. (14.30), is equal to

$$\Delta Q_5 = \frac{e^2}{2\pi^2} \int d^3x dt \; \boldsymbol{E} \cdot \boldsymbol{H} . \tag{14.32}$$

Problem. Discuss a diagrammatic interpretation of the 2D anomaly

---

¶To see this in the particular case $\boldsymbol{E} \| \boldsymbol{H}$, which corresponds to our simplified kinematics, one has to substitute Eq. (14.25) into Eq. (14.29), expand the left side over q and compare it with

$$\langle \partial^\mu j_{\mu5}(0) \rangle_{\boldsymbol{E}\|\boldsymbol{H}} = ie \int \langle T\{j_{\mu5}(0)j_0(x)\}\rangle_H (-Ex^3) d^4x .$$

(12.27) in the Schwinger model.

Solution. The right side of Eq. (12.27) is linear in field, which means that the anomaly shows up in the vacuum 2-point correlator

$$T_{\mu\nu}^{2D}(q) = -i \int \langle T\{j_{\mu 5}(x)j_\nu(0)\}\rangle \, e^{iq\cdot x} d^2x \,. \tag{14.33}$$

Thus, instead of the graphs in Fig. 14.2, we have just one anomalous diangle. In spite of the fact that the corresponding Feynman integral converges in the ultraviolet (it diverges logarithmically by power counting, but the leading term $\propto \text{Tr}\{\gamma_\mu\gamma_5\slashed{p}\gamma_\nu\slashed{p}\}$ vanishes after integration), it still has to be regularized (otherwise, the property $q^\nu T_{\mu\nu} = 0$ required by gauge invariance would not hold). We will choose Pauli–Villars regularization. Subtracting the heavy fermion loop, we obtain for small $m^2 \ll |q^2|$

$$
\begin{aligned}
T_{\mu\nu}(q) &= -i \lim_{M\to\infty} \int \frac{d^2p}{(2\pi)^2} \left[\text{Tr}\left\{\gamma_\mu\gamma_5 \frac{\slashed{p}}{p^2}\gamma_\nu \frac{\slashed{p}+\slashed{q}}{(p+q)^2}\right\} \right. \\
&\qquad\qquad \left. - \frac{M^2}{(p^2-M^2)^2}\text{Tr}\{\gamma_\mu\gamma_5\gamma_\nu\} \right] \\
&= \frac{1}{2\pi}\left[-\frac{(\epsilon_{\mu\alpha}q_\nu + \epsilon_{\nu\alpha}q_\mu)q^\alpha}{q^2} + \epsilon_{\mu\nu} \right] = -\frac{q_\mu\epsilon_{\nu\alpha}q^\alpha}{\pi q^2},
\end{aligned} \tag{14.34}
$$

where the conventions (N.17) and the kinematic identity $\eta_{\alpha\beta}\epsilon_{\mu\nu} + \eta_{\beta\mu}\epsilon_{\nu\alpha} + \eta_{\nu\beta}\epsilon_{\alpha\mu} = 0$ were used. The result (14.34) coincides with Eq. (14.28) up to the change $\bar\epsilon_{\nu\alpha} \to \epsilon_{\nu\alpha}$ and without the factor $-eH/2\pi$. The relations (14.29)–(14.31) also have their exact two-dimensional counterparts. In addition, the interpretation is the same: an external electric field brings about the change of axial charge due to the creation of soft massless e^+e^- pairs.

As a side remark, note that in two dimensions vector and axial currents are related as $j_{\mu 5} = \epsilon_{\mu\nu}j_\nu$, and the singularity of the correlator (14.33) means also the singularity of the vector polarization operator $\Pi_{\mu\nu} \propto (\eta_{\mu\nu} - q_\mu q_\nu/q^2)$, i.e. nonvanishing $\Pi(0)$. As a result, the Schwinger photon acquires the mass,

$$\mu^2 = e^2/\pi \,. \tag{14.35}$$

Actually (we leave it now without proof), the only physical states in the massless Schwinger model are *free bosons* with mass (14.35). And that means that the fermion fields entering the Lagrangian of the model are confined! We will return to the discussion of this issue in the last lecture.

14.4   Chiral Symmetry Breaking and Confinement.

Look again at Eq. (12.21). The axial current entering the left-hand side is an external current in a sense that no dynamical field is coupled directly to $j_{\mu 5}$. But the fields entering on the right-hand side are dynamical gluon fields present in the QCD Lagrangian.

In the chiral (left-right asymmetric) gauge theories like the standard electroweak model, both vector and axial currents are coupled directly to the physical gauge fields. Anomalous divergence of such current would mean explicit breaking of the gauge invariance, which is not nice. Therefore, in chiral theories one should always take care that such purely *internal anomalies* would cancel out at the end of the day. In the Standard Model, they do.

Let us discuss, however, purely *external anomalies* in QCD which are not related to breaking of any symmetry but only mean that certain correlators involving external currents are singular.[||]

As a simplest nontrivial example, consider the theory with two massless flavors and look at the correlator

$$K_{\mu\nu}^{ab,\ \mathcal{H}}(q) = -i \int \langle T\{j_{\mu 5}^a(x) j_\nu^b(0)\}\rangle_{\mathcal{H}}\ e^{iq\cdot x} d^4 x \ , \qquad (14.36)$$

where a, b are flavor indices and $\mathcal{H}$ is the external flavor-singlet "magnetic field".[**] The correlator (14.36) is nothing but a three-point vacuum expectation value (14.27) in kinematics in which one of the external momenta associated with the vector current is set to zero.

The one-loop calculation of the corresponding graph displays a singularity,

$$K_{\mu\nu}^{ab,\ \mathcal{H}}(q) = \frac{\mathcal{H}}{2\pi^2} \frac{q_\mu \tilde{\epsilon}_{\nu\alpha} q^\alpha}{q^2} \times N_c \times \frac{1}{2}\delta^{ab} \ , \qquad (14.37)$$

where the last two factors come from the color and flavor trace. The imaginary part of this amplitude is also singular, $\sim \delta(p^2)$, which can be re-

----

[||]The anomaly (12.21) is related to the correlator (14.27) involving both internal (j_μ^a) and external (j_λ^5) currents and can be called *mixed* in this setting.

[**]The quotation marks distinguish $\mathcal{H}$ that couples to the baryon charge from the physical magnetic field, which has the matrix structure $\mathrm{diag}(2/3, -1/3)$ and is a mixture of isotriplet and isosinglet. But we are not interested in dynamics of electromagnetic or weak interactions here. In QCD proper, all color-singlet currents are external. $\mathcal{H}$ is just a source of such vector flavor-singlet current.

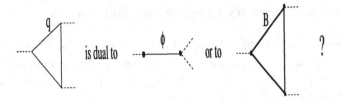

Fig. 14.3   Saturating the external anomaly.

lated to the masslessness of quarks [40]. However, the quarks (in contrast to electrons in QED) do not exist as physical particles due to confinement and one can ask where does the singularity in the imaginary part Im $K_{\mu\nu}^{ab,\,\mathcal{H}}(q) \propto \delta(q^2)$ come from? This is a good question, the answer is better still: the singularity $\sim 1/q^2$ comes from the propagator of a massless Goldstone boson, which appears due to the spontaneous chiral symmetry breaking and which is directly coupled to the axial current $j_{\mu 5}$ (see the middle graph in Fig. 14.3).

Let us ask now: can one reproduce the singularity in Eq. (14.37) *without* Goldstone bosons and without spontaneous chiral symmetry breaking, but in some other way?

As far as the theory with two light quarks is concerned, the answer is positive: the singularity of the correlator above can be reproduced, in principle, if *massless baryons* are present. Protons and neutrons represent, like quarks, a flavor SU(2) doublet. There are $N_c = 3$ quark doublets and only one baryon doublet $|P\rangle = |uud\rangle$ and $|N\rangle = |udd\rangle$. The absence of the overall N_c factor is compensated, however, by the fact that the baryon charge of nucleon is 3 times larger than that of the quark, and the vertex involving the "magnetic field" $\mathcal{H}$ is 3 times larger for baryons.

Thus, this purely algebraic *anomaly matching* argument due to 't Hooft does not rule out a dynamical scenario where the physical spectrum in the theory with just two massless quark flavors would not involve massless pions but, instead, the massless proton and neutron. It is rather remarkable that, in the theories with $N_f \geq 3$, the scenario with massless baryons *is* ruled out. Suppose that, instead of the octet of massless Goldstone fields, we have an octet of massless baryons. The contribution of the corresponding triangle graph in (14.36) would have the same structure as in Eq. (14.37), but with the factor

$$\text{Tr}\,\{T^a T^b\} = C_8 \delta^{ab} \tag{14.38}$$

instead of $\delta^{ab}/2$, where T^a are the flavor generators in the octet representation. To find the *Dynkin index* of the octet representation C_8, it is sufficient to assume that $a = b = 1$ and decompose the octet with respect to the SU(2) flavor subgroup: $\mathbf{8 = 3 + 2 + 2 + 1}$. The contribution of each doublet to C_8 is $1/2$ and the contribution of the triplet is 2. Adding it together, we obtain $C_8 = 3 \neq 1/2$ and the required result (14.37) is *not* reproduced. Also, a massless decuplet and all other possible color-singlet baryon representations would give the coefficient in front of the singularity much larger than that in Eq. (14.36), and the anomaly matching condition would not be fulfilled. Therefore, massless baryons do not exist.[tt]

We have arrived at a remarkable result. In QCD with three massless quarks, the assumption of confinement, allowing the existence of only colorless states in the physical spectrum, *and* the anomaly matching condition lead to the conclusion that massless Goldstone states *should* appear and chiral symmetry *must* be spontaneously broken. If the latter is not true, the only possibility to saturate the anomaly is to assume that massless quarks still exist as physical states in the spectrum and there is no confinement!

In the real world confinement and spontaneous chiral symmetry breaking in the limit of massless u, d, and s quarks are experimental facts. Whether or not these phenomena take place in hypothetic theories with $N_f \geq 4$ is an open question. It is quite possible that starting, say, from $N_f \overset{?}{=} 6$, the small eigenvalues in the Dirac operator spectrum are punished by the determinant factor so strongly that $\rho(0) = 0$ and, in view of the Banks–Casher relation (14.18), the quark condensate vanishes and the symmetry is not broken. The anomaly matching argument tells then that

---

[tt]To be quite precise, one could, in principle, saturate the anomaly with *several* baryon multiplets with positive and negative baryon charges. This possibility is so unaesthetic, however, that it can be rejected simply by that reason.

there is no confinement in this case.[‡‡]

We have called this result — that the chiral symmetry breaking and confinement go together — remarkable. It is also somewhat mysterious. Even though we do not understand well dynamical reasons for confinement to occur, we still expect that *the same* mechanism which works for the theory with $N_f = 3$ works also for the theory with $N_f = 2$. But 't Hooft's argument works only for $N_f \geq 3$...

[‡‡]We know, of course, that if the number of the quark flavors is *very* high $N_f > 16$, the asymptotic freedom is lost and we cannot expect confinement. We are almost sure that quarks and gluons are not confined at $N_f = 16$ or $N_f = 15$, in which case the theory is asymptotically free, but has the infrared fixed point (9.38) at a small value of α_s, and the coupling constant never grows large. The argument about suppression of the small Dirac eigenvalues and the results of some numerical simulations indicate that confinement might be lost at a smaller value of N_f, not just $N_f = 15$.

Lecture 15
Mesoscopic QCD

In this lecture, we derive a set of exact relations for QCD placed in a finite Euclidean volume. The importance of these results is that they can be compared with numerical lattice calculations. Like the exact theorems of the previous lecture, they rely heavily on the ideas of quark–hadron duality.

15.1 Partition Function: $N_f = 1$.

Consider first the theory with just one light quark flavor in a finite Euclidean box whose size L is much larger than the characteristic scale μ_{hadr}^{-1}. We assume that the quark mass m is complex so that the mass term in the Lagrangian has the form

$$\mathcal{L}_m^{N_f=1} = m\bar{\psi}_R\psi_L + \bar{m}\bar{\psi}_L\psi_R , \qquad (15.1)$$

and the vacuum angle θ is nonzero. The partition function of the theory can be written in the form

$$Z(\theta) = \exp\{-V\epsilon(\theta, m, L)\} , \qquad (15.2)$$

where $V = L^4$ and $\epsilon(\theta, m, L)$ becomes the vacuum energy density $\epsilon_{\text{vac}}(\theta, m)$ in the thermodynamic limit $L \to \infty$. It is important that the finite volume corrections $\epsilon(\theta, m, L) - \epsilon_{\text{vac}}(\theta, m)$ are exponentially suppressed $\propto \exp\{-\mu_{\text{hadr}}L\}$. Indeed, the corrections appear due to modification of the spectrum of excitations in the finite box by the Casimir mechanism. At the technical level, they are due to modification of the corresponding Green's functions by finite volume effects. But in the theory with $N_f = 1$, all physical (meson and baryon) excitations are massive and their Green's functions decay as $\exp\{-\mu_{\text{hadr}}R\}$ at large distances. Also, the modification of the Green's functions at coinciding points due to boundary effects has the order $\exp\{-\mu_{\text{hadr}}L\}$.

The vacuum energy depends on quark mass m and vacuum angle θ. It turns out that ϵ_{vac} is actually a function of just one complex parameter, which is $me^{i\theta}$. To see that, present $Z(\theta)$ as a series

$$Z(\theta) = \sum_{q=-\infty}^{\infty} e^{-iq\theta} Z_q , \qquad (15.3)$$

where Z_q is given by the functional integral over quark and gluon fields in the sector of definite topological charge q [cf. Eq. (5.25)]. Consider some particular q and suppose for definiteness that $q > 0$. One of the factors entering the integrand of Z_q is the quark determinant

$$\det \left\| -i\not{D}^E + m\frac{1-\gamma^5}{2} + \bar{m}\frac{1+\gamma^5}{2} \right\|$$

$$= \bar{m}^q \prod_k{}' \det \left\| \begin{matrix} \bar{m} & -i\lambda_k \\ -i\lambda_k & m \end{matrix} \right\| = \bar{m}^q \prod_k{}' (\lambda_k^2 + \bar{m}m) , \qquad (15.4)$$

where the product $\prod'$ runs over nonzero λ_k's and the factor $\bar{m}^q$ reflects the presence of q right-handed fermion zero modes, as the index theorem (12.34) requires [cf. Eq. (4.32) written earlier for the case of real mass]. If $q < 0$, the factor $\bar{m}^q$ should be substituted by m^{-q}. Averaging the determinant (15.4) over the gluon field configurations with a given q and substituting it in Eq. (15.3), we obtain that each term in the sum depends, indeed, only on the combinations $z = me^{i\theta}$ and $\bar{z}$ (being real, the partition function and $\epsilon_{\text{vac}} \propto \ln Z$ cannot, of course, be holomorphic in z).

Next we assume that $|m| \ll \mu_{\text{hadr}}$ and that the dependence of ϵ_{vac} on mass is analytic. This allows us to expand

$$\epsilon_{\text{vac}} = \epsilon_0 - \Sigma \text{Re}\left(me^{i\theta}\right) + o(m) , \qquad (15.5)$$

where Σ is a real constant, which, for $\theta = 0$ and real m, coincides with the variation of the vacuum energy with respect to mass, i.e. with the quark condensate $\langle \psi\bar{\psi} \rangle_{\text{vac}}$.* Substituting Eq. (15.5) into Eq. (15.2) and comparing

---

*A finite nonzero value of the quark condensate was exactly our assumption when we have written the expansion (15.5).

it with the Fourier expansion (15.3), we can derive

$$Z_q \propto \frac{1}{2\pi} \int_0^{2\pi} d\theta e^{iq\theta} e^{V\Sigma m \cos\theta + o(m)} = I_q(m\Sigma V) \qquad (15.6)$$

$[I_q(x) = I_{-q}(x)$ stands for the Bessel function which grows exponentially for large real x].

It is very instructive now to consider two limits: *(i)* $m\Sigma V \gg 1$ and *(ii)* $m\Sigma V \lesssim 1$ (keeping the assumption $L\mu_{\text{hadr}} \gg 1$). The limit *(i)* is a genuine thermodynamic limit. The sum (15.3) is saturated by many terms with large values of $|q|$. Actually, one easily derives (for any $m\Sigma V$)

$$\langle q^2 \rangle_{\text{vac}} = m\Sigma V . \qquad (15.7)$$

The quantity $\chi = \langle q^2 \rangle_{\text{vac}}/V$ can also be expressed via the correlator of topological charge densities,

$$\chi = \int d^4x \left\langle \frac{1}{16\pi^2} \text{Tr}\{F_{\mu\nu}\tilde{F}_{\mu\nu}\}(x) \; \frac{1}{16\pi^2}\text{Tr}\{F_{\mu\nu}\tilde{F}_{\mu\nu}\}(0) \right\rangle , \qquad (15.8)$$

and is called the topological susceptibility of the vacuum. We have already met it in Lecture 5. Eq. (5.6) relates χ to the curvature of the function $\epsilon_{\text{vac}}(\theta)$ at $\theta = 0$. The region

$$\mu_{\text{hadr}}^{-1} \ll L \lesssim (m\Sigma)^{-1/4} \qquad (15.9)$$

can be called *mesoscopic*. The word is borrowed from condensed matter physics, where mesoscopic system is defined as a system involving a large enough number of particles for the statistical description to be adequate, but which is not sufficiently large for the boundary effects to be irrelevant. The same is true in our case. In particular, when $m\Sigma V \ll 1$, the partition function is well approximated by just the first term with $q = 0$ and, if we are interested in the mass dependence, by the terms with $q = \pm 1$ corresponding to the instanton and anti-instanton sectors.

If $m\Sigma V \ll 1$, the fermionic condensate is provided by instanton-like configurations supporting a single fermion zero mode. If $m\Sigma V \gg 1$, the condensate appears due to the finite average density of small but nonzero eigenvalues of the Dirac operator, according to the Banks and Casher formula (14.18). Quite remarkably, the *value* of the condensate is the same, however, and does not depend on the parameter $m\Sigma V$.

Note that if m is real and $\theta = 0$ [more generally, if the combination $\theta_{\text{phys}} = \theta + \arg(m)$ on which everything depends vanishes], all terms in the partition function (15.3) are positive and the path integral for $Z(\theta)$ has a probabilistic interpretation. And if not — then not.

15.2   Partition Function: $N_f \geq 2$.

The spectrum of the theory with several light flavors involves light pseudo-Goldstone states. If the length of our box is comparable with the Compton wavelength of pseudo-Goldstone particles, finite volume corrections in $\epsilon(\theta, m, L)$ in Eq. (15.2) are not exponentially small and should be taken into account. We will be interested here in the region

$$\mu_{\text{hadr}}^{-1} \ll L \ll \frac{1}{M_\pi} \sim \frac{1}{\sqrt{m\mu_{\text{hadr}}}} . \tag{15.10}$$

The scale $\sim m^{-1/4}\mu_{\text{hadr}}^{-3/4}$ entering Eq. (15.9) and lying in the middle of the newly defined mesoscopic interval (15.10) is relevant in the multiflavor case too. The formulae we are going to derive in this section will be universally valid, however, for any value of $m\Sigma V$ as long as the condition (15.10) is satisfied.

The main idea is to present the partition function as the path integral in the *effective theory* (12.57) and (12.61), and notice that, when the condition $LM_\pi \ll 1$ is fulfilled, only the zero Fourier harmonics of the pseudo-Goldstone fields are relevant.[†] Then we can disregard the kinetic term and keep only the mass term (12.61) assuming $U(x) = const$. The path integral is reduced to an ordinary one. We want to calculate the partition function at arbitrary θ and need to know the form of the effective potential at arbitrary θ. In the case $N_f = 1$, all physical quantities depended on the combination $me^{i\theta}$, but not on m and θ separately. By the same token, for $N_f \geq 2$, they depend on the combination $\mathcal{M}e^{i\theta/N_f}$ (where $\mathcal{M}$ is the $N_f \times N_f$ complex mass matrix, $\mathcal{L}_\mathcal{M} = \bar{\psi}_R \mathcal{M} \psi_L + \bar{\psi}_L \mathcal{M}^\dagger \psi_R$). The proof

---

[†]This can be observed by looking at the pseudo-Goldstone Green's function in finite volume:

$$G_\pi(x) = \frac{1}{V} \sum_{\{n_\mu\}} \frac{e^{ipx}}{M_\pi^2 + \left(\frac{2\pi n_\mu}{L}\right)^2} .$$

is quite analogous and is based on the multiflavor version of Eq. (15.4),

$$
\det \left\| -i\not{D}^E + \mathcal{M}\frac{1-\gamma^5}{2} + \mathcal{M}^\dagger\frac{1+\gamma^5}{2} \right\| =
$$

$$
[\det \|\mathcal{M}^\dagger\|]^q \prod_k{}' \det \left\| \begin{matrix} \mathcal{M}^\dagger & -i\lambda_k \\ -i\lambda_k & \mathcal{M} \end{matrix} \right\|
$$

$$
= [\det \|\mathcal{M}^\dagger\|]^q \prod_k{}' \det \|\lambda_k^2 + \mathcal{M}^\dagger\mathcal{M}\| \,. \tag{15.11}
$$

Similar to the case $N_f = 1$, we see that the physical vacuum angle on which all physical quantities depend is

$$
\theta_{\text{phys}} = \theta + \arg(\det \|\mathcal{M}\|) \,, \tag{15.12}
$$

rather than just θ. In particular, the statement at the end of Lecture 5 that the vacuum angle in QCD is either zero or very small refers to the combination (15.12).

Thus, we can write

$$
Z_{N_f}(\theta) = \int dU \exp\left\{ V\Sigma\text{Re}\left[\text{Tr}\{Me^{i\theta/N_f}U\}\right] \right\} \,, \tag{15.13}
$$

where the integral is done over the group $SU(N_f)$ with the proper Haar measure. Integrating it further over $d\theta$ as in Eq. (15.6), we find the partition function Z_q in the sector with a definite topological charge. In the simple case $\mathcal{M} = m\mathbb{1}$, a beautiful analytic expression for Z_q can be derived (see **Problem 1** below):

$$
Z_q = \det \left\| \begin{matrix} I_q(\kappa) & I_{q+1}(\kappa) & \cdots & I_{q+N_f-1}(\kappa) \\ I_{q-1}(\kappa) & I_q(\kappa) & \cdots & I_{q+N_f-2}(\kappa) \\ \vdots & \vdots & \vdots & \vdots \\ I_{q-N_f+1}(\kappa) & \cdots & \cdots & I_q(\kappa) \end{matrix} \right\| \,, \tag{15.14}
$$

where $\kappa = |m\Sigma V|$.

Let us assume $\theta = 0$ and m real and study the quantity

$$
\langle\psi\bar\psi\rangle_{m,V} = \frac{1}{N_f V}\frac{\partial}{\partial m}\ln Z \,, \tag{15.15}
$$

the quark condensate at finite mass and finite volume. In the limit $\kappa \to \infty$, the integral in Eq. (15.13) has the main support in the region $U \sim 1$, and we obtain

$$Z_{N_f} \propto \frac{e^{N_f \kappa}}{\kappa^{(N_f^2-1)/2}} \; , \tag{15.16}$$

so that the condensate (15.15) tends to a constant Σ as it should. [Note the presense of the power-like pre-exponential factor in Eq. (15.16). It reflects the fact that, while the condition $LM_\pi \ll 1$ is satisfied and higher Fourier harmonics of pion fields are decoupled, we are not in the true thermodynamic limit yet. In the latter, the partition function is given by a simple exponential as in Eq. (15.2), and the finite volume corrections to $\epsilon(\theta, m, L)$ are suppressed as $\exp\{-M_\pi L\}$.]

In the opposite limit $\kappa \ll 1$, $Z_{q \neq 0}$ are suppressed compared to Z_0 as $\kappa^{|q|N_f}$, reflecting the presence of $N_f|q|$ fermion zero modes in the sector with topological charge q. The condensate (15.15) is also suppressed $\propto \kappa$. This is due to the fact that, in the multiflavor case, the condensate plays a role of the order parameter signaling the spontaneous breaking of the chiral symmetry. But, strictly speaking, spontaneous symmetry breaking never occurs in quantum systems at finite volume in a sense that the true ground state wave function is always symmetric even though the gap separating it from excited asymmetric states may be very (exponentially) small.[‡] Therefore, for $N_f > 1$, the condensate should vanish in the chiral limit $m \to 0$ no matter how large a fixed volume V is. And it does.

15.3 Spectral Sum Rules.

Let us return to a simple case, $N_f = 1$. Assume that m is real and q is positive, and rewrite the result (15.6) as

$$\left\langle m^q \prod_k{}' (\lambda_k^2 + m^2) \right\rangle_q = CI_q(\kappa) \; , \tag{15.17}$$

where the averaging is done over the gluon field configurations with a given topological charge q. We can expand now the two sides of Eq. (15.17) in m and compare the coefficients of the expansion. The first nontrivial relation (one of the so called *Leutwyler–Smilga sum rules*) is obtained when the

---

[‡]Never say never: the statement above is not correct for supersymmetric systems. But QCD is not supersymmetric and we may forget this subtlety.

terms $\propto m^q$ and $\propto m^{q+2}$ are compared. We obtain

$$\left\langle\!\!\left\langle \sum_k{}' \frac{1}{\lambda_k^2} \right\rangle\!\!\right\rangle_q = \frac{(\Sigma V)^2}{4(q+1)}, \tag{15.18}$$

where the sum runs over only nonzero eigenvalues and the symbol $\langle\!\langle\cdots\rangle\!\rangle$ means averaging with the weight $\exp\{-S_g\}\prod_k{}' \lambda_k^2$ (the second factor is the fermionic determinant in the limit $m \to 0$, with discarded zero modes).

Expanding Eq. (15.17) up to the order $\sim m^{q+4}$, we obtain the next sum rule

$$\left\langle\!\!\left\langle \sum_{k\neq l}{}' \frac{1}{\lambda_k^2 \lambda_l^2} \right\rangle\!\!\right\rangle_q = \frac{(\Sigma V)^4}{16(q+1)(q+2)}. \tag{15.19}$$

Similar relations can be derived in the multiflavor case. Expanding Equation (15.14) in mass and comparing the terms $\sim m^{N_f q}$, $\sim m^{N_f q+2}$, and $\sim m^{N_f q+4}$, we obtain the same relations (15.18) and (15.19), with q being substituted by $q + N_f - 1$. We can, however, extract more information by considering the expression for the partition function with *different* quark masses. Two different sum rules can be derived by considering the terms $\sim m^{N_f q+4}$ of the expansion (see **Problem 2**):

$$\left\langle\!\!\left\langle \left(\sum_k{}' \frac{1}{\lambda_k^2} \right)^2 \right\rangle\!\!\right\rangle_q = \frac{(\Sigma V)^4}{16\left[(q+N_f)^2 - 1\right]},$$

$$\left\langle\!\!\left\langle \sum_k{}' \frac{1}{\lambda_k^4} \right\rangle\!\!\right\rangle_q = \frac{(\Sigma V)^4}{16(q+N_f)\left[(q+N_f)^2 - 1\right]}. \tag{15.20}$$

Subtracting one sum rule in Eq. (15.20) from the other and setting $N_f = 1$, we reproduce Eq. (15.19), but the relations (15.20) are valid only for $N_f \geq 2$. Actually, the average $\langle\!\langle\sum_k{}'(1/\lambda_k^4)\rangle\!\rangle_q$ diverges at $N_f = 1$. For $N_f = 2$, there are two sum rules on each level with higher inverse powers of λ. For $N_f = 3$, three different sum rules can be derived starting from the terms $\sim m^{N_f q+6}$ of the expansion, etc.

The left side of the sum rule (15.18) can be presented as the integral

$$V \int \frac{\rho_q(\lambda)}{\lambda^2} \, d\lambda,$$

where $\rho_q(\lambda)$ is the microscopic spectral density (14.12) in the sector with a given topological charge q. The sum rules (15.19) and the first sum rule in Eq. (15.20) are not expressed in terms of $\rho(\lambda)$ only, but also via a certain integral involving the correlation function $\rho(\lambda_1, \lambda_2)$. The sum rule for

$$\langle\!\langle \sum_{k \neq l \neq p}' (1/\lambda_k^2 \lambda_l^2 \lambda_p^2) \rangle\!\rangle_q$$

is expressed in terms of an integral of the correlator $\rho(\lambda_1, \lambda_2, \lambda_3)$, etc. Using ingenious arguments which are beyond the scope of our discussion here, J. Verbaarschot, I. Zahed, and D. Toublan managed to determine the functional form of $\rho_q(\lambda)$ and of all correlators. For example,

$$\rho_{N_f,q}(\lambda) = \frac{\Sigma^2 V |\lambda|}{2} \left[J_{N_f+|q|}^2(\Sigma V \lambda) \right.$$
$$\left. - J_{N_f+|q|+1}(\Sigma V \lambda) J_{N_f+|q|-1}(\Sigma V \lambda) \right] , \qquad (15.21)$$

where $J_\nu(x)$ are standard oscillating Bessel functions. The expression is valid for $\lambda \ll \mu_{\text{hadr}}$, and the function is "alive" in the region of very small eigenvalues $\lambda \sim 1/\Sigma V$. In the thermodynamic limit $\lambda \Sigma V \to \infty$, the function (15.21) tends to the constant Σ/π in agreement with the theorem (14.18).

The spectral sum rules are well adapted to be confronted with the numerical lattice calculations. The main interest here is not so much to "confirm" these exact theoretical results by computer, but rather to test lattice methods. This was a challenge for lattice people for some time. By now reasonably good numerical data were obtained, they agree well with theory. One can expect that the accuracy will grow substantially if the algorithms based on the lattice fermion action with exact chiral symmetry (as was discussed in details in Lecture 13) are implemented.

15.4   Instanton Gas and Instanton Liquid.

Let us look again at the exact result (15.6). Remarkably enough, its form coincides exactly with Eq. (5.19) obtained for the quantum pendulum problem in the framework of the dilute instanton gas picture. For the quantum pendulum, the result was exact and the picture was exactly correct if the action of an individual instanton $S_I = 8\omega$ was large. Let us formulate now the dilute instanton gas model in QCD and discuss its applicability.

Consider the Euclidean path integral for the partition function in the sector $q = 1$. Let us calculate it in the semi-classical approximation assuming that characteristic field configurations saturating the integral are close to the classical instanton solution (2.18). Eq. (2.18) is written for the instanton centered at $x = 0$. Of course, the solutions centered at any $x = x_0$ are also admissible and they enter on the equal footing. Also, the size ρ of the solution can acquire any value (that is in contrast to the quantum pendulum model, where the characteristic Euclidean time extension of the instanton was fixed $\rho_{\text{pend}} \sim \omega^{-1}$). At classical level, the instanton action does not depend on ρ. But, when the Gaussian functional integral over the fluctuations around the classical instanton solution is actually calculated (this was done by 't Hooft), the classical action $8\pi/g_0^2$ is substituted by the effective action

$$S_I^{\text{eff}} = \frac{2\pi}{\alpha_s(\rho)}, \tag{15.22}$$

where $\alpha_s(\rho)$ is the running coupling constant. Thus, the analog of Eq. (5.15) in QCD with one flavor is

$$\begin{aligned} \frac{Z_1}{Z_0} &= \int \frac{d^4 x_0 \, d\rho}{\rho^5} F(m\rho) \exp\left\{ -\frac{2\pi}{\alpha_s(\rho)} \right\} \\ &= V \int \frac{d\rho}{\rho^5} F(m\rho) \exp\left\{ -\frac{2\pi}{\alpha_s(\rho)} \right\}, \end{aligned} \tag{15.23}$$

where the factor ρ^5 downstairs appears just by dimensional reasons. $F(m\rho)$ is dimensionless. In the QCD with one flavor, $F(m\rho) = Cm\rho$ (it should be so because $Z_1 \propto m$ in the chiral limit). To leading order, the running charge $\alpha_s(\rho)$ is given by Eq. (9.34) (with ρ^{-1} being substituted for μ). Substituting it in Eq. (15.23), we find

$$\frac{Z_1}{Z_0} \propto mV (\mu_{\text{hadr}})^{b_0} \int \rho^{b_0 - 4} d\rho \tag{15.24}$$

with $b_0 = (11N_c - 2)/3$. The integral (15.24) diverges for large ρ, which signals the *breakdown* of semi-classical approximation. Let us assume, however, that the integral over ρ is effectively cut off at the confinement scale μ_{hadr}^{-1}, so that $Z_1/Z_0 \sim mV\mu_{\text{hadr}}^3$ (we know from the exact analysis at the beginning of this lecture that it *is* true, indeed, if $mV\mu_{\text{hadr}}^3 \ll 1$).

The main assumption of the instanton gas model is that, even for large $m\Sigma V$, characteristic field configurations in the Euclidean path integral represent a superposition of some number of instantons and anti-instantons. Both the characteristic distance between instantons and their characteristic size are of order μ_{hadr}^{-1}. The model assumes, however, that the former is several times larger than the latter. As a result, the form of individual instantons is not so much distorted and correlations between different instantons and anti-instantons are not so much important. As this model gives at least one thing [the behavior of $Z_q(m\Sigma V)$ in the whole range of $m\Sigma V$] quite correctly, it is bound to give reasonable predictions also for other physical quantities. And it does.

For several light flavors, the situation is less clear. For small ρ, $F(m\rho) \sim (m\rho)^{N_f}$, and the integral for Z_1/Z_0 still diverges, but what is worse is that the assumption that the integral is cut off at some $\rho \sim \mu_{\text{hadr}}^{-1}$ is wrong here. The exact results (15.14) tells us that $Z_1/Z_0 \sim (m\Sigma V)^{N_f}$ at small $m\Sigma V$. If we want to interpret it in the framework of Eq. (15.23) [whose right side now has the form $m^{N_f} V (\mu_{\text{hadr}})^{b_0} \int \rho^{b_0-5+N_f} d\rho$], we should say that the characteristic size of instanton grows with the length of the box or maybe even speculate that instanton effectively "dissociates" into N_f bunches of fractional topological charge $q = 1/N_f$ ...

We are at the frontier of the QCD studies now and it is better to stop the discussion at this point. We just mention that a modified *instanton liquid* model developped by D. Diakonov, V. Petrov, E. Shuryak, and J. Verbaarschot exists. It takes into account correlations between individual instantons due to "determinant interaction" (i.e. the Dirac determinant evaluated in certain approximation is taken into account in the measure) and is compatible with Eq. (15.14). This model successfully describes the behavior of various quark correlators in the real QCD and provides a simple qualitative explanation of *why* spontaneous breaking of chiral symmetry occurs in QCD. It does not explain confinement, however...

Problem 1. Derive Eq. (15.14).

Solution. It is convenient to rewrite the Fourier component of the partition function (15.13) in the form

$$Z_q = \int \mathcal{D}\tilde{U} \left(\det \|\tilde{U}\| \right)^q \exp \left(\text{Re} \left[\text{Tr}\{\mathcal{X}\tilde{U}\} \right] \right) , \qquad (15.25)$$

where $\mathcal{X} = V\Sigma\mathcal{M}$, $\tilde{U} = U e^{i\theta/N_f} \in U(N_f)$, and the integral is done over the whole $U(N_f)$ group. Any $U(N_f)$ matrix can be brought to the diagonal form $\text{diag}(e^{i\phi_1}, \ldots, e^{i\phi_{N_f}})$ by a conjugation $U \to RUR^\dagger$. Note that, for

$\mathcal{M} = m\mathbb{1}$, the integrand is invariant with respect to such a conjugation. This allows us to use the Weyl formula (A.7) for the group measure [without $2\pi\delta\left(\sum_i^N \phi_i\right)$ in the integrand and with an extra factor $1/N_f!$ introduced to normalize the measure to $\int_{U(N_f)} DU = 1$] and write

$$Z_q =$$

$$\frac{1}{N_f!} \int_0^{2\pi} \prod_{i=1}^{N_f} \frac{d\phi_i}{2\pi} \left| \prod_{k<l}^{N_f} \left(e^{i\phi_k} - e^{i\phi_l}\right) \right|^2 \exp\left\{ \sum_{k=1}^{N_f} (\kappa \cos\phi_k + iq\phi_k) \right\},$$

$$(15.26)$$

where $\kappa = |m\Sigma V|$.

If $P = \prod_{k<l}^{N_f} \left(e^{i\phi_k} - e^{i\phi_l}\right)$ is multiplied out, we obtain a set of terms for which the integral factorizes into N_f independent one-dimensional integrals, each of them representing a Bessel function. To check that the combinatorics of these terms indeed gives rise to the determinant (15.14), we note that P is a linear combination of factors of the form $\exp[i(n_1\phi_1 + \cdots +n_{N_f}\phi_{N_f})]$. Since the result must be odd under the interchange of any two angles ϕ_i, ϕ_k, only those terms survive for which the integers $(n_1, \ldots, n_{N_f})$ are mutually different. Hence the set of integers must represent a permutation of the numbers $(0, 1, 2, \ldots, N_f - 1)$; i.e. P consists of $N_f!$ terms which only differ by a permutation of $\phi_1, \ldots, \phi_{N_f}$ and by a sign if the permutation is odd. Now the integrand in Eq. (15.26) involves the factor $\bar{P}P$. In view of the permutation symmetry of the remainder, we can replace the quantity $\bar{P}$ by a single term, say, $(n_1, \ldots, n_{N_f}) = (0, 1, \ldots, N_f - 1)$, dropping the factor $1/N_f!$ in front of the integral. One then easily checks that the $N_f!$ permutations occuring in P precisely give rise to the determinant (15.14).

Problem 2. Derive the sum rules (15.20).

Solution. This is another nice exercise in group theory. Let us look at Eq. (15.25) for a generic $\mathcal{M}$. Exploiting the invariance of the Haar measure, one can deduce that

$$Z_q(R_1 \mathcal{X} R_2) = \left(\det \|R_1^\dagger R_2^\dagger\|\right)^q Z_q(\mathcal{X}).$$

$$(15.27)$$

This implies that the expansion of $Z_q(\mathcal{X})$ at small $\mathcal{X}$ has the form

$$
\begin{aligned}
Z_q \;=\;& \mathcal{N}_q \left(\det \|\mathcal{X}^\dagger\| \right)^q \left[1 + a_q \mathrm{Tr}\{\mathcal{X}^\dagger \mathcal{X}\} \right. \\
& \left. + \; b_q \left(\mathrm{Tr}\{\mathcal{X}^\dagger \mathcal{X}\} \right)^2 + c_q \mathrm{Tr}\{\mathcal{X}^\dagger \mathcal{X} \mathcal{X}^\dagger \mathcal{X}\} + \ldots \right].
\end{aligned}
\tag{15.28}
$$

Decompose the matrix $\mathcal{X}$ as

$$
\mathcal{X} \;=\; \sum_{a=1}^{N_f^2} \mathcal{X}^a t^a ,
\tag{15.29}
$$

where t^a are the generators of $U(N_f)$ (with $t^0 = 1/\sqrt{2 N_f}$) satisfying the relations

$$
\mathrm{Tr}\{t^a t^b\} \;=\; \frac{1}{2}\delta^{ab}, \quad \text{and} \quad \sum_a t^a t^a = \frac{1}{2} N_f .
$$

Using the identity

$$
\sum_a \mathrm{Tr}\{t^a A\} \mathrm{Tr}\{t^a B\} \;=\; \frac{1}{2}\mathrm{Tr}\{AB\} ,
\tag{15.30}
$$

one readily checks that the integral (15.25) obeys the second-order differential equation

$$
\sum \frac{\partial^2}{\partial \mathcal{X}^a \partial \bar{\mathcal{X}}^a} Z_q \;=\; \frac{N_f}{8} Z_q .
\tag{15.31}
$$

This relation generalizes the familiar differential equation obeyed by the Bessel functions, which corresponds to the case $N_f = 1$. Inserting the expansion (15.28), one obtains $a_q = 1/4(N_f + |q|)$ as well as the relation involving a particular combination of the coefficients b_q and c_q:

$$
(N_f^2 + |q| N_f + 1) b_q + (2 N_f + |q|) c_q \;=\; \frac{N_f}{32(N_f + |q|)} .
\tag{15.32}
$$

To separately determine b_q and c_q, one may use the fourth-order differential equation

$$
\sum_{abcd} \mathrm{Tr}\{t^a t^b t^c t^d\} \frac{\partial^4}{\partial \mathcal{X}^a \partial \bar{\mathcal{X}}^b \partial \mathcal{X}^c \partial \bar{\mathcal{X}}^d} Z_q \;=\; \frac{N_f}{16} Z_q ,
\tag{15.33}
$$

which also follows from the representation (15.25) upon repeated use of Equation (15.30). Inserting the expansion (15.28) and using the identity

$\sum_a t^a A t^a = (1/2)\text{Tr}A$ which holds for any $N_f \times N_f$ matrix A, this gives

$$(2N_f + |q|)b_q + (N_f^2 + |q|N_f + 1)c_q = \frac{1}{32(N_f + |q|)} . \qquad (15.34)$$

Solving the linear system (15.32, 15.34), we finally obtain

$$b_q = \frac{1}{32(k^2 - 1)}, \qquad c_q = -\frac{1}{32k(k^2 - 1)}, \qquad (15.35)$$

with $k \equiv |q| + N_f$. This should be compared with the expansion of Eq. (15.11). Using the relation

$$\det \|1 + \epsilon\| = 1 + \text{Tr}\,\epsilon + \frac{(\text{Tr}\,\epsilon)^2 - \text{Tr}\,\epsilon^2}{2} + o(\epsilon^2) ,$$

we finally reproduce the sum rule (15.18), with $q + 1$ being replaced by k, and the sum rules (15.20).

Lecture 16
Fairy QCD

We know that in the real world the gauge group is $SU(3)$ and we have six fundamental quarks with some particular mass values. We also know that θ_{phys} defined in (15.12) is very close to zero. It presents a considerable theoretical interest, however, to study what happens in the same theory, but with other values of the parameters. Though these "fairy" * variants of the theory do not have a direct relation to reality, in order to understand *well* the physics of our world, it is important to also understand how it varies if the rules of the game are changed.

16.1   Finite θ.

One of the ways to modify the theory is to assume a nonzero θ_{phys}. We will not assume that the quark masses coincide with their experimental values, but will keep them (or rather some of them) small enough for the chirality considerations to be relevant. The expressions for the partition function in finite volume as a function of θ were written in the previous lecture. In this lecture, we will concentrate, however, not on the mesoscopic regime $m\Sigma V \sim 1$, but on dynamics of the theory in the thermodynamic limit $m\Sigma V \gg 1$.

The vacuum energy density at finite θ (its mass-dependent part) is ob-

---

*This word was coined by chess problemists. A fairy chess problem is a problem referring to a game with modified rules or on a nonstandard board.

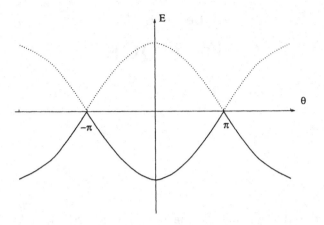

Fig. 16.1 $N_f = 2$: Minima (solid line) and maxima (dotted line) of the effective potential for different θ.

tained by minimizing the chiral effective potential over U,

$$\epsilon_{\text{vac}}(\theta) = -\Sigma \min_U \left[\text{Re Tr} \left\{ \mathcal{M} e^{i\theta/N_f} U \right\} \right] . \qquad (16.1)$$

In the two-flavor case, a nice analytic expression for $\epsilon_{\text{vac}}(\theta)$ can be written for a generic mass matrix $\mathcal{M}$,

$$\epsilon_{\text{vac}}(\theta) = -\Sigma \sqrt{ \text{Tr}\{\mathcal{M}^\dagger \mathcal{M}\} + e^{i\theta} \det \|\mathcal{M}\| + e^{-i\theta} \det \|\mathcal{M}^\dagger\| } . \qquad (16.2)$$

In the general case, it is a continuous smooth function of θ. It is, of course, also periodic: $\epsilon_{\text{vac}}(\theta + 2\pi) = \epsilon_{\text{vac}}(\theta)$. The case of degenerate masses is more subtle. Substituting $\mathcal{M} = m\mathbb{1}$ (with real m) in Eq. (16.2), we obtain

$$\epsilon_{\text{vac}}(\theta) = -2m\Sigma \left| \cos \frac{\theta}{2} \right| . \qquad (16.3)$$

This function is still periodic in θ, but obviously involves a singularity at $\theta = \pi$. A mathematical reason for this is rather simple. Consider the chiral potential as a function of U. It has two stationary points: a minimum and a maximum. At $\theta = \pi$, they fuse together. The maximum and minimum of energy are plotted in Fig. 16.1. We see that two smooth curves $\propto \pm \cos(\theta/2)$ cross at the point $\theta = \pi$ and, after passing this point, the maximum and minimum are interchanged. While $\epsilon_{\text{vac}}(\theta)$ is determined by the curve $-2m\Sigma \cos(\theta/2)$ at $\theta < \pi$, it is determined by the curve $2m\Sigma \cos(\theta/2)$ for $\theta > \pi$.

Physically, this is the situation of the second-order phase transition: the energy is a continuous function of the parameter θ, but its derivative is not. A more detailed analysis shows, however, that the conclusion of the existence of the second-order phase transition is an artifact of the approximation used. Indeed, for $N_f = 2$, the potential in Eq. (16.1) just vanishes identically at the point $\theta = \pi$. This implies that the pion excitations become exactly massless and chiral symmetry is restored.[†]

It is clear from physical considerations that, no matter what the value of θ is, the symmetry (12.52) is broken explicitly if the quarks are massive. For $\theta = \pi$, the leading term vanishes, but the breaking is still there being induced by the terms of *higher order in mass* in the chiral effective potential. The relevant term has the structure $\propto m^2 \sin^2(\theta/2)\mathrm{Tr}\{U^2\}$.

One can show that taking this into account modifies the curves in Fig. 16.1 in the vicinity of $\theta = \pi$ so that, at the point $\theta = \pi$, one has two exactly degenerate vacuum states separated by a low but nonvanishing barrier. If θ is slightly less than π, the degeneracy is lifted, and we have a metastable vacuum state on top of the true vacuum. When θ slightly exceeds π, the roles of these two states are reversed: the one which was stable before becomes metastable, and the metastable one becomes stable. And this is exactly the physical picture of the *first-order* phase transition with superheated water, supercooled vapor, and all other familiar paraphernalia.

Consider now the theory with three degenerate light flavors. By a conjugation $U \to RUR^\dagger$, any unitary matrix U can be rendered diagonal,

$$U = \mathrm{diag}(e^{i\alpha}, e^{i\beta}, e^{-i(\alpha+\beta)}).$$

When $\mathcal{M} = m\mathbb{1}$, the conjugation does not change the potential in Eq. (16.1). For diagonal U, the latter acquires the form

$$V(\alpha,\beta) = -m\Sigma \left[\cos\left(\alpha + \frac{\theta}{3}\right) + \cos\left(\beta + \frac{\theta}{3}\right) + \cos\left(\alpha + \beta - \frac{\theta}{3}\right) \right].$$

$$(16.4)$$

[†]Another way to understand it is to look at the Gell-Mann–Oakes–Renner relation (12.62). Equation (12.62) was originally written for $\theta = 0$, but a generalization for arbitrary θ is rather straightforward. In our case, we even do not need it: the point $\theta = \pi$, $m_u = m_d$ is equivalent to the point $\theta = 0$, $m_u = -m_d$ by virtue of (15.12), and we see that the pion mass vanishes.

The function U has six stationary points,

$$\mathbf{I}: \alpha = \beta = 0 \ , \qquad \mathbf{II}: \alpha = \beta = -\frac{2\pi}{3} \ , \qquad \mathbf{III}: \alpha = \beta = \frac{2\pi}{3} \ ,$$

$$\mathbf{IV}: \alpha = \beta = \frac{2\theta}{3} + \pi \ , \qquad \mathbf{IV}a: \alpha = -\alpha - \beta = \frac{2\theta}{3} + \pi \ ,$$

$$\mathbf{IV}b: \beta = -\alpha - \beta = \frac{2\theta}{3} + \pi \ . \tag{16.5}$$

The points $\mathbf{IV}$a and $\mathbf{IV}$b are obtained from $\mathbf{IV}$ by the Weyl permutations $\alpha \leftrightarrow \beta$, $\alpha \leftrightarrow -\alpha - \beta$, etc., and their physical properties are the same. Actually, we have here not three distinct stationary points, but, rather, a four-dimensional manifold $SU(3)/[SU(2) \times U(1)]$ of the physically equivalent stationary points related to each other by conjugation. The values of the potential at the stationary points are

$$\epsilon_{\mathrm{I}} = -3m\Sigma \cos\frac{\theta}{3} \ , \qquad \epsilon_{\mathrm{II}} = -3m\Sigma \cos\frac{\theta - 2\pi}{3} \ ,$$

$$\epsilon_{\mathrm{III}} = -3m\Sigma \cos\frac{\theta + 2\pi}{3} \ , \qquad \epsilon_{\mathrm{IV}} = m\Sigma \cos\theta \ . \tag{16.6}$$

Studying the expressions (16.6) and the matrix of the second derivatives of the potential (16.4) at $\theta = \pi$, one can readily see that *(i)* the points $\mathbf{I}$ and $\mathbf{II}$ are degenerate minima separated by a barrier, *(ii)* the point $\mathbf{III}$ is a maximum, and *(iii)* the points $\mathbf{IV}$ are saddle points. The physical picture we arrive at is that of the first-order phase transition. The situation is even simpler than in the two-flavor case, since the barrier between the minima appears in the leading order and we need not bother about the subleading terms $\propto m^2$ in the potential.

Figure 16.2 illustrates how the stationary points of the potential move when the vacuum angle is changed. At $\theta = \pi/2$, $\theta = 3\pi/2$, etc., the metastable minima coalesce with the saddle points and disappear. No trace of them is left at $\theta = 0$. One can show that, for the physical values of the quark masses, the metastable vacua are also absent at $\theta = 0$.

If the quark masses are not degenerate but their values are close, nice curves in Fig. 16.2 are distorted a little bit, but physics remains largely the same: the first-order phase transition is robust and cannot be destroyed by small variations of parameters. If the masses are essentially different, the phase transition may disappear. This is true, in particular, for physical values of the quark masses. In this case, the vacuum energy depends on

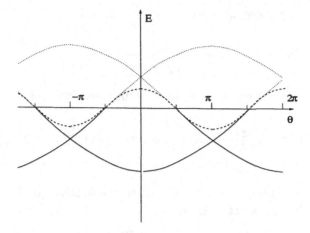

Fig. 16.2 $N_f = 3$: Stationary points of $V(\alpha, \beta)$ for different θ. The solid lines are minima, the dotted lines are maxima, and the dashed lines are saddle points.

θ analytically, as in Eq. (16.2). Unfortunately, no such simple formula for $\epsilon_{\text{vac}}(\theta)$ can be written for a generic mass matrix $\mathcal{M}$ for three or more flavors.

What we can do easily and quite generally is to determine the behavior of the function $\epsilon_{\text{vac}}(\theta)$ at small θ and the topological susceptibility (5.7). Let us do it for arbitrary N_f still assuming, for simplicity, that $\mathcal{M}$ is real and diagonal. Then U can be taken to be diagonal too,

$$U = \text{diag}\left(e^{i\alpha_1}, \ldots e^{i\alpha_{N_f}}\right), \qquad \sum_{j=1}^{N_f} \alpha_j = 0 .$$

For $\theta \ll 1$, we anticipate that the vacuum values of α_j will also be small. We can write the effective potential as

$$V^{\text{eff}}(U) = -\Sigma \sum_{j=1}^{N_f} m_j \cos\left(\alpha_j + \frac{\theta}{N_f}\right)$$

$$= \text{const} + \frac{\Sigma}{2} \sum_{j=1}^{N_f} m_j \left(\alpha_j + \frac{\theta}{N_f}\right)^2 + \cdots . \qquad (16.7)$$

Adding here the Lagrange multiplier term $\lambda \sum_{j=1}^{N_f} \alpha_j$ and minimizing the expression thus obtained over α_j and λ, we find

$$\alpha_j = \theta \left[\frac{1}{m_j \sum_{j=1}^{N_f} m_j^{-1}} - \frac{1}{N_f} \right]. \tag{16.8}$$

Substituting this result for α_j in Eq. (16.7), we finally obtain

$$\chi = m_{\mathrm{eff}}^{-1} = \frac{\partial^2 \mathcal{E}_{\mathrm{vac}}(\theta)}{\partial \theta^2} \bigg|_{\theta=0} = \frac{1}{V} \langle q^2 \rangle_{N_f} = \Sigma \left[\frac{1}{m_1} + \cdots + \frac{1}{m_{N_f}} \right]^{-1}. \tag{16.9}$$

Roughly speaking, χ is proportional to the lightest quark mass (lightest quark masses). In actual QCD

$$\chi_{\mathrm{QCD}} \approx \frac{m_u m_d}{m_u + m_d} \Sigma \, ; \tag{16.10}$$

the strange and other quarks are too heavy to be relevant. If $\mathcal{M} = m\mathbb{1}$, Eq. (16.9) is reduced to $\chi = m\Sigma/N_f$. The latter result can be easily reproduced by substituting the asymptotics (15.16) for Z_q in the definition

$$\langle q^2 \rangle = \frac{\sum_q q^2 Z_q}{\sum_q Z_q}.$$

Problem 1. Find the surface energy density of the *domain wall* separating the degenerate vacua at $\theta = \pi$ in the 3-flavor case [41].

Solution. We have to find a stationary solution to the classical equations of motion in effective chiral theory which does not depend on two spatial coordinates (y, z) and satisfies the boundary conditions $U(-\infty) = e^{-2\pi i/3}\mathbb{1}$, $U(\infty) = \mathbb{1}$. Assuming the ansatz

$$U(x) = e^{-i\pi/3}\mathrm{diag}\left(e^{i\gamma(x)}, \ e^{i\gamma(x)}, \ -e^{-2i\gamma(x)} \right)$$

and substituting it in the energy functional, we obtain for the surface energy density of the wall

$$\sigma = \min_{\gamma(x)} 3F_\pi^2 \int_{-\infty}^{\infty} dx \left[\frac{1}{2}\gamma'^2 + \frac{M_\pi^2}{3} \left(\cos\gamma - \frac{1}{2} \right)^2 \right],$$

$$\gamma(\pm\infty) = \pm\frac{\pi}{3}. \tag{16.11}$$

The field configuration minimizing Eq. (16.11) is

$$\cos\gamma = \frac{E^2 + 4E + 1}{2(E^2 + E + 1)} \quad , \quad E = \exp\left\{\frac{M_\pi x}{\sqrt{2}}\right\} . \tag{16.12}$$

We obtain

$$\sigma = 3\sqrt{2}\left(1 - \frac{\pi}{3\sqrt{3}}\right) M_\pi F_\pi^2 . \tag{16.13}$$

Remark. If θ does not coincide with π but is close to it and one of the vacua becomes metastable, one can estimate the probability of its decay. For almost all θ in the interval $-\pi/2 < \theta < \pi/2$, this probability turns out to be negligibly small, and the Universe could exist in a metastable state for billions of years and more awaiting the eventual cataclysm. It is a real pity that such a beautiful possibility is not realized in Nature.

16.2 Large N_c.

The multicolored version of QCD is a rather popular object of studies. One of the reasons is that the number of quark colors in the real world $N_c = 3$ can sometimes be treated as large (in a sense that the expansion in the parameter $1/N_c$ is not altogether stupid).

Before proceeding further, we should warn the reader that we are now going to lower the standards of rigor in our discussion and to make it somewhat more heuristic. This is just the level of understanding we are at now. In spite of the fact that the results to be presented below do not have the status of mathematical theorems, there is little doubt, however, that they are correct.

We want to send N_c to infinity while keeping $\mu_{\text{hadr}} \sim \Lambda_{\text{QCD}}$ fixed. A glance at Eqs. (9.34), (9.27) tells us that, at a fixed energy scale (say, at $\mu = 3\Lambda_{\text{QCD}}$), the effective coupling constant is supressed $\propto 1/N_c$. Thus, the relevant limit is

$$N_c \to \infty, \quad g^2 \to 0 \qquad \text{with} \quad g^2 N_c \quad \text{fixed} . \tag{16.14}$$

μ_{hadr} determines the mass scale of the lightest hadrons, which are *mesons*. The baryon masses grow $\propto N_c$. This is most simply seen in the framework of the *constituent quark model:*[‡] the mass grows with the number of

_____
[‡]This is a phenomenological model assuming that hadrons are made of quarks (in a more

constituents. The same result follows from the *skyrmion model* (see **Problem 2** below) .

Consider a correlator of two quark currents, $C(x) = \langle T\{\bar{\psi}\psi(x)\,\psi\psi(0)\}\rangle$. Let us discuss the order of magnitude of some individual contributions in the perturbative series for $C(x)$. To leading order (see Fig. 16.3a), $C^{(1)}(x) \sim$ $\text{Tr}\{1\} = N_c$. At the 2-loop level (Fig. 16.3b and two other similar graphs),

$$C^{(2)}(x) \sim g^2\text{Tr}\{t^a t^a\} = g^2 c_F N_c \sim (g^2 N_c)N_c \;,$$

which is of the same order in N_c as $C^{(1)}(x)$. Compare now three different graphs in Figs. 16.3c–16.3e contributing in $C(x)$ at the 3-loop level. The contribution of Fig. 16.3c is $\sim g^4\text{Tr}\{t^a t^b t^b t^a\} \sim (g^2 N_c)^2 N_c \sim N_c$. But the contribution of Fig. 16.3d is

$$\sim g^4\text{Tr}\{t^a t^b t^a t^b\} \sim g^4 c_F(c_F - c_V/2)N_c \sim g^4 N_c \;,$$

which is suppressed, compared to that of Fig. 16.3a, as $\sim 1/N_c^2$. The contribution of Fig. 16.3e involving a quark loop is $\sim g^4\text{Tr}\{t^a t^b\}\text{Tr}\{t^a t^b\} \sim$ $g^4 N_c^2$ and is suppressed, compared to the leading term, as $\sim 1/N_c$.

This reasoning can, of course, be made more precise and done in a more general way for any graph. The conclusions are: *(i)* internal quark loops are not relevant to leading order in N_c, with each quark loop providing a suppression factor $\sim 1/N_c$; *(ii)* to leading order in N_c, only the *planar* graphs involving gluon exchanges like in Fig. 16.3c survive. Crossing the gluon legs costs at least $1/N_c^2$.

One of the physical conclusions which follow from this analysis is that the width of the meson states tends to zero in the limit $N_c \to \infty$. Indeed, in this limit, all meson states can be subdivided into the classes: *(i)* pure glueballs, *(ii)* mesons involving one $\bar{q}q$ pair, *(iii)* mesons with two such pairs, etc. These states do not mix because their mixture would mean the creation or annihilation of $\bar{q}q$ pairs, which is suppressed at large N_c. Actually, a difficulty to create pairs renders the constituent quark model in the large N_c limit *more* reasonable and justified than in actual QCD. Note now that, for a $\bar{q}q$ meson to decay into two other mesons of the same class, an extra $\bar{q}q$ pair should be created, and such a transition is suppressed. For a $\bar{q}q$ meson to decay into a $\bar{q}q$ meson and a glueball, a system of gluons which

---

sophisticated version — also of gluons) roughly in the same way as atoms are made of their constituents. It is, of course, too naïve, but some features of the hadron spectrum are described very well in this framework.

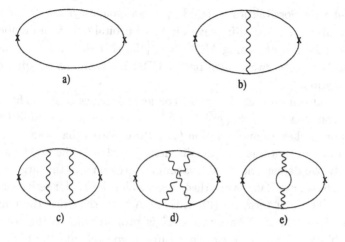

Fig. 16.3   Contributions to the quark correlator.

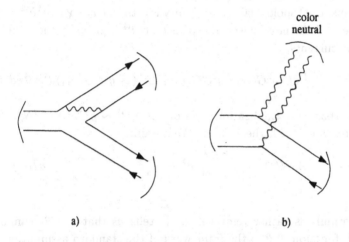

Fig. 16.4   Decay of mesons at large N_c.

is neutral in color should be "separated" from the diagram (see Fig. 16.4). But this is also suppressed: a probability to find a system of several colored gluons in a globally color-singlet state is small $\lesssim 1/N_c^2$.

The mesons are very narrow, but there are a lot of them: the density of states grows exponentially with mass. The explanation of this comes from the constituent model combinatorics. The higher the limit M, the more is the number of possibilities to build up a compound system of constituent quarks and gluons with different radial and orbital quantum numbers and with the mass not exceeding M. In Ref. [43], this assertion was confirmed *numerically* in the two-dimensional QCD with fermions in adjoint color representation.

Les us also note that the quark condensate scales $\propto N_c$. That simply follows from the fact that $\langle \psi\bar{\psi} \rangle = \sum_{i=1}^{N_c} \langle \psi_i\bar{\psi}_i \rangle$. From this and from the Gell-Mann–Oakes–Renner relation (12.62), it follows that also $F_\pi^2 \propto N_c$.

We want to continue now the discussion started in the previous section and study the dependence of the vacuum energy on θ in the large N_c limit. Consider first pure Yang–Mills theory. Consider the Taylor series for $\epsilon_{\text{vac}}(\theta)$ at $\theta = 0$. The first term $\epsilon_{\text{vac}}(0)$ scales as N_c^2, which is just the number of degrees of freedom.§ The term $\propto \theta^2$ is proportional to the topological susceptibility (15.8). To make an estimate, go over into the perturbative normalization of the field so that $F_{\mu\nu}$ is multiplied by g and the factor g^4 in front of the integral in Eq. (15.8) appears. The gluon loop gives the factor N_c^2. Thus, the topological susceptibility is estimated as $\chi \propto g^4 N_c^2 \sim 1$. The coefficient of the next power of expansion $\propto \theta^4$ is given by the correlator of the four currents,

$$\chi^{(4)} \sim \int \langle T\{g^2 G\tilde{G}(x)\, g^2 G\tilde{G}(y)\, g^2 G\tilde{G}(z)\, g^2 G\tilde{G}(0)\} \rangle d^4x\, d^4y\, d^4z \ ,$$

in perturbative normalization. We obtain $\chi^{(4)} \sim g^8 N_c^2 \sim 1/N_c^2$. Similar estimates work for higher orders. We obtain:

$$\epsilon_{\text{vac}}(\theta) = A_0 N_c^2 + A_2\theta^2 + \frac{A_4}{N_c^2}\theta^4 + \frac{A_6}{N_c^4}\theta^6 + \cdots = N_c^2 f(\theta/N_c) \ .$$

$$(16.15)$$

This formula is rather remarkable. It tells us that $\epsilon_{\text{vac}}(\theta)$ cannot be a smooth function of θ in the framework of the standard assumption that it is a periodic function of θ with the period 2π. A kind of phase transition in θ, presumably at $\theta = \pi$, is bound to take place in this case. Little is known about the nature of this phase transition. Assuming that it occurs at $\theta = \pi$ and is of the first order, we arrive at the picture with domain walls separating two degenerate vacua.

---

§ $\epsilon_{\text{vac}}(0)$ also drastically diverges in ultraviolet, but we are concerned here only with color factors. Anyway the ultraviolet divergence is absent in the higher terms of the expansion in θ.

Another possibility is that the original assumption that $\epsilon_{vac}(\theta + 2\pi) = \epsilon_{vac}(\theta)$ is *wrong* and the true period of $\epsilon_{vac}(\theta)$ is $2\pi N_c$ rather than 2π. Then would-be degenerate vacua at $\theta = \pi$ correspond actually to *different sectors* in the Hilbert space with different values of θ (say, $\theta = \pi$ and $\theta = -\pi$, which are no longer equivalent). These sectors do not talk to each other and, in particular, there is no physical domain wall which separates them.

A longer interval of periodicity for $Z(\theta)$ would necessarily imply the presense of fractional topological charges $q \propto 1/N_c$ in the Euclidean path integral for pure Yang–Mills theory [see Eq. 15.3)]. The presense of such configurations in a finite box with the so-called twisted boundary conditions was established 20 years ago by 't Hooft. The problem is that these *toron* configurations are essentially delocalized, and it is not clear whether they do or do not give an essential contribution into the path integral in the large volume limit. The same refers to the meron configuration (5.28), whose status and relevance for the physics of QCD is not clear by now. If you would force me to lay my bets, I would opt for the second possibility (with fractional topological charges and without domain walls). But a scientific resolution of this very interesting problem has not yet been obtained.

We want to emphasize again that this controversy refers *only* to pure Yang–Mills theory. In a theory with quarks, fractional charges are *not* allowed (one simply cannot define the Dirac operator for fundamental fermions in a gauge background with fractional q), $\epsilon(\theta + 2\pi) = \epsilon(\theta)$, there is a first-order phase transition at $\theta = \pi$ with the physical domain walls.

Consider now the theory with light quarks in the large N_c limit. The low energy dynamics is modified, compared to the theory with finite N_c, due to the fact that the effects associated with the explicit breaking of the singlet axial symmetry by anomaly become irrelevant. Anyway, as this symmetry is spontaneously broken together with nonsinglet axial symmetries, due to formation of quark condensate, so the flavor-singlet pseudoscalar meson, the η', reveals now the pseudo-Goldstone nature and becomes massless in the chiral limit.

Heuristically, it is related to the fact that, in the limit $N_c \to \infty$, the dependence of the vacuum energy on θ in the interval $0 \leq \theta \leq 2\pi$ becomes flat, which means that the instanton effects (which are responsible for the phenomenological implementation of the $U_A(1)$ anomaly) become irrelevant in the limit. To be more quantitative and more formal, let us *assume* that η' becomes light at large N_c and then *justify* it by studying the corresponding effective Lagrangian. The form of the latter is similar to what we had before, only the field $U(x)$ describing the fluctuations of the order parameter $\langle \psi_L^f \bar{\psi}_R^g \rangle$ is now not an $SU(N_f)$, but a general $U(N_f)$ matrix. The phase of its determinant describes the η' field.

The singlet axial transformation

$$\psi_L \rightarrow e^{i\alpha/(2N_f)}\psi_L, \qquad \psi_R \rightarrow e^{-i\alpha/(2N_f)}\psi_R \qquad (16.16)$$

is realized at the level of the effective Lagrangian as $U(x) \rightarrow e^{i\alpha/N_f}U(x)$. On the other hand, the transformation (16.16) is equivalent to multiplying the mass matrix by a phase: $\mathcal{M} \rightarrow e^{i\alpha/N_f}\mathcal{M}$. That adds to θ_{phys} as given by Eq. (15.12) the constant α, which can also be realized by leaving $\mathcal{M}$ intact but shifting $\theta \rightarrow \theta + \alpha$. Thereby, the effective chiral Lagrangian should be invariant under the transformation $U(x) \rightarrow e^{i\alpha/N_f}U(x)$, $\theta \rightarrow \theta - \alpha$, which dictates to us the following form of the effective Lagrangian to leading order in momenta, $\mathcal{M}$, and N_c:

$$\mathcal{L}_{\text{eff}} = \frac{F_\pi^2}{4}\text{Tr}\left\{\partial_\mu U \partial_\mu U^\dagger\right\} + \Sigma \, \text{Re}\left[\text{Tr}\{\mathcal{M}U\}\right] - \frac{\chi}{2}(\phi + \theta)^2 \, , \quad (16.17)$$

where $\phi = \arg[\det\|U\|]$ and χ is the topological susceptibility (15.8) calculated in *pure* Yang–Mills theory [indeed, when $\mathcal{M} \rightarrow \infty$ and the quark fields and hence the chiral fields decouple, Eq. (16.17) is reduced to $\epsilon_{\text{vac}}(\theta) = \frac{\chi}{2}\theta^2$].

The mass of the η' meson described by the field $\frac{F_\pi}{\sqrt{2N_f}}\phi(x)$ can now be readily found. When $\mathcal{M} = m\mathbb{1}$, we obtain

$$M_{\eta'}^2 = \frac{2}{F_\pi^2}(N_f\chi + \Sigma m) \, . \qquad (16.18)$$

In the chiral limit $m \rightarrow 0$, $M_{\eta'}^2 = 2N_f\chi/F_\pi^2$ (the *Witten–Veneziano formula*), which is suppressed $\propto 1/N_c$ Q.E.D.

In the limit $N_c \rightarrow \infty$ with fixed m, the second term in Eq. (16.18) dominates, and η' becomes degenerate with other pseudogoldstones. Finding the minimum of the effective potential in Eq. (16.17) and differentiating it twice with respect to θ, we find

$$\frac{\langle q^2 \rangle}{V}\bigg|_{\text{with quarks}} = \frac{\chi \Sigma m}{N_f\chi + \Sigma m} \, , \qquad (16.19)$$

which tends to χ in the limit $N_c \to \infty$ with fixed finite m and to $\Sigma m/N_f$ in the chiral limit, as it should [see Eq. (16.9)].

Problem 2. Estimate the mass of baryons in the large N_c limit using the effective chiral Lagrangian (12.57) with account of the higher derivative terms (12.59).

Solution. As was mentioned, F_π^2 scales as N_c. The same concerns the coefficient $m\Sigma$ of the mass term, the coefficients $L_{1,2,3}$ in Eq. (12.59), and also all other coefficients in $\mathcal{L}_{\text{eff}}$: at $p \sim \mu_{\text{hadr}} \propto N_c^0$ all the terms should be of the same order.

Now note that the nonlinear effective Lagrangian $\mathcal{L}_{\text{eff}}^{(2)} + \mathcal{L}_{\text{eff}}^{(4)}$ admits a solitonic solution called *skyrmion*. To see this, assume $N_f = 2$ and, following Skyrme, choose the Lagrangian as

$$\mathcal{L}_{\text{eff}} = \frac{F_\pi^2}{4}\text{Tr}\left\{\partial_\mu U \partial^\mu U^\dagger\right\} + \frac{1}{32e^2}\text{Tr}\left\{[U^\dagger\partial_\mu U, U^\dagger\partial_\nu U]^2\right\} . \quad (16.20)$$

The second "Skyrme" term corresponds to a particular choice of L_i in Eq. (12.59). Take the ansatz

$$U(x) = \exp\left\{i\sigma^a\frac{x^a}{r}f(r)\right\} . \quad (16.21)$$

If $f(0) = 0$ and $f(\infty) = \pi$, then $U(0) = 1$ and $U(\infty) = -1$, and the function (16.21) describes a topologically nontrivial map of S^3 ($\equiv R^3$ with the infinity added) onto the group $SU(2)$ (cf. the discussion at the beginning of Lecture 2). The degree of this map is $q = 1$. There are, of course, the maps with all other integer values of q. The skyrmion solution is the configuration in the class $q = 1$ minimizing the energy functional

$$E = 4\pi \int_0^\infty r^2 dr \left\{\frac{F_\pi^2}{2}\left(f'^2 + 2\frac{\sin^2 f}{r^2}\right)\right.$$
$$\left. + \frac{1}{2e^2}\frac{\sin^2 f}{r^2}\left(2f'^2 + \frac{\sin^2 f}{r^2}\right)\right\} . \quad (16.22)$$

One can observe that, if only the term $\propto F_\pi^2$ were present, the minimum of energy would be just zero. The minimum would be realized by the functions $f(r)$ which are equal to their asymptotic value $f(\infty) = \pi$ practically everywhere, but fall rapidly to zero in the region of very small r. The presence of the Skyrme term does not allow for the solution to collapse down to the origin, and a nonsingular configuration realizing the minimum of energy exists. Its exact form can be found numerically.

The solution exists also for $N_f > 2$. To construct it explicitly, one has to consider a map of the distant sphere onto a certain $SU(2)$ subgroup of $SU(N_f)$ in the same way as it is done for instantons. Also the particular choice (16.20) of the higher derivative terms is not crucial. The solution exists in the wide range of parameters L_i.

As all the terms in the effective Lagrangian and Hamiltonian scale as N_c, the energy of the classical solution, i.e. the mass of the soliton, also scales as N_c, while the characteristic size of the solution does not depend on N_c. This soliton can be associated with the baryon and there are a number of arguments why this is reasonable. In particular, the degree of map discussed above is naturally associated with the baryon charge. The accurate quantization of the classical solution displays that the corresponding quantum state is an isotopic doublet and has spin 1/2. Still this argumentation is of Babylonian quality because, strictly speaking, we can only apply the effective Lagrangian philosophy to describe the interactions of pseudo–Goldstone particles at low energy, and not their collective state of large mass...

ITEP Sum Rules: the Duality Festival

17.1 The Method.

In previous lectures, we have derived some exact results for QCD using nonperturbative methods. Unfortunately, most of these results [like the Banks and Casher theorem (14.18)] refer to the quantities not directly measurable in a physical experiment. The exact quantitative results of QCD that *can* be tested in experiment are obtained in a straightforward way for the processes with large characteristic momentum transfer, where the effective coupling constant is small and perturbation theory works.

The vast majority of experimental data on strong interaction physics refer, however, to the low energy region, where coupling constant is large and no analytic theoretical results can be obtained. Certain predictions for low energy hadron properties can be obtained numerically by calculating the corresponding path integrals. These calculations are, on one hand, very difficult (so that the "honest" calculations involving evaluation of the fermionic determinant have become possible only recently) and, on the other hand, they always leave an uneasy feeling. OK, a numerical result for, say, the nucleon mass is obtained. But *why* nature prefers this particular value for the mass and not some other?

Another possible approach is to determine the masses and other hadron properties in the framework of some phenomenological model: the *constituent quark model* and the *instanton liquid model* mentioned above, or the *MIT bag model* and some other models which we did not mention. All these models are "QCD-inspired", but have an unclear theoretical status: one cannot really derive the constituent quark model from the first QCD principles.

Fortunately, there are two remarkable methods which allow one to describe the properties of hadrons *quantitatively* in the framework of QCD. We have already mentioned the first method, chiral perturbation theory, which can be used to determine the properties of the lightest pseudoscalar states and their interactions at low energies. The chiral perturbation theory in its standard form is not capable of saying anything, however, on the properties of ρ meson, nucleon, and other massive hadrons. This is possible in the framework of the so-called ITEP (*Institute of Theoretical and Experimental Physics* in Moscow) or SVZ (*Shifman–Vainshtein–Zakharov*) *sum rules* [44].

The method relies heavily on the philosophy of quark–hadron duality, which we already exploited in the previous lectures. It allows one to determine the mass and other properties of any hadron h which represents the *lightest* state in a channel with given quantum numbers (the method does not work for excited states). The basic algorithm is the following.

(1) Pick up a local gauge-invariant Heisenberg operator (current) $J(x)$ carrying the quantum numbers of h, i.e. with the nonzero residue $\langle 0|J|h\rangle$.

(2) Consider the correlator $\langle J(\tau)J^\dagger(0)\rangle_0$ and choose the Euclidean time τ large enough so that nonperturbative corrections to the perturbative result (valid at very small τ) are already important. On the other hand, τ should be small enough so that only leading nonperturbative corrections, which depend on the bulk characteristics of the QCD vacuum state, like the quark condensate or the gluon condensate, would give an essential contribution.

(3) Write the spectral decomposition

$$\langle J(\tau)J^\dagger(0)\rangle_0 \;=\; \sum_n \langle 0|J|n\rangle\langle n|J^\dagger|0\rangle e^{-E_n\tau} \,, \qquad (17.1)$$

where $J(\tau)$ is the Euclidean Heisenberg operator (4.29). It is our great luck that, in the region of τ defined above, the sum is "focused" in most cases on the first term with $|n\rangle = |h\rangle$ (the integration over the momenta and the summation over the polarizations of the hadron h is, of course, assumed) and the contribution of the excited states is more or less suppressed.

(4) We do not, however, simply neglect the contribution of the ex-

cited states (that would give a bad accuracy), but estimate it using quark–hadron duality (see below).

(5) Comparing the left side and the right side in Eq. (17.1), we determine the mass M_h of the lowest state and the residue $\langle 0|J|h \rangle$.

17.2 Nucleon Mass and Residue.

Let us now explain how it works in the nucleon case. Consider the current

$$\eta(x) = \epsilon^{ABC} \left[u^A(x) C \gamma_\mu u^B(x) \right] \gamma_5 \gamma^\mu d^C(x) , \tag{17.2}$$

where $u(x)$ and $d(x)$ are the up- and down-quark fields, A, B, C are the fundamental color indices, and C is the charge conjugation matrix ($C^2 = 1$, $C\gamma_\mu C = -\gamma_\mu^T$). The spinorial fermionic current (17.2) has the nonzero residue,

$$\langle 0|\eta|P \rangle = \lambda u_P \tag{17.3}$$

$[(\not{p} - m_P)u_P = 0; \; \bar{u}_P u_P = 2m_P]$, in the proton state.[*] There is another such current,

$$\tilde{\eta} = \epsilon^{ABC} \left[u^A(x) C \sigma_{\mu\nu} u^B(x) \right] \gamma_5 \sigma^{\mu\nu} d^C(x) ,$$

orthogonal to η, but the residue $\langle 0|\tilde{\eta}|P \rangle$ happens to be relatively small, and the corresponding sum rule is not "focused" on the proton and is useless.

Consider the Euclidean correlator $\Pi_{ij}(\tau) = \langle \eta_i(\tau)\bar{\eta}_j(0) \rangle_0$. The leading contribution at small τ is given by the graph in Fig. 17.1a. A simple calculation gives

$$[\Pi_{ij}(\tau)]^{\text{pert}} = \frac{24i(\gamma_0)_{ij}}{\pi^6 \tau^9} \{1 + O[\alpha_s(\tau)]\} . \tag{17.4}$$

(γ_0 is the Euclidean γ-matrix). Perturbative corrections come from the graphs with gluon exchange like in Fig. 17.1b. The coefficient of $\alpha_s(\tau)$ turns out to be comparatively small and, in the region of interest of the parameter τ, the perturbative corrections to the leading-order result can be disregarded.

---

[*] The microconventions for this lecture: "P" is proton, "p" is momentum, "N" is nucleon, and "n" is neutron.

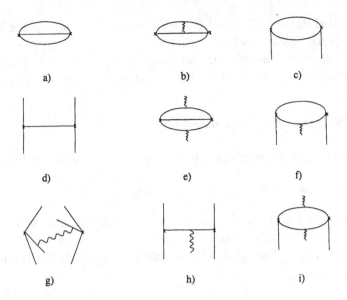

Fig. 17.1   Some QCD graphs contributing to $\Pi_{ij}(\tau)$.

There are, however, essential nonperturbative corrections. The most important ones are associated with the nonzero quark condensate Σ. Technically, the presence of the latter can be taken into account by writing the quark Green's function in the form

$$\langle q_i^A(\tau)\bar{q}_j^B(0)\rangle = \delta^{AB}\left[\frac{i(\gamma_0)_{ij}}{2\pi^2\tau^3} + \frac{\Sigma}{12}\delta_{ij}\right] + \ldots , \qquad (17.5)$$

where the dots stand not only for the perturbative corrections $\propto \alpha_s^n(\tau)$, but also for the higher nonperturbative corrections like $\propto \langle\bar{q}\sigma_{\alpha\beta}G_{\alpha\beta}q\rangle_0$ (G is the gluon field), etc. It is important that the higher nonperturbative corrections involve extra factors $\propto \tau^k$ with positive k. When τ is small, they are all suppressed. Substituting the propagator (17.5) into the loop in Fig. 17.1a, we obtain

$$\Pi_{ij}(\tau) = \frac{24i(\gamma_0)_{ij}}{\pi^6\tau^9}\left[1 + \frac{\pi^4\Sigma^2\tau^6}{72}\right] + \frac{2\Sigma}{\pi^4\tau^6}\delta_{ij} . \qquad (17.6)$$

The contributions $\propto \Sigma$ and $\propto \Sigma^2$ can be described by the graphs in Fig. 17.1c,d with disconnected quark lines. No contribution $\propto \Sigma^3$ appears at this level: it would be described by a completely disconnected graph, which has no imaginary part and, as we will shortly see, is irrelevant by that reason.

The estimate (17.6) takes into account the most essential *power corrections* (they are suppressed at small τ as a power of $\tau\mu_{\text{hadr}}$) to the leading order result (17.4) and is already reasonable enough. It can be improved, however.

(i) As was mentioned above, perturbative corrections coming from the graphs like in Fig. 17.1b with the standard gluon propagator are not so important. But the gluon propagator involves, by the same token as the quark one, also nonperturbative contributions. The leading contribution is associated with the *gluon condensate*. Choosing the fixed point gauge $x_\alpha A_\alpha^a(x) = 0$, we can express $A_\alpha^a(x) = -(1/2)x_\beta G_{\alpha\beta}^a + \cdots$ [cf. Eq. (12.24)] and write

$$\langle A_\alpha^a(x) A_\beta^b(y) \rangle = -\frac{\delta^{ab}\delta_{\alpha\beta}}{4\pi^2(x-y)^2} + \frac{\delta^{ab}\langle G_{\mu\nu}^c G_{\mu\nu}^c \rangle}{384}[\delta_{\alpha\beta}(xy) - y_\alpha x_\beta] + \ldots$$

$$(17.7)$$

The contribution of the second term in Eq. (17.7) to the nucleon polarization operator is usually represented by the graphs with a disconnected gluon line like in Fig. 17.1e. The gluon condensate $\langle G^2 \rangle$ is the universal parameter characterizing the vacuum wave function. It enters not only the nucleon sum rules discussed here, where it gives a comparatively small correction, but also the ITEP sum rules for all other hadron channels. The value of the gluon condensate is best determined from the sum rules in the charmonium $(\bar{c}c)$ and bottomonium $(\bar{b}b)$ channels. The result is

$$\left\langle \frac{\alpha_s}{\pi} G_{\mu\nu}^c G_{\mu\nu}^c \right\rangle_0 \approx 0.012 \text{ GeV}^4 \qquad (17.8)$$

with the uncertainty $\approx 30 - 40\%$.

(ii) Another correction is brought about by the nonzero vacuum expectation value

$$\langle g_s \bar{q} i \sigma_{\alpha\beta} G_{\alpha\beta}^a t^a q \rangle_0 \equiv m_0^2 \Sigma \,, \qquad (17.9)$$

where the convention (N.11) for $\sigma_{\alpha\beta}$ is chosen. The parameter $m_0^2 \approx 0.8 \text{ GeV}^2$ was evaluated independently from the analysis of certain meson sum rules. As was already mentioned, the average (17.9) gives the next subleading term in the expansion (17.5) of the quark propagator, and the corresponding coefficient can be exactly determined. Another contribution

$\sim m_0^2\Sigma$ is due to a nonperturbative piece in the 3-point function $\langle q\bar{q}A\rangle$ (it is described by the graphs like in Fig. 17.1f obtained from the graphs of Fig. 17.1b by breaking apart the gluon line *and* one of the quark lines to which it is attached). In our case, the two contributions $\sim m_0^2\Sigma$ happen to cancel, however.

(iii) Besides, there is a contribution $\propto \alpha_s\Sigma^3$ coming from the graphs like in Fig. 17.1g, which is obtained from the graph in Fig. 17.1b by cutting 3 quark lines, the contribution $\propto m_0^2\Sigma^2$ (see Fig. 17.1h), and the contribution $\propto \Sigma\langle G^2\rangle_0$ (see Fig. 17.1i). Putting everything together, we obtain

$$\Pi_{ij}(\tau) = \frac{24i(\gamma_0)_{ij}}{\pi^6\tau^9}\left[1 + \frac{b\tau^4}{3\cdot 2^8} + \frac{\pi^4\Sigma^2\tau^6}{72} - \frac{\pi^4 m_0^2\Sigma^2\tau^8}{9\cdot 2^7}\right]$$
$$+ \frac{2\Sigma}{\pi^4\tau^6}\delta_{ij}\left[1 - \frac{b\tau^4}{3\cdot 2^8}\right] - \frac{68}{81}\delta_{ij}\left(\frac{\alpha_s}{\pi}\right)\Sigma^3\ln(\tau\mu_{\text{hadr}}) , \quad (17.10)$$

where $b = 4\pi^2\langle(\alpha_s/\pi)G^2\rangle_0$.

The underlying philosophy behind Eq. (17.10) is the Wilsonian *operator product expansion*. Treating $\eta(x)$ as a Heisenberg operator, we can write

$$\langle T\{\eta(x)\bar{\eta}(0)\}\rangle_0 = \sum_n c_n(x)\,\langle O_n(0)\rangle_0 , \quad (17.11)$$

where O_n is a set of local operators with nonzero vacuum expectation values, like $q\bar{q}$, $\alpha_s G^2$, etc., and x is assumed to be spacelike. Naïvely,

$$c_n(x) \sim \left(\sqrt{-x^2}\right)^{d_n - 2d_\eta} , \quad (17.12)$$

where d_n, d_η are the canonical dimensions of the corresponding operators. Going over into Euclidean space, we reproduce the expansion (17.10).

Equation (17.11) has an exact meaning, however, only in the framework of the Wilsonian procedure with careful separation of scales so that $\langle O_n\rangle_0$ is associated with the small momenta region $p < \mu$, and the *coefficient functions* $c_n(x)$ — with the region $p > \mu$. Both $\langle O_n\rangle_0$ and $c_n(x)$ would then depend on μ. Speaking of Eq. (17.10), it is a clever approximation for $\Pi(\tau)$ describing it well at not so large τ, but it does not have the status of an *exact* result of QCD:

- First, there are terms of still higher order in τ depending on the operators of still higher dimension.
- When deriving Eq. (17.10), we assumed the factorization of vacuum expectation values $\langle(q\bar{q})^2\rangle_0 = \langle q\bar{q}\rangle_0^2$, etc. We are lucky that this

approximation works well in most cases (this can be justified by a special reasoning).

- Nice power ordering is already broken by the last term in (17.10) involving a logarithm. Logarithms also appear in perturbative corrections to all other terms. To leading order, they bring about anomalous dimension factors [cf. Eq. (9.56)]. The perturbative corrections and also the mixed perturbative + power corrections can be tackled. (They can be handled to every finite order in α_s and one can also sum up the leading and even subleading logarithms to all orders.) What is more difficult to handle are the so-called *direct instanton* contributions, which are suppressed $\sim (\tau\mu_{\text{hadr}})^{11-2N_f/3}$ but, in contrast to the condensate contributions, come from the small distance region. The *renormalon* corrections coming from the resummation of a certain infinite set of perturbative graphs and which are suppressed only as $(\tau\mu_{\text{hadr}})^2$ are even less tractable. It is our luck again that in most cases the corrections due to direct instantons and renormalons are numerically small.[†]

Our next task is to evaluate the correlator $\Pi_{ij}(\tau)$ by saturating it by the physical states in the spectral decomposition (17.1). The most important such state is the proton. We have $\Pi^P(\tau) = \lambda^2 G_m(\tau)$, where

$$[G_m(\tau)]_{ij} = \frac{m^2}{4\pi^2\tau}[i(\gamma_0)_{ij}K_2(m\tau) + \delta_{ij}K_1(m\tau)] \qquad (17.13)$$

is the Euclidean Green's function of a massive free fermion ($m \equiv m_P$). [$K_n(x)$ are the exponentially decreasing Bessel functions.]

Besides, there is a contribution of excited states. There are two types of such states: the states with the same parity as a nucleon, whose coupling to the current (17.2) has the same form as in Eq. (17.3), and the states with $I(J^P) = \frac{1}{2}\left(\frac{1}{2}^-\right)$ involving the factor γ^5 in the residue. Bearing in mind that $(\gamma^5 u)^\dagger\gamma^0 = -\bar{u}\gamma^5$ and that $-\gamma^5(\not{p}+m)\gamma^5 = \not{p}-m$, the negative parity states can be viewed as the states of positive parity but with negative mass. The whole polarization operator can be presented as a sum of two spectral

---

[†]It is not always so. A classical example where both effects are large so that the ITEP sum rules do *not* work is the correlator $\langle\bar{q}\gamma^5 q(x)\ \bar{q}\gamma^5 q(0)\rangle_0$ of light quark pseudoscalar currents.

integrals,

$$\Pi^{\text{phys}}(\tau) = \int_0^\infty G_{\sqrt{s}}(\tau)\, \rho_+(s) ds + \int_0^\infty G_{-\sqrt{s}}(\tau)\, \rho_-(s) ds \, , \quad (17.14)$$

where $\rho_\pm(s)$ are the spectral densities in the channels $\frac{1}{2}^+$ and $\frac{1}{2}^-$, correspondingly. Substituting here the Green's function (17.13), we obtain

$$\Pi^{\text{phys}}(\tau) = \int_0^\infty ds\, \rho_1(s) \left[\frac{i\gamma_0 s}{4\pi^2 \tau} K_2(\sqrt{s}\tau) \right]$$

$$+ \int_0^\infty ds\, \rho_2(s) \left[\frac{\sqrt{s}}{4\pi^2 \tau} K_1(\sqrt{s}\tau) \right] \, , \quad (17.15)$$

where $\rho_1(s) = \rho_+(s) + \rho_-(s)$ and $\rho_2(s) = \sqrt{s}\,[\rho_+(s) - \rho_-(s)]$ are the chirality odd and chirality even structures in the imaginary part of the polarization operator in momentum space:

$$\frac{1}{\pi} \text{Im}[i\Pi(p)] \equiv \frac{1}{\pi} \text{Im}\left[i \int \Pi(x) e^{ip\cdot x} d^4 x \right] = \not{p}\rho_1(s) + \rho_2(s) \, ,$$

$$s = p^2 \geq m^2 \quad (17.16)$$

(We are back in Minkowski space now). What can we say about $\rho_1(s)$ and $\rho_2(s)$? To begin with, there is the nucleon contribution $\rho_2^N(s) = m\rho_1^N(s) = m\lambda^2 \delta(s - m^2)$. To understand what happens for $s > m^2$, consider first a simpler problem, the spectral density of the polarization operator of two electromagnetic currents, which is nothing but the total cross section of annihilation $e^+e^- \to$ hadrons well known from experiment.

At $s \approx 0.5$ GeV^2, this cross section has a sharp narrow peak associated with the ρ and ω mesons. For s slightly exceeding m_ρ^2, the cross section is very small. Then it rises again and exhibits a resonance behavior (much less distinct than in the ρ meson case) at $s \approx m_{\rho^*}^2$, where ρ^* is the meson with the same quantum numbers as ρ, the "radial excitation" of the latter ($m_{\rho^*} \approx 1450$ MeV). There are also still heavier resonances in the ρ channel, but the larger the mass is, the broader are the states and the less the distance between neighboring resonances is. As a result, for $s \gtrsim 2$ GeV^2, the adjacent Breit–Wigner peaks overlap and the behavior of $\rho^{\text{em}}(s) \propto \sigma_{e^+e^- \to \text{hadrons}}(s)$ is smooth.

When s is large, $\sigma_{e^+e^- \to \text{hadrons}}(s)$ coincides with the quark cross section $\sigma_{e^+e^- \to \bar{q}q}(s)$. This duality is the central point of the whole QCD philosophy: when the characteristic energy of the process is large, one can calculate its cross section in the framework of QCD with quarks and gluons and be sure that the result is correct even though we see hadrons and not quark and gluons in the final state. In terms of spectral density, this means that, for large s, the physical spectral density coincides with the perturbative quark spectral density. If perturbative gluon corrections are neglected, it coincides with the flavor sum of the imaginary parts of the simplest quark loop diagrams.

Let us go back to the nucleon channel. The function $\rho_1(s)$ can be related to the physical cross section $e^+ + \gamma \to hadrons$.[‡] Even though we do not have any experimental information on $\sigma_{e^+\gamma \to \text{hadrons}}$, we can expect that it behaves exactly in the same way as $\sigma_{e^+e^- \to \text{hadrons}}$: a very sharp peak at $s = m^2$, a gap, and, after minor oscillations associated with the Roper resonance $N^*(1440)$ and other "radial excitations" of nucleon, $\rho_1(s)$ should approach the quark spectral density calculated by the graph in Fig. 17.1a.

The situation with $\rho_2(s)$ is more subtle. First, it is not positive definite and probably oscillates rather significantly at $s \approx 2 - 2.5$ GeV^2. Second, its value at large s is related not to the total cross section itself, but to a small correction to $\sigma_{e^+\gamma \to \text{hadrons}}$ proportional to the mass m_e of the initial electron or, better to say, to the sensitivity of the cross section to m_e in the massless quark limit:

$$\rho_2(\text{large } s) \propto \frac{d}{dm_e}\sigma_{e^+\gamma \to \text{hadrons}}(\text{large } s) . \tag{17.17}$$

The quantity (17.17) is also physical, however. We can deduce that, like $\rho_1(\text{large } s)$ can be calculated by the perturbative quark diagram Fig. 17.1a, so $\rho_2(\text{large } s)$ is determined by the imaginary part of the graph in Fig. 17.1c:

$$\rho_1(\text{large } s) \sim \rho_1^{\text{quark}}(s) = \frac{s^2}{64\pi^4} , \qquad \rho_2(\text{large } s) \sim \rho_2^{\text{quark}}(s) = \frac{\Sigma s}{4\pi^2} . \tag{17.18}$$

The detailed behavior of $\rho_1(s)$ and $\rho_2(s)$ is not known but, for an estimate of the contribution of excited states in the integrals (17.15), we do not need

---

[‡]Such a process involving baryon number nonconservation is predicted by grand unified theories. Its cross section is very very small, but it is not relevant for us, neither it is relevant whether this process goes in Nature, indeed, or not. Also the electromagnetic interactions responsible for the process $e^+e^- \to hadrons$ are *external* to QCD. The only important point is that $\rho_1(s)$ can *in principle* be observed and has therefore a status of physical quantity.

it and can adopt a rough but sufficient (and efficient) *resonance+continuum model*:

$$\rho_1(s) = \lambda^2 \delta(s - m^2) + \rho_1^{\text{quark}}(s)\theta(s - W_1^2) ,$$
$$\rho_2(s) = m\lambda^2 \delta(s - m^2) + \rho_2^{\text{quark}}(s)\theta(s - W_2^2) , \qquad (17.19)$$

where $W_{1,2}$ are certain thresholds, free parameters of the model. The simplest assumption is $W_1 = W_2$ and we will stick to it. Actually, in view of the discussion above, one could rather expect that, effectively, $W_2 > W_1$.

We substitute now Eq. (17.19) into Eq. (17.15) and transfer the continuum contribution to the left side. Two spinor structures give rise to two sum rules:

$$\frac{96}{\pi^4 \tau^8} \Delta_1(W_1) + \frac{b}{8\pi^4 \tau^4} + \frac{4\Sigma^2}{3\tau^2} - \frac{m_0^2 \Sigma^2}{12} = \lambda^2 m^2 K_2(m\tau) ,$$

$$\frac{8\Sigma}{\pi^2 \tau^5} \Delta_2(W_2) - \frac{\Sigma b}{96\pi^2 \tau} + \frac{272}{81}\pi^2 \left(\frac{\alpha_s}{\pi}\right)\tau\Sigma^3 \ln(\tau\mu_{\text{hadr}})$$
$$= \lambda^2 m^2 K_1(m\tau) , \qquad (17.20)$$

where

$$\Delta_1(W) = \frac{\tau^8}{3 \cdot 2^{11}} \int_0^{W^2} s^3 ds K_2(\sqrt{s}\tau)$$

$$= 1 - \frac{(\tau W)^7}{3 \cdot 2^{10}} K_3(\tau W) - \frac{(\tau W)^6}{3 \cdot 2^8} K_4(\tau W) - \frac{(\tau W)^5}{3 \cdot 2^7} K_5(\tau W) ,$$

$$\Delta_2(W) = \frac{\tau^5}{32} \int_0^{W^2} s^{3/2} ds K_1(\sqrt{s}\tau)$$

$$= 1 - \frac{(\tau W)^4}{16} K_2(\tau W) - \frac{(\tau W)^3}{8} K_3(\tau W) . \qquad (17.21)$$

The task now is to fit three parameters m, λ, W to achieve the best agreement between the left side and right side of the sum rules (17.20) in the *fiducial interval* of τ such that both the uncertainty due to unknown higher power corrections and the uncertainty brought about by our rough model (17.19) for the spectral densities $\rho_{1,2}$ are reasonably small.

The enthusiastic reader who followed us faithfully up to this point and would rush immediately to his PC to perform such a fit might get slightly disappointed, because the accuracy of the predictions for the residue λ and for the nucleon mass which follow from the analysis of the sum rules (17.20) appears to be reasonable but not remarkable. The point is, however, that the sum rules (17.20) based on the analysis of the polarization operator Π as a function of Euclidean time τ are not exactly what is usually called the ITEP sum rules.

Following the pioneers, people usually consider *not* $\Pi(\tau)$, but another quantity, the *Borel transform* of the polarization operator,

$$\hat{\mathcal{B}}_{M^2}\Pi(p) = \frac{1}{\pi}\int_0^\infty ds\, e^{-s/M^2}[\not{p}\rho_1(s) + \rho_2(s)] \,. \tag{17.22}$$

Qualitatively, M^{-1} plays the same role as the Euclidean time interval τ in the spectral decomposition (17.15): when M is large, the power corrections on the theoretical side of the sum rules are suppressed, when M is small, the contribution of the excited states in the phenomenological side is suppressed, and there is a certain fiducial interval of M when both are under control. Borel sum rules have somewhat simpler analytical form (exponentials instead of Bessel functions) and they give better numerical fits.[§] On the other hand, Euclidean time τ is a more natural quantity than the Borel parameter M, looking somewhat artificial, and we have chosen to start our discussion with the Euclidean time sum rules mostly for pedagogical purposes.

To derive the Borel sum rules, we have to calculate the Fourier image of the coordinate space polarization operator (17.10), take its imaginary part in the timelike region $s = p^2 \ge 0$,

$$\begin{aligned}\frac{1}{\pi}\mathrm{Im}\,\Pi(p) = {}& \not{p}\left[\frac{2s^2+b}{128\pi^4} + \frac{2\Sigma^2}{3}\delta(s) - \frac{m_0^2\Sigma^2}{6}\delta'(s)\right] \\ & + \frac{\Sigma}{4\pi^2}\left[s - \frac{b}{24}\delta(s)\right] + 32\pi^2\frac{17}{81}\left(\frac{\alpha_s}{\pi}\right)\Sigma^3\delta'(s) \,, \quad (17.23)\end{aligned}$$

substitute it in the integral (17.22), and compare it with the phenomenological contribution to the same integral estimated using the resonance+continuum model (17.19) for the physical spectral density. We obtain

$$M^6 E_2(W^2/M^2)L^{-4/9} + \frac{4}{3}a^2 L^{4/9} + \frac{1}{4}bM^2 E_0(W^2/M^2)L^{-4/9} - \frac{1}{3}a^2\frac{m_0^2}{M^2}$$
$$= \tilde{\lambda}^2 \exp(-m^2/M^2) ,$$

$$2aM^4 E_1(W^2/M^2) + \frac{272}{81}\left(\frac{\alpha_s}{\pi}\right)\frac{a^3}{M^2} - \frac{ab}{12} = m\tilde{\lambda}^2 \exp(-m^2/M^2) ,$$

$$(17.24)$$

where $a = 4\pi^2\Sigma$, $\tilde{\lambda}^2 = 32\pi^4\lambda^2$, $E_0(x) = 1 - e^{-x}$,
$E_1(x) = 1 - (1+x)e^{-x}$, $E_2(x) = 1 - (1 + x + x^2/2)e^{-x}$,

$$L = \frac{\ln(M/\Lambda_{\text{QCD}})}{\ln(\mu_{\text{hadr}}/\Lambda_{\text{QCD}})} , \quad \Lambda_{\text{QCD}} \approx 150 \text{ MeV} , \quad \text{and} \quad \mu_{\text{hadr}} \approx 500 \text{ MeV} ,$$

and we have assumed $W_1 = W_2 = W$. We have included here the anomalous dimension factors $\propto L^{\pm 4/9}$ mentioned above. They do not provide an important numerical effect, however. The factor $E_0(W^2/M^2)$ multiplying the term $\propto b$ in the first sum rule corresponds to a certain refinement of the model (17.19) for the spectral density, but it does not affect the results either.

The nucleon mass carries dimension and it is determined by the dimensionful input parameters in the sum rules (17.20) or (17.24), of which the quark condensate Σ plays the most important role. To get a rough idea of what happens here, let us play the following game:

(1) Neglect the continuum contributions in Eqs. (17.24) and set $E_0 = E_1 = E_2 = 1$.

(2) Set $L = 1$ and neglect the terms $\propto b$, $\propto m_0^2$, and $\propto \alpha_s a^3$.

(3) Set $M = m$ and compare the two sum rules. We obtain the equation

$$m^6 + \frac{4}{3}a^2 = 2am^3 . \tag{17.25}$$

The equation (17.25) has no real solutions for m (a real solution appears, however, if taking into account the neglected terms on the left side of the sum rules), but the best agreement (a rather close one) is achieved at

$$m \approx a^{1/3} = (4\pi^2\Sigma)^{1/3} \approx 0.82 \text{ GeV} . \tag{17.26}$$

This is the famous *Ioffe formula*. Due to many approximations involved in its derivation, the formula cannot, of course, be treated as an exact result, but it is beautiful, and also the value of mass extracted in a more

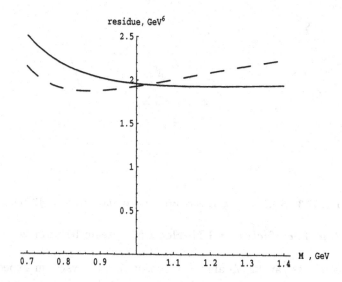

Fig. 17.2   The nucleon residue $\tilde{\lambda}^2$ extracted from the first (solid line) and the second (dashed line) sum rule in Eq. (17.24).

rigorous analysis is not very much different from Eq. (17.26) and from the experimental value.

Speaking of the more rigorous analysis, its results are presented in Fig. 17.2. We have adopted here the experimental value $m = .94$ GeV for the nucleon mass and plotted the residue $\tilde{\lambda}^2$ determined from the sum rules (17.24) as a function of M. The continuum threshold $W = 1.5$ GeV is chosen such that the fluctuations of the residue in the fiducial interval of M be minimal. If we took the values of mass, say, $m = 1.2$ GeV or $m = .75$ GeV, the fit would be considerably worse no matter what value of W is chosen. The extracted value $\tilde{\lambda}^2 \approx 2$ GeV^2 (with an estimated accuracy $\approx 15\%$) can be used to calculate the proton decay rate in grand unified models.

We discussed in details the nucleon channel, but the same method can be applied to extract the masses and the residues of the lowest hadron states in all other channels. And in almost *all* cases, the results agree well with experiment.

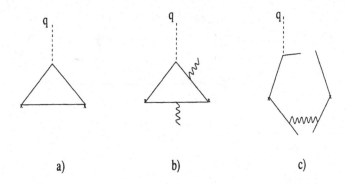

Fig. 17.3   Some QCD graphs contributing to the correlator (17.27).

17.3   Pion Formfactor and Nucleon Magnetic Moments.

All these results can be obtained by considering the vacuum expectation values of the proper 2-point correlators. But nothing prevents us to study some other matrix elements of these correlators or consider 3-point correlators (mostly, the vacuum expectation values thereof), etc. A lot of nontrivial predictions on *dynamical* characteristics of hadrons can be obtained in this way.

Consider the correlator

$$T_{\mu\nu\lambda}(p,p') = -\int d^4x\,d^4y\; e^{i(p\cdot x - p'\cdot y)} \langle A_\mu^+(x) J_\lambda^{\text{em}}(0) A_\nu^-(y) \rangle_0 \;, \quad (17.27)$$

where $A_\mu^+ = \bar{u}\gamma_\mu\gamma_5 d$, $A_\nu^- = \bar{d}\gamma_\nu\gamma_5 u$, and $J_\lambda^{\text{em}} = Z_u \bar{u}\gamma_\mu u + Z_d \bar{d}\gamma_\mu d$ $(Z_u = 2/3, Z_d = -1/3)$, in the Euclidean region p^2, $p'^2 < 0$, and $q^2 = (p - p')^2 < 0$. When $|p^2|$ and $|p'^2|$ are large (and $Q^2 = -q^2$ is not too small), we can calculate $T_{\mu\nu\lambda}(p,p')$ in the framework of QCD. The leading contribution is provided by the triangle quark graph (see Fig. 17.3a). There are graphs with perturbative gluon exchange whose contribution grows with Q^2 and which *determine* the formfactor at asymptotically large Q^2, but which are irrelevant in the region $Q^2 \sim$ several GeV^2. In addition, there are nonperturbative power corrections due to the gluon condensate (see Fig. 17.3b) and the (square of) the quark condensate as in Fig. 17.3c. On the other hand, the correlator (17.27) is saturated by the physical hadron states with nonzero coupling residue to the axial current. The pion is the lightest of such states.

One can focus on the pion contribution either by directly considering the Euclidean correlator $\langle A_\mu^+(x) J_\lambda^{em}(0) A_\nu^-(y) \rangle_0$ or, as it is usually done, by performing the *double* Borel transform and considering the quantity

$$\frac{1}{\pi} \int \int ds ds' e^{-(s+s')/M^2} \rho(s, s'; Q^2) , \qquad (17.28)$$

where $\rho(s, s'; Q^2)$ is the double spectral density of the invariant amplitude $T(p^2, p'^2; Q^2)$ multiplying the structure $(p+p')_\mu (p+p')_\nu (p+p')_\lambda$ in $T_{\mu\nu\lambda}$. The sum rule associated with this structure gives better results than the others by the reasons which we will not discuss here.

The quark (perturbative and nonperturbative) spectral density is calculated by the diagrams in Fig. 17.3 but, to write down the sum rule, we should adopt also some model for the physical spectral density. In this case, the resonance+continuum model looks as follows:

$$\rho^{phys}(s, s'; Q^2) \sim \frac{1}{4} F_\pi^2 f_\pi(Q^2) \delta(s) \delta(s')$$
$$+ \left[1 - \theta(W^2 - s)\theta(W^2 - s')\right] \rho^{quark}(s, s'; Q^2) , \quad (17.29)$$

where F_π is the pion decay constant and $f_\pi(Q^2)$ is the electromagnetic formfactor of the pion, which we are set to determine. The pion is assumed to be massless here, and we will also consistently neglect all other effects due to small nonzero masses m_u and m_d. Eq. (17.29) assumes the presence of the gap at small enough nonzero s, s'. For $s > W^2$ or $s' > W^2$ (the threshold W^2 being chosen as the fitted value of $W^2 \approx 1.2$ GeV^2 in the sum rule for the 2-point axial current correlator), $\rho^{phys}(s, s'; Q^2)$ is assumed to coincide with the perturbative quark spectral density (the double cut of the graph in Fig. 17.3a):

$$\rho^{quark}(s, s'; Q^2)$$
$$= \frac{3Q^4}{16\pi^2 \lambda^{7/2}} \left[3\lambda(x + Q^2)(x + 2Q^2) - \lambda^2 - 5Q^2(x + Q^2)^3\right] ,$$
$$x = s + s' , \quad \lambda = (s + s' + Q^2)^2 - 4ss' . \quad (17.30)$$

As a result, we obtain the following approximate expression for the pion formfactor:

$$f_\pi(Q^2) = \frac{4}{F_\pi^2} \left[\int_0^{W^2} \int_0^{W^2} ds ds' e^{-(s+s')/M^2} \rho^{quark}(s, s'; Q^2) \right.$$
$$\left. + \frac{1}{48M^2} \left\langle \frac{\alpha_s}{\pi} G^2 \right\rangle_0 + \frac{52\pi}{81M^4} \alpha_s \Sigma^2 \left(1 + \frac{2Q^2}{13M^2}\right) \right] . \quad (17.31)$$

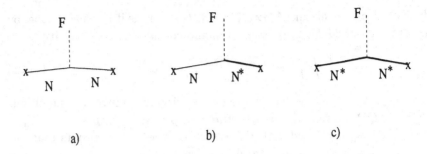

Fig. 17.4   Physical contributions to $\Pi_{\mu\nu}(p)$.

We can believe in this result in the intermediate Q^2 region $0.5\ \text{GeV}^2 \lesssim Q^2$ $\lesssim 3\ \text{GeV}^2$. When Q^2 is too small or too large, there is no reliable fiducial interval in the Borel parameter M^2. In the quoted interval of Q^2 the formfactor (17.31) practically does not depend on M^2 in the interval $0.7\ \text{GeV}^2 \lesssim M^2 \lesssim 1.7\ \text{GeV}^2$ and agrees well with experiment.

The same method can be (and was) applied to determine the electromagnetic formfactors of other mesons and baryons, the rates of semileptonic decays like $D^+ \to e^+\nu + hadrons$, etc.

To give the last example, consider the polarization operator of the proton currents (17.2) and their neutron counterparts in the presence of a weak (electro)magnetic field $F_{\mu\nu}$. (For the vacuum to be stable, the condition $|\boldsymbol{H}| > |\boldsymbol{E}|$ should be fulfilled and, by a Lorentz transformation, the electric components of $F_{\mu\nu}$ can be brought to zero.) The first terms in the expansion of $\Pi_F(p)$ in F are

$$\Pi_F(p) \;=\; \Pi_0(p) \;+\; \sqrt{4\pi\alpha}\,F_{\mu\nu}\Pi_{\mu\nu}(p) \;+\ldots \qquad (17.32)$$

We will be interested in the quantity $\Pi_{\mu\nu}(p)$. The leading phenomenological contribution to $\Pi_{\mu\nu}(p)$ depicted in Fig. 17.4a is proportional to the proton or neutron magnetic moment. Writing down and analysing the corresponding sum rules, the values of μ_P and μ_n can be determined.

This problem is somewhat more refined and complicated than the analysis of the mass sum rules. The basic novelty is the appearance of new expectation values proportional to the strength of the external field. The most important such parameter χ is defined as

$$\langle i\bar{q}\sigma_{\mu\nu}q\rangle_F \;=\; -\sqrt{4\pi\alpha}\,Z_q\chi\Sigma F_{\mu\nu}\,. \qquad (17.33)$$

[The assumption $\langle \bar{q}\sigma_{\mu\nu}q\rangle_F \propto Z_q$ is violated by the contributions described by the graphs including closed quark loops in nonperturbative gluon background. Their smallness can be justified by the same (hidden from the reader!) arguments which justify the suppression of direct instanton contributions in the sum rules under discussion.] The parameter χ can be called *magnetic susceptibility* of the vacuum. Actually, there is an infinite set of such susceptibilities:

$$\langle g_s \bar{q} t^a G^a_{\mu\nu} q\rangle_F = -\sqrt{4\pi\alpha} Z_q \kappa \Sigma F_{\mu\nu} ,$$

$$\epsilon_{\mu\nu\lambda\sigma}\langle g_s \bar{q}\gamma^5 t^a G^a_{\lambda\sigma} q\rangle_F = -i\sqrt{4\pi\alpha} Z_q \xi \Sigma F_{\mu\nu} , \qquad (17.34)$$

etc. Like the vacuum expectation values, these generalized susceptibilities are (more or less) ordered according to their canonical dimensions so that, in the fiducial interval of the sum rules we are going to derive, the contribution of the higher susceptibilities is irrelevant. Still, the presence of three new parameters χ, κ, ξ is a nuissance which could have prevented us to determine the physical quantities μ_P, μ_n.

Fortunately, we have at our disposal not just one but four different sum rules: there are two sum rules for the proton corresponding to the chirality odd $\propto \not{p}\sigma_{\mu\nu} + \sigma_{\mu\nu}\not{p}$ and chirality even $\propto (p_\mu\gamma_\nu - p_\nu\gamma_\mu)\not{p}$ tensor structures in $\Pi_{\mu\nu}(p)$ and two such sum rules for the neutron. We are lucky enough to be able to choose two different combinations of the four sum rules where the parameters $\chi, \kappa,$ and ξ *do* not enter.

In principle, there is also the third structure $\propto \sigma_{\mu\nu}$, but the corresponding sum rules are worse than for the structure $\propto (p_\mu\gamma_\nu - p_\nu\gamma_\mu)\not{p}$ because the nucleon contribution is not so well separated there from continuum. (The rule of thumb here is the following: always try to choose the structures which involve as many powers of external momenta as possible — they focus on the lowest state better.) This example and the examples discussed above show that *selecting* proper sum rules is a necessary preliminary step for *any* evaluation of hadron parameters with this method. Wrong selection gives at best bad accuracy and at worst wrong results...

Another complication is a somewhat more intricate model for the phenomenological spectral density which we have to adopt. We are interested in the graph in Fig. 17.4a with double nucleon pole. Its contribution to

$\Pi_{\mu\nu}(p)$ is [cf. (12.41)]

$$\Pi_{\mu\nu}(p) = -\frac{1}{4}\frac{\lambda^2}{(p^2-m^2)^2}\{\mu_N(\not{p}\sigma_{\mu\nu}+\sigma_{\mu\nu}\not{p})$$
$$+ 2\mu_N\sigma_{\mu\nu}m + 2\mu_N^{\mathrm{a}}(p_\mu\gamma_\nu - p_\nu\gamma_\mu)\frac{\not{p}}{m}\}\,, \qquad (17.35)$$

where we have retained only the terms $\propto (p^2-m^2)^{-2}$. In the above expression, λ is the nucleon residue defined in Eq. (17.3), μ_N is the full nucleon magnetic moment expressed in nuclear magnetons $e\hbar/(2mc)$, and μ_N^{a} is the anomalous magnetic moment ($\mu_P^{\mathrm{a}} = \mu_P - 1$ and $\mu_n^{\mathrm{a}} = \mu_n$). The imaginary part of Eq. (17.35) involves the derivative of delta function $\delta'(s-m^2)$. Besides, there is a contribution of Fig. 17.4b describing the transitions of nucleon to excited continuum states in the presence of a magnetic field. This contribution is proportional to $\delta(s-m^2)$. Finally, there is a contribution $\propto \langle continuum|continuum\rangle_F$ depicted in Fig. 17.4c. For large s, its imaginary part coincides with the perturbative quark spectral density. Thus, the "resonance+continuum" model of the mass sum rules should be modified to the "resonance + $\langle resonance|continuum\rangle_F$ + continuum" model:

$$\rho_{\mathrm{odd}}^{\mathrm{phys}}(s) = \mu_N\frac{\lambda^2}{4}\delta'(s-m^2) + A_N\delta(s-m^2) + \rho_{\mathrm{odd}}^{\mathrm{quark}}(s)\theta(s-W^2)\,,$$

$$\rho_{\mathrm{even}}^{\mathrm{phys}}(s) = \mu_N^{\mathrm{a}}\frac{\lambda^2}{4}\delta'(s-m^2) + B_N\delta(s-m^2) + \rho_{\mathrm{even}}^{\mathrm{quark}}(s)\theta(s-W^2)\,,$$
$$(17.36)$$

where the spectral densities are defined according to

$$-\frac{1}{\pi}\,\mathrm{Im}[\Pi_{\mu\nu}(p)] = (\not{p}\sigma_{\mu\nu} + \sigma_{\mu\nu}\not{p})\rho_{\mathrm{odd}}^{\mathrm{phys}}(s) + 2\rho_{\mathrm{even}}^{\mathrm{phys}}(s)(p_\mu\gamma_\nu - p_\nu\gamma_\mu)\not{p}\,.$$
$$(17.37)$$

Now, A_N and B_N are "parasitic" parameters contaminating the sum rules. Fortunately, the terms $\propto \delta'(s-m^2)$ and $\propto \delta(s-m^2)$ in the spectral density correspond to the terms with different dependence of the Borel transform $\Pi_{\mu\nu}(M^2)$ on M^2. This allows one to separate the signal from noise and to determine μ_P and μ_n with surprisingly good accuracy. The theoretical predictions $\mu_P \approx 3.0$, $\mu_n \approx -2.0$ agree well with experiment ($\mu_P^{\mathrm{exp}} = 2.79$, $\mu_n^{\mathrm{exp}} = -1.91$).

Comparing the Borel sum rules for the magnetic moments with the Borel mass sum rules at $M = m$ (which is in the center of the fiducial interval) and neglecting some terms, one can derive nice approximate formulae

$$\mu_P = \frac{8}{3}\left(1 + \frac{2\pi^2\Sigma}{3m^3}\right), \text{ and } \mu_n = -\frac{4}{3}\left(1 + \frac{8\pi^2\Sigma}{3m^3}\right). \quad (17.38)$$

These estimates are even closer to the experimental values than the results of the accurate analysis. Their theoretical status is the same as that of the Ioffe formula (17.26) for the nucleon mass.

The described external field method can be and was used to determine the magnetic moments and electromagnetic radii of other hadrons, the couplings $g_{\pi\bar{N}N}$, $g_{\pi D^{\bullet}D}$, etc.

Lecture 18
Hot and Dense QCD

Up to now, we discussed only the properties of the QCD vacuum state and of its elementary excitations. This lecture is devoted to the properties of the QCD *heat bath*, a statistical system characterized by the thermal density matrix $\exp\{-\beta H\}$ or the grand canonical density matrix $\exp\{-\beta(H - \mu Q_B)\}$, where Q_B is the operator of the conserved baryon charge, and the constant μ is the *chemical potential*. Nonzero μ means a finite baryon number density. Let us first assume $\mu = 0$. Gross features of the dynamics of hot QCD as a function of temperature are as follows:

At low temperatures, the system may be regarded as a gas of colorless hadron states, the eigenstates of the QCD Hamiltonian at zero temperature. When the temperature is small, this gas is composed mainly of pions — other mesons and baryons have higher masses and their admixture in the medium is exponentially small $\sim \exp\{-M/T\}$. At small temperatures, also the pion density is small — the gas is rarefied and pions practically do not interact with each other.

However, as the temperature increases, the pion density grows, their interactions become strong, and also other strongly interacting hadrons appear in the medium. For temperatures of order $T \gtrsim 150$ MeV, the interaction is so strong that the hadron states do not provide a convenient basis to describe the properties of the medium anymore, and no analytic calculation is possible.

On the other hand, when the temperature is very high, much higher than the characteristic hadron scale $\mu_{hadr} \sim 0.5$ GeV, theoretical analysis becomes possible again. Only in this range, the proper basis is not given by the hadron states but, rather, by quarks and gluons — the elementary fields entering the QCD Lagrangian. One can say that, at high temperatures, hadrons get "ionized" to their basic compounds. In the 0th approximation, the system is a heat bath of freely propagating colored particles. For sure, quarks and gluons interact with each other, but at high temperatures the effective coupling constant is small, $\alpha_s(T) \ll 1$, and the effects due to the interaction can be taken into account perturbatively. This interaction has a long-distance Coulomb nature, and the properties of the system are in many respects very similar to the properties of an ordinary nonrelativistic plasma involving charged particles with weak Coulomb interactions. The only difference is that quarks and gluons carry color rather than electric charge. Hence the name: *quark–gluon plasma*.

Thus, the properties of the system at low and at high temperatures have nothing in common. A set of natural questions arises: What is the nature of the transition from the low-temperature hadron gas to the high-temperature quark–gluon plasma? Is it a *phase* transition? If yes, what is its order? I want to emphasize that this question is highly nontrivial. A drastic change in the properties of the system in a certain temperature range does not guarantee the presence of a phase transition *point* where the free energy of the system or its specific heat is discontinuous. Recall that there is no phase transition between an ordinary gas and ordinary plasma.

There are at least four reasons why the question of the phase transition and the related questions of the low temperature and high temperature dynamics of QCD are interesting to study:

(1) They are very amusing from a purely theoretical viewpoint. The results can only be obtained by using an appropriate spicy mixture of methods of quantum field theory and statistical physics.

(2) Theoretical conclusions can be checked in lattice numerical experiments. Scores of papers devoted to lattice studies of thermal properties of QCD have been published.

(3) *Some* theoretical results can be confronted with the results of experiments done in the *laboratory* with heavy ion colliders.

(4) There might be cosmological implications. During the first second of its evolution, our Universe passed through the stage of high-T quark–gluon plasma which later cooled down to a hadron gas (and eventually to dust and stars, of course). We will argue later that this transition was not a *phase* transition and, contrary to early expectations, did not bring about significant inhomogenuities or

other observable effects for the Universe we now live in. But it is important to understand why.

We will touch here on just a few selected issues of the physics of hot QCD. See the review [45] for more details.

There is also a related question — what are the properties of relatively cold but very dense matter, and whether there is a phase transition when the chemical potential (corresponding to the baryon charge), rather than temperature, is increased. From experimental viewpoint this question is no less interesting and important: in realistic heavy ion experiments we have *both* finite temperature and finite baryon number density. Though theoretical predictions are less certain here, we have recently acquired some qualitative understanding of high density physics; I will discuss it at the end of the lecture.

18.1   Lukewarm Pion Gas. Restoration of Chiral Symmetry.

To begin with, consider the thermodynamics of the "lukewarm" QCD system, a rarefied pion gas where the pion interactions can be taken into account perturbatively in the framework of chiral perturbation theory. When $T \ll \mu_{hadr}$, interactions can be neglected altogether and the eigenstates of the Hamiltonian are simple combinations of individual pion states. The free energy of such a system of free pions has the form

$$F^{\text{free pion}} = -T \ln Z = -T \ln \left[\prod_{p=\frac{2\pi}{L}m} \left(\sum_{n=0}^{\infty} e^{-\beta n E_p} \right)^3 \right]$$

$$= 3T \sum_{p} \ln \left(1 - e^{-\beta E_p} \right) , \qquad (18.1)$$

where $E_p = \sqrt{p^2 + M_\pi^2}$, L is the length of the spatial box introduced as an infrared regulator, and the components of the vector m are integers. The factor "3" is just the number of pion species. Trading the sum for the integral, we obtain for the volume density of the free energy

$$\frac{F^{\text{free pion}}}{V} = 3T \int \frac{dp}{(2\pi)^3} \ln \left(1 - e^{-\beta\sqrt{p^2+M_\pi^2}} \right) . \qquad (18.2)$$

An alternative way to obtain this result is to calculate the simplest pion "vacuum loop" diagram in Fig. 18.1a. A very important corollary of this simple formula (it is nothing but the Stephan–Boltzman formula for massive scalar particles) is the fact that the order parameter of the spontaneously broken chiral symmetry, the quark condensate, decreases with temperature. The temperature-dependent condensate $\Sigma(T)$ is defined as the derivative of free energy with respect to the quark mass

$$\Sigma(T) = -\frac{1}{2}\frac{1}{V}\frac{\partial F(T)}{\partial m}, \tag{18.3}$$

where we endowed both the u and d quarks with the same mass m, to be set to zero after differentiation. In actuality, $m_q \neq 0$, but let us consider first a theoretically clean case when the quark masses are very small so that not only is $m_q \ll T$, but also $M_\pi \ll T$. The expansion of Eq. (18.2) in pion mass in the region $M_\pi \ll T$ reads

$$\frac{F^{\text{free pion}}}{V} = 3\left[-\frac{\pi^2 T^4}{90} + \frac{T^2 M_\pi^2}{24} + O(M_\pi^3 T)\right]. \tag{18.4}$$

(Note that the expansion of F in M_π^2 is not analytic. We will capitalize on this fact later on.) Substituting the free energy (18.4) in Eq. (18.3) (with the added vacuum energy density) and taking into account the Gell-Mann–Oakes–Renner relation (12.62), we obtain

$$\Sigma(T) = \Sigma(0)\left[1 - \frac{T^2}{8F_\pi^2} + o\left(\frac{T^2}{\mu_{\text{hadr}}^2}\right)\right]. \tag{18.5}$$

The important result obtained from this analysis is that the chiral condensate Σ, the order parameter associated with the chiral symmetry breaking, *decreases* with temperature.

Higher-order terms in the parameter T^2/μ_{hadr}^2 can be determined by taking into account the pion interactions in the framework of chiral perturbation theory. Gerber and Leutwyler calculated the two- and three-loop graphs in Fig. 18.1b–e. (They did it for nearly massless pions, i.e. only the terms linear in mass were taken into account, with the contributions due to other hadrons being disregarded.) The result is

$$\Sigma_T = \Sigma_0\left[1 - \frac{T^2}{8F_\pi^2} - \frac{T^4}{384F_\pi^4} - \frac{T^6}{288F_\pi^6}\ln\frac{\Lambda}{T} + \ldots\right], \tag{18.6}$$

where all effects from the higher-derivative terms in the effective chiral Lagrangian are described by the constant Λ. Experimental data on pseudo-Goldstone interactions give the value $\Lambda \sim 500 \pm 100$ MeV.

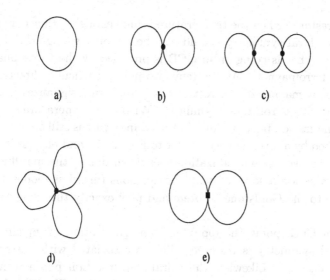

Fig. 18.1 Pion loop contributions to the free energy. Black circles are the pion vertices following from the leading-order chiral Lagrangian (12.57), (12.61). The black box stands for the vertices involving four derivatives like in Eq. (12.59), or the vertices with two derivatives, but proportional to small quark masses.

It is remarkable that *all* calculated terms in the expansion (18.6) are negative. The quark condensate does fall off, and a simple extrapolation of the dependence (18.6) leads us to the conclusion that the condensate goes to zero at roughly

$$T_c \approx 190 \text{ MeV} . \tag{18.7}$$

At this point, a phase transition occurs and chiral symmetry is restored.

To be exact, the result (18.6) was derived in the theory with two massless quarks and even there it has, of course, a limited range of validity. The expansion in the parameter $\sim T^2/8F_\pi^2$ makes sense when this parameter is small, which is not the case at $T \sim T_c$. We cannot, strictly speaking, extrapolate the dependence (18.6) to the very point of the phase transition. But the conclusion that the phase transition exists in the *massless* theory, as well as the estimate (18.7), is rather solid.

The symmetry restoration phenomenon just observed is very natural. Almost *any* physical symmetry that is broken spontaneously at zero temperature is restored when the temperature is high enough.* Hence, in massless QCD the chiral symmetry should also be restored at high T. The exact analogy of what is going on in QCD is provided by the Curie phase transition in ferromagnets. At low temperatures, a particular direction of the spontaneous magnetization vector in a ferromagnet signalizes the spontaneous breaking of rotational symmetry. When the temperature is increased, the magnetization falls. As long as the temperature is still small, this effect is described by a formula very similar to Eq. (18.5), and physics is also very similar: the average magnetization goes down due to thermal fluctuations of magnetization associated with the magnons (massless modes which appear due to the Goldstone theorem and play exactly the same role as the pions).

At the Curie point, the spontaneous magnetization disappears, and the rotational symmetry is restored. This is associated with a *second-order* phase transition. Likewise, the chirality restoration phase transition in the theory with two massless quarks is second-order. We will substantiate this assertion by a reasoning which may look somewhat heuristic, but can actually be made quite precise in the framework of the general theory of phase transitions, which is beyond the scope of this book. Consider the theory in the vicinity of the phase transition point. The description of the system in terms of the pion fields and the chiral effective Lagrangian makes little sense there. One can, instead, introduce composite colorless fields $\Phi^{fg} = \psi_L^f \bar{\psi}_R^g$ ($f, g = 1, \ldots, N_f$) and write down an effective Lagrangian depending on Φ^{fg}. In condensed matter physics, this trick is well known and is called the *mean field* or the *Ginsburg–Landau* approach. For $N_f = 2$, it is convenient to consider the combinations

$$\Phi_\mu = \Phi^{fg}(\sigma_\mu)_{fg} = \psi_L \sigma_\mu \bar{\psi}_R , \qquad \sigma_\mu = (i, \boldsymbol{\sigma}) , \qquad (18.8)$$

which transform as 4-vectors under $O(4) \equiv SU_L(2) \times SU_R(2)$. In the broken phase, $\langle \Phi_\mu \rangle \neq 0$ and the $O(4)$ symmetry is broken down to $O(3)$ [or $SU_L(2) \times SU_R(2) \to SU_V(2)$]. If you will, QCD with two massless flavors is an "$O(4)$-magnet".

A general form of the $O(4)$ invariant Ginsburg–Landau effective poten-

---

*This is not the case for supersymmetry. Also, there are some rather artificial counterexamples involving ordinary symmetries remaining spontaneously broken at high temperatures, which we will not discuss here.

tial is

$$g_1 \Phi_\mu^2 \; + \; g_2 \left(\Phi_\mu^2\right)^2 \; + O(\Phi^6) \,, \tag{18.9}$$

where the operators $\propto \Phi^6$, etc. are "irrelevant" and can be discarded. The coefficients $g_{1,2}$ depend on temperature. A stability argument tells us that g_2 is always non-negative. Assume it to be positive. The phase transition occurs at the point where $g_1(T_c) = 0$. For $T > T_c$, $g_1(T)$ is positive, while for $T < T_c$ it is negative, and the average $\langle \Phi^2 \rangle_T \propto -g_1(T)/[2g_2(T)]$ and also the average $\langle \Phi_\mu \rangle_T = \sqrt{-g_1(T)/[2g_2(T)]} \, n_\mu$, signaling the spontaneous breaking of the $O(4)$ symmetry, appear. At $T \sim T_c$, the dependence of $\langle \Phi^2 \rangle_T$ on temperature is smooth (no discontinuity) and one has a second-order phase transition.

If the same logic is applied to a system with three or more massless flavors, one finds that a first-order phase transition with a finite discontinuity of the quark condensate at the phase transition point is not excluded and would probably occur, indeed, if we had three massless flavors. In reality, however, all quarks have finite masses. If, for example, we had three quarks with very small masses $m_u \sim m_d \sim m_s \sim 1$ MeV, we would still expect a first-order phase transition even though the exact chiral symmetry of the Lagrangian would be lost because a finite discontinuity in energy density and other thermodynamic quantities cannot disappear at once when the parameters of the Lagrangian are only slightly changed.

An analysis using certain phenomenological information shows, however, that the strange quark mass $m_s \approx 150$ MeV is large in this context, and the first-order phase transition, even if it took place at $m_u = m_d = m_s = 0$, is no longer there in the real world. Speaking of the second-order phase transition, it is always violated whenever the symmetry whose spontaneous breaking brings about the phase transition is violated explicitly. There *is* no phase transition in a ferromagnet when an external magnetic field is applied. There *is* no phase transition in the real QCD with two light but not exactly massless quarks.

But the *crossover*, an essential change of the dynamic properties of the system in a comparatively narrow temperature range, is still there.

18.2   Quark–Gluon Plasma.

Let us discuss now the properties of the high temperature QCD system and, to begin with, let us calculate its free energy. In the lowest order in $\alpha_s(T)$, the result is

$$\frac{F^g}{V} = -\frac{\pi^2 T^4}{90}\left[2(N_c^2 - 1) + \frac{7}{8}(4N_c N_f)\right] , \qquad (18.10)$$

where the first term corresponds to the free gluon loop [and the factor $2(N_c^2 - 1)$ is just the number of gluon degrees of freedom] and the second term corresponds to the free quark loop [$4N_c N_f$ is the number of quark degrees of freedom and the factor 7/8 appears due to the Fermi statistics: for quarks, we have to integrate over momenta the expression $\ln(1 + e^{-\beta E_p})$ rather than $\ln(1 - e^{-\beta E_p})$ as we did in the boson case].

Perturbative corrections to this quantity were subject to intense study during the last 20 years. Their structure is rich and interesting. The first correction $\sim \alpha_s T^4$ comes from the two-loop diagrams and presents no particular complications. It turns out, however, that starting from the 3-loop level, we cannot restrict ourselves to a given order of perturbation theory but have to sum over a certain class of graphs. A general reason for this is that the infrared behavior of thermal graphs is worse than at $T = 0$. Indeed, the recipe (6.17) implies that, whenever we have a logarithmic integral such as $\int d^4p/p^4$ at zero temperature, the lowest finite T Matsubara mode (with $n = 0$) gives rise to a power infrared divergence $\int d^3p/p^4 \sim \mu_{\text{infr}}^{-1}$, etc.

The value of μ_{infr} depends on the problem under consideration. Heuristically, μ_{infr} is just a mass which a gluon acquires due to interaction with the heat bath. At $T \neq 0$, Lorentz invariance is lost, and the notion of mass is not uniquely defined. To begin with, the gluon polarization operator $\Pi_{\mu\nu}(\omega, \boldsymbol{k})$ satisfying the Ward identity $k_\mu \Pi_{\mu\nu}(\omega, \boldsymbol{k}) = 0$ (it holds also at finite T) admits now two different (transverse and longitudinal) tensor structures:

$$\Pi_{00} = \Pi_l(\omega, k) ; \qquad \Pi_{l0} = \frac{k_i \omega}{k^2}\Pi_l(\omega, k) ;$$

$$\Pi_{ij} = -\Pi_t(\omega, k)\left(\delta_{ij} - \frac{k_i k_j}{k^2}\right) + \frac{\omega^2}{k^2}\frac{k_i k_j}{k^2}\Pi_l(\omega, k) \qquad (18.11)$$

($k \equiv |\boldsymbol{k}|$). It makes sense now to study the poles of the gluon propagator

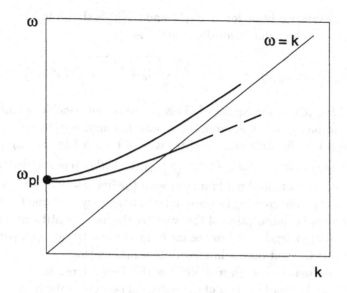

Fig. 18.2 Plasmon spectrum.

obtained as the solutions to the dispersive equation

$$\det \| g_{\mu\nu}(\omega^2 - k^2) - \Pi_{\mu\nu}(\omega, k) \| = 0 . \tag{18.12}$$

The equation splits up in three:

$$\begin{aligned}
\omega^2 - k^2 - \Pi_t(\omega, k) &= 0 , \\
k^2 + \Pi_l(\omega, k) &= 0 , \\
\omega^2 - k^2 &= 0 .
\end{aligned} \tag{18.13}$$

The solutions of the first and the second equation correspond to the transverse $\omega_\perp(k)$ and the longitudinal $\omega_\|(k)$ branches of collective excitations with the gluon quantum numbers.[†] Physics is exactly the same as in an ordinary plasma, where the transverse and longitudinal collective excitations (the propagating electromagnetic waves) were known for a long time. They are called plasmons. The branch $\omega^2 - k^2 = 0$ describes the unphysical gauge modes, which are not affected by the medium.

[†] By the same token, one can also study the collective excitations with the quark quantum numbers corresponding to the poles of the quark propagator, but we will not do it here.

The qualitative behavior of $\omega_\perp(k)$ and $\omega_\|(k)$ is shown in Fig. 18.2. At $k = 0$, the two branches coincide. The value

$$\omega_\perp(0) = \omega_\|(0) \equiv \omega_{\text{pl}} = \frac{gT}{3}\sqrt{N_c + \frac{N_f}{2}} + O\left(g^2 \ln g\right) \qquad (18.14)$$

is called the plasmon frequency.[‡] This is one of the possible definitions of the gluon mass. When k grows, the branches split apart and their fate at large k is quite different. The transverse branch has the asymptotics $\omega_\perp(k \gg \omega_{\text{pl}}) \approx k^2 + \frac{3}{2}\omega_{\text{pl}}^2$ (thus, $\sqrt{\frac{3}{2}}\omega_{\text{pl}}$ is *another* possible definition of the mass). The longitudinal branch crosses the line $\omega = k$ at some $k_* \sim gT$ and soon after the crossing acquires a large imaginary part due to so-called *Landau damping* [absorption of the wave by the hard constituents carrying the (colored) charge] and disappears from the spectrum. The pattern is very similar to what occurs in an ordinary plasma [46].

Another well-known plasma effect is the *Debye screening* — the exponential fall-off $\sim \exp\{-r/r_D\}$ of the potential between static heavy external charges immersed into the plasma. By the same token, such a screening shows up in the quark–gluon plasma. The parameter r_D is called the Debye screening length, and its inverse the Debye screening mass. The latter coincides with the pole position of the gluon propagator $D_{00}(\omega, k)$ at zero frequency ω (we are talking about the *static* potential) and imaginary k. To leading order in g, it can be determined as

$$m_D^2 = \lim_{k \to 0} \Pi_l^{1\ \text{loop}}(0, k) = 3\omega_{\text{pl}}^2 . \qquad (18.15)$$

Returning to the free energy calculations, the infrared divergences which we mentioned earlier show up only at zero frequency and have a static nature. The relevant infrared regulator is exactly the Debye screening mass.

Breaking of the naïve perturbation expansion due to infrared effects results in the terms in the expansion of $F(\alpha_s)$ which are not analytic in the coupling constant α_s. The term $\sim \alpha_s T^4$ is followed by the term $\sim m_D^3 T$ [cf. the last term in Eq. (18.4)!], which is proportional to $\alpha_s^{3/2}$. The next term is $\sim \alpha_s^2$ but also $\sim \alpha_s^2 \ln(\alpha_s)$, and the next after is $\sim \alpha_s^{5/2}$. Not going into details of the calculations (they are not so difficult for the term $\propto \alpha_s^{3/2}$, but the calculation of the coefficient of $\propto \alpha_s^{5/2}$ is a rather involved task),

---

[‡]The result (18.14) and the whole picture in Fig. 18.2 follow from the one-loop finite T calculations of $\Pi_{\mu\nu}(\omega, \boldsymbol{k})$. See Ref. [45] for more details.

we cannot resist the temptation to quote the result:

$$F = -\frac{8\pi^2 T^4}{45}\left[F_0 + F_2\frac{\alpha_s(\mu)}{\pi} + F_3\left(\frac{\alpha_s(\mu)}{\pi}\right)^{3/2} + \right.$$

$$\left. F_4\left(\frac{\alpha_s}{\pi}\right)^2 + F_5\left(\frac{\alpha_s}{\pi}\right)^{5/2} + O(\alpha_s^3\ln\alpha_s)\right], \qquad (18.16)$$

where

$$F_0 = 1 + \frac{21}{32}N_f, \quad F_2 = -\frac{15}{4}\left(1 + \frac{5}{12}N_f\right), \quad F_3 = 30\left(1 + \frac{N_f}{6}\right)^{3/2},$$

$$F_4 = 237.2 + 15.97N_f - 0.413N_f^2 + \frac{135}{2}\left(1 + \frac{N_f}{6}\right)\ln\left[\frac{\alpha_s}{\pi}\left(1 + \frac{N_f}{6}\right)\right]$$

$$\qquad - \frac{165}{8}\left(1 + \frac{5}{12}N_f\right)\left(1 - \frac{2}{33}N_f\right)\ln\frac{\mu}{2\pi T},$$

$$F_5 = \left(1 + \frac{N_f}{6}\right)^{1/2}\left[-799.2 - 21.96N_f - 1.926N_f^2\right.$$

$$\qquad \left. + \frac{495}{2}\left(1 + \frac{N_f}{6}\right)\left(1 - \frac{2}{33}N_f\right)\ln\frac{\mu}{2\pi T}\right], \qquad (18.17)$$

where μ is the normalization scale (*not* the chemical potential, *not* the infrared cutoff μ_{infr}, and *not* the characteristic hadron scale μ_{hadr}), and the value $N_c = 3$ is assumed. Such coefficients as 237.2 are not a result of numerical integration but are expressed via certain special functions. One expected nice feature of the result (18.16) and (18.17) is its renormalization-group invariance. The coefficients F_4 and F_5 involve a logarithmic μ dependence in such a way that the whole sum does not depend on μ [cf. Eqs. (9.45), (9.46) and the discussion thereabout]. The series (18.16) converges reasonably well at high enough temperatures $T \gtrsim 10$ GeV. In the region $T \sim 1$ GeV accessible for experimental studies, the effective coupling is not small enough yet and a perturbative description does not apply.

Let us assume, however, that the temperature is so high that $\alpha_s(T)$ is *really* small. A very important remark is that the term $\sim \alpha_s^{5/2}$ in the expansion (18.16) is the absolute limit beyond which no perturbative calculation is possible. At the level $\sim \alpha_s^3$, we are running into the so-called *magnetic mass* problem, in other words the infrared divergences in the graphs become so bad that no prescription for their resummation can be suggested.

In contrast to the Debye mass and plasmon frequency, a nonzero magnetic mass (or a finite magnetic screening length) is a specifically non-

Abelian phenomenon; it does not occur in an ordinary Abelian plasma. Magnetic fields cannot be screened by ordinary charges, only magnetic monopoles (which are not abundant in Nature) could do the job. The absence of the monopoles in ordinary QED is technically related to one of the Maxwell equations, $\partial_i B_i = 0$. In the non-Abelian case, the corresponding equation reads $\mathcal{D}_i B_i = 0$, where $\mathcal{D}$ is the covariant derivative. Thus, gluon field configurations with local color magnetic charge density $\rho_m \sim \partial_i B_i$ (with the ordinary derivative) are admissible. The presence of such configurations in the gluon heat bath results in screening of chromomagnetic fields.

The value of the magnetic mass can be estimated by considering the gluon loop contributions in the *transverse* polarization operator $\Pi_t(0, k)$. At 1-loop level, we have

$$\Pi_t^{1\ \text{loop}}(0, k) \sim g^2 k^2 T \int \frac{d\boldsymbol{p}}{p^2(\boldsymbol{p} - \boldsymbol{k})^2} \sim g^2 T k \ . \tag{18.18}$$

A similar estimate for higher loop graphs gives $\Pi_t^{2\ \text{loop}}(0, k) \sim (g^2 T)^2$, $\Pi_t^{3\ \text{loop}}(0, k) \sim (g^2 T)^3/k$, etc. We see that, at $k \sim g^2 T$, all contributions are of the same order and no analytic calculation of the polarization operator at $k \lesssim g^2 T$ is possible. *If* the limit $\lim_{k \to 0} \Pi_t(0, k)$ exists, it can be naturally associated with the magnetic screening length. The crux of the issue, however, is the breakdown of perturbation theory at large distances in thermal QCD no matter how large the temperature is and how small $\alpha_s(T)$ is.

There is an alternative way to see this. Consider the finite T partition function (4.24) of a pure glue system. When β is small, the Euclidean time dependence of the fields may be disregarded. In addition, the effects due to $A_0^a(\tau, \boldsymbol{x})$ can be disregarded: the temporal components of gluon field acquire a comparatively large mass $m_D \sim gT \gg g^2 T$ and decouple. We are left with the expression

$$Z = \int \prod_{\boldsymbol{x}} dA_i^a(\boldsymbol{x}) \exp \left\{ -\frac{1}{2g^2 T} \int d\boldsymbol{x}\ \text{Tr}\{F_{ij} F_{ij}\} \right\} \ . \tag{18.19}$$

The partition function (18.19) describes 3-dimensional Yang–Mills theory with the dimensional coupling $g_3^2 = g^2 T$. This theory involves a spectrum of glueball states with masses of order $g^2 T$, but it is impossible to *calculate* these masses (which determine the rate of exponential decay of different gauge-invariant correlators at large spatial distances) by perturbative methods.

As was mentioned before, the magnetic mass effects bring in corrections $\sim \mu_{mag}^3 T \sim \alpha_s^3(T) T^4$ to the free energy which cannot be calculated analytically.

18.3 Finite Baryon Density. Color Superconductivity.

In the actual experimental situation, not only the temperature, but also the baryon number density is high: the colliding nuclei are not only heated, but also squeezed. What do theorists know about the properties of the hadronic systems with a finite baryon number density n?

First of all, the problem can be solved exactly while the density is still small and the medium presents a nuclear liquid. In particular, by calculating the nucleon loop contribution in the free energy at finite chemical potential, and differentiating it with respect to the quark mass, one can derive a formula describing the suppression of the quark condensate in the dense cold hadronic matter [47]:

$$< \bar{\psi}\psi >_n = < \bar{\psi}\psi >_0 \left[1 - \frac{2n\sigma_{\pi N}}{F_\pi^2 M_\pi^2} + o(n) \right] . \tag{18.20}$$

Here $\sigma_{\pi N} \approx 45$ MeV is the so-called sigma-term — the contribution to the nucleon mass due to the nonvanishing light quarks masses (thus $\sigma_{\pi N} \propto m_q$ and the ratio $\sigma_{\pi N}/M_\pi^2$ is constant in the chiral limit). The correction turns out to be surprisingly high. For the nuclear density $n_0 \approx .15(\text{fm})^{-3}$, the drop of the condensate constitutes about $\sim 30\%$! One should thereby expect a drastic change in the properties of the system, including probably a phase transition in the region $n \approx 3n_0$. Based on the analogy between (18.20) and (18.6) and the previous discussion for the finite temperature case, one could guess that the phase transition is also second-order.

This is not true, however, and the phase transition in this case is *first-order*. Actually, in an imaginary world without electromagnetic interactions, there would be *two* different first-order phase transitions in density. The existence of the first such phase transition can be qualitatively explained in a very simple way. It is well known that, if the protons had no electric charge and did not repel each other, energetically the most favorable state for an ensemble of a large number of nucleons would be the uniform nuclear matter state. It has a definite density n_0. Suppose we want to study the system with zero temperature and an overall density $n < n_0$. The system would then create a set of nuclear matter clusters hovering in empty space. This is what is called the *mixed* inhomogeneous phase characteristic for the systems with first-order phase transition (just imagine boiling water). The interval of the densities $0 \leq n \leq n_0$ where the mixed phase can exist corresponds to *one and the same* value for the chemical potential μ_c. If we increase the temperature, nucleons would evaporate from the clusters. In addition, the pion excitations would show up. As long as the temperature is small, a system can still exist in the mixed phase, but it would be less inhomogeneous and the interval Δn of average densities allowing the existence of the mixed phase would be smaller than at $T = 0$. At a certain critical temperature T_c, the allowed density interval shrinks to zero and no inhomogenuity in nucleon distribution is possible if the temperature becomes still higher.

But nuclear matter itself is also an inhomogeneous system: it consists of nucleons, where the density of matter is higher than in relatively empty regions of space between them. Qualitatively, one should expect a first-order phase transition associated with squeezing the system to the point that individual nucleons unite in homogeneous clusters carrying large baryon charges. Again, in some density interval the system would exist in a mixed phase: dense homogeneous clusters floating in normal nuclear matter or maybe clusters of the rarified nuclear matter immersed in the dense homogeneous medium, depending on what the net density is. Again, the density interval where the mixed phase can exist shrinks with temperature and disappears at some critical point.

Reliable theoretical calculations of the exact position of this second critical point are not possible now, but reasonable estimates show that the critical density is $\sim (2.5 - 3.0)n_0$. Model calculations show that the quark condensate vanishes in the homogeneous phase. They also show that, at least in the theory with two massless flavors, it is probably not a critical point, but a *tricritical* point, i.e. the line of first-order phase transitions does not end there but continues as a line of the second-order phase transitions yielding to the finite T second-order phase transition at $\mu = 0$.

An exciting recent theoretical finding is that, although the standard quark condensate $< \bar{\psi}\psi >$ vanishes in the high density phase, another condensate, the colored diquark condensate $< \psi\psi >$ which was absent at zero density, *appears*. This phenomenon was called *color superconductivity*.

A heuristic physical picture for very high densities is rather clear. Remember what happens in an ordinary superconductor. The electrons on the surface of the Fermi sphere attract each other due to phonon exchange and form the Cooper pairs. This is associated with the appearance of the charged order parameter $\langle ee \rangle$, the Meißner effect, etc. If hadron system is very dense, it makes no sense to speak of individual hadrons anymore, the system is better described in terms of quarks.§ The colored quarks interact with each other. Consider first their perturbative interaction due to the one-gluon exchange. In contrast to electrons which always repel each other, the quarks repel if their color wave function is symmetric (the diquark is in the sextet representation), but attract if it is antisymmetric (the antitriplet representation). Thus, there is a way to form the Cooper pairs and they are formed.

This possibility was realized already some time ago [48]. At the same time, it was realized that the perturbative quark interaction is rather weak. The characteristic values of the superconducting gap were estimated to be just several MeV, which could not lead to any observable effect. According to an important recent observation [49], there exists also a nonperturbative mechanism of forming the Cooper pairs in the attractive antitriplet channel, due to instantons. This gives rise to rather large values of the mass gap and the diquark condensate when density is high (above the phase transition) but not asymptotically high.¶ A possible form of the net phase diagram for the theory with two massless flavors in the (T, μ) and (T, n) planes suggested by K. Rajagopal is shown in Fig. 18.3. Solid lines mark the second-order phase transitions, and dashed lines the first-order phase transitions. The line of the trivial first-order phase transition "nucleon gas → nuclear matter" is not displayed in these plots. The values of T and μ achieved in real heavy ion collision experiments are expected to be close to

§ Temperature is still zero, so we have only quarks and no antiquarks, neither do we have gluons.

¶ For academically high values of the chemical potential μ such that

$$\frac{\alpha_s(\mu)}{2\pi} \ln^2 \frac{\mu}{\mu_{\text{hadr}}} \gg 1 ,$$

a semi-exact theoretical analysis is possible and the mass gap is estimated as [50]

$$\Delta \sim \mu \exp \left\{ -\frac{C}{\sqrt{\alpha_s(\mu)}} \right\}$$

with $C = 3\pi^2/\sqrt{8\pi}$.

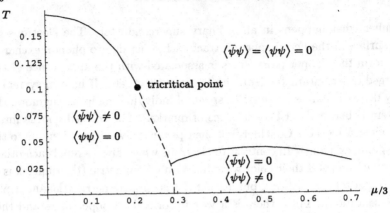

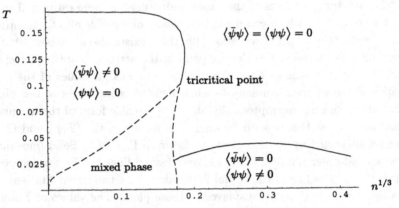

Fig. 18.3 Phase diagram of hot and dense QCD ($N_f = 2, m_q = 0$).

the position of the upper tricritical point and this may lead to some distinct observable effects.

Lecture 19
Confinement

As was repeatedly mentioned before, the hypothesis of confinement in QCD, i.e. the assertion that the physical spectrum of the theory has little to do with the quanta of the underlying quark and gluon fields, but involves colorless hadron states, is not proven yet. Bearing this in mind, one of the possible options would be not to discuss here this subject whatsoever and leave it for a future author who will write a "final" textbook on QCD when the edifice of this theory is covered with a roof and plastered.

At the moment, it is not clear, however, *when* exactly this will happen. A generation of theorists have broken their teeth trying to prove confinement in QCD, and it is not unconceivable that this problem will keep bothering us for a couple of generations to come. One is reminded of the problem of turbulence which was posed more than a century ago and is still not satisfactorily solved. Remember also the great Fermat problem, which took 350 years to be resolved. Therefore, we have chosen to acquaint the impatient reader, who might not be prepared to wait for centuries, with at least some current ideas on how confinement might be realized.

The first remark in order is the following. Confinement was defined above as the absence of colored states in the spectrum. But *any* admissible state is *in a sense* colorless because it is annihilated by the generator G^a of gauge transformations. Even properly defined asymptotic electron states which definitely exist in QED are all neutral in this sense.* Putting the sys-

---

*This requires a comment. Though usual free asymptotic electron states form a convenient basis for perturbative calculations, they are not annihilated by the operator $G(x) = \partial_i E_i(x) - e\bar\psi(x)\gamma^0\psi(x)$ and are not "properly defined" in this sense. Asymptotic electron states satisfying $G(x)|\Psi\rangle = 0$ represent coherent states involving an

tem into a finite three-dimensional box with periodic boundary conditions and integrating, we obtain

$$\int G(x)dx|\Psi\rangle = Q|\Psi\rangle = 0 , \qquad (19.1)$$

where $Q = e \int \bar{\psi}(x)\gamma^0\psi(x) \, dx$ is the operator of net electric charge. We see, however, that the cancellation (19.1) occured due to boundary effects (actually, imposing periodic boundary conditions for a Coulomb field amounts to putting a compensating charge on the boundary). If no box and no boundary conditions are introduced, Coulomb states carry electric charge associated with a nonzero electric field flux at spatial infinity. Saying that all physical states in QCD are colorless just means that, in non-Abelian case, such a long-distance color-electric Coulomb tail is absent.

19.1 Weak Confinement and Strong Confinement. Wilson Criterium.

Besides confinement in the weak sense, the absence of free quarks and gluons, one also often discusses confinement in the strong sense: the statement that the potential between static colored sources rises indefinitely when the separation between the sources increases. Suppose we have a color-neutral system formed by a very heavy quark and antiquark. If the potential between the constituents rises with the distance, it means that the energy of a "liberated" quark is infinite, i.e. that such a quark is not present in the spectrum and therefore confined. The converse is not true, however. Colored states may well be absent from the spectrum, but the interquark potential may not rise. In particular, we are sure that the potential does not rise in QCD: if the separation between the sources becomes large enough, a light quark–antiquark pair is created which screens the sources. Physically, this describes the decay of a heavy quarkonium excited state $(\bar{Q}Q)^*$ into light–heavy mesons $(\bar{Q}q)$ and $(\bar{q}Q)$ (see Fig. 19.1).

There are still cases when the potential rises, however. If we have pure Yang–Mills theory (without quarks) and probe it by heavy sources in the fundamental color representation, screening cannot occur because adjoint gluon fields presumably cannot form an object which is fundamental in

---

indefinite number of longitudinal Coulomb photons. Further discussion of this question can be found in Ref. [42].

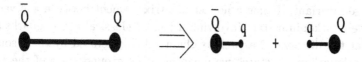

Fig. 19.1 Breaking the string in QCD.

color so that it could screen the sources.[†]

Confinement can occur in this case only if the potential grows at infinity. People believe[‡] that the growth is linear,

$$V_{\bar{Q}Q}(R) \sim \sigma R \, , \tag{19.2}$$

where σ is a dimensionful constant of order Λ_{YM}^2 which is called *string tension*. A picture which is kept in mind here is the following. When the separation R between static quarks is large, chromoelectric field acting between them is concentrated in a tube of finite width $a \sim \Lambda_{YM}^{-1} \ll R$, with the field strength $\boldsymbol{E}$ and the volume energy density, $\epsilon \sim \mathrm{Tr}\{\boldsymbol{E}^2\}/g^2$, being roughly independent of the position along the tube axis. The total energy stored in the tube $\equiv$ string is $\epsilon S_{\mathrm{eff}} R$ ($S_{\mathrm{eff}} \sim \pi a^2$) and hence $\sigma \sim \epsilon S_{\mathrm{eff}}$. These strings are depicted as the lines connecting quarks in Fig. 19.1.

Consider now the operator

$$M(\boldsymbol{R}) = \bar{Q}(\boldsymbol{R})S(\boldsymbol{R})Q(0) = \bar{Q}(\boldsymbol{R})P\exp\left\{i\int_0^{\boldsymbol{R}} A_k(\boldsymbol{x})dx_k\right\}Q(0) \tag{19.3}$$

(the integral is performed along the straight line connecting the points $\boldsymbol{R}, 0$) in the theory involving gauge fields and also a very heavy quark field $Q(\boldsymbol{x})$. The operator (19.3), involving the P-ordered exponential (1.11),

---

[†]Surprisingly, this natural assumption sometimes fails. In particular, massless 2D Majorana fermions in the adjoint color representation *can* form a collective state which is fundamental in color [51] and which can screen fundamental probe charges. The possibility that our 4D gluons also behave like this and screen quark sources seems to be ruled out by experiment, however, since no colorless states with fractional baryon charges were observed.

[‡]It is worth remembering that, while confinement in QCD is an experimental fact, pure Yang–Mills theories are not observed in Nature and confinement there is an educated guess. Most probably, it is true, but an exotic scenario, where the theory involving light quarks is confining while a theory without quarks is not, has not been completely excluded.

is gauge-invariant. It has a nonzero matrix element between the vacuum and the quarkonium state involving a pair of classical heavy quarks separated at the distance R as well as gluon fields generated by these sources. By the "quarkonium state" we mean here the ground state of the Born–Oppenheimer Hamiltonian involving only relatively light gluon degrees of freedom, while the heavy quark degrees of freedom, as well as the distance between the quarks, are frozen. The quark–antiquark potential $V_{\bar{Q}Q}(R)$ is then defined as the energy $E(R)$ of such a state modulo $2M_Q$.

We can write

$$\langle M(\boldsymbol{R},T)M^\dagger(\boldsymbol{R},0)\rangle_0 \sim \exp\{-E(R)T\}, \qquad (19.4)$$

where $M(\boldsymbol{R},T) = \exp\{\hat{H}T\}M(\boldsymbol{R},0)\exp\{-\hat{H}T\}$, and the Euclidean time T is large.

The heavy quark propagators entering Eq. (19.4) can be considered free [a graph involving n vertices describing the interaction with the gluon field is suppressed as $(\Lambda_{YM}/M_Q)^n$]. At large T, the propagators fall off as $\exp\{-M_Q T\}$. We obtain

$$\langle \mathrm{Tr}\,\{S(\boldsymbol{R},T)S^\dagger(\boldsymbol{R},0)\}\rangle_0 =$$

$$\left\langle \mathrm{Tr}\left(P\exp\left\{i\int_0^R A_k(\boldsymbol{x},T)dx_k\right\}P\exp\left\{i\int_R^0 A_k(\boldsymbol{x},0)dx_k\right\}\right)\right\rangle_0$$

$$\propto \exp\left\{-V_{\bar{Q}Q}(R)T\right\}. \qquad (19.5)$$

This was written in the Hamilton gauge $A_0 = 0$. The same expression in a generic gauge acquires the form

$$W(C) = \frac{1}{N_c}\left\langle \mathrm{Tr}\left(P\exp\left\{i\oint_C A_\mu(x)dx_\mu\right\}\right)\right\rangle_0 \propto \exp\left\{-V_{\bar{Q}Q}(R)T\right\}, \qquad (19.6)$$

where C is a rectangular contour depicted in Fig. 19.2 and the factor $1/N_c$ normalizes the Wilson loop operator $W(C)$ to unity for small contours [see Eq. (1.16)].

If $V_{\bar{Q}Q}(R) \sim \sigma R$ at large R, it follows that

$$W(C) \sim \exp\{-\sigma S\} \qquad (19.7)$$

for a large rectangle, where S is its area. Since the theory involves mass gap and a finite correlation length, the property $W(C_1 \cup C_2) = W(C_1)W(C_2)$

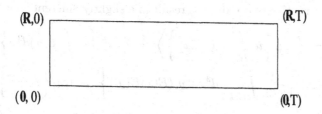

Fig. 19.2   Canonical rectangular Wilson loop.

should hold for large smooth contours (for the contours C_1 and C_2 having a common part, the unification $C_1 \cup C_2$ is defined such that the common part is removed), which implies that the *area law* (19.7) is true not only for the rectangle in Fig. 19.2, but for smooth flat contours of any other shape as well. Actually, a contour need not even be flat; the area law should hold in the framework of our assumptions for all smooth (with curvature much less than Λ_{YM}^2) contours, where S is the area of the soap film having the wire of shape C as a boundary.

19.2   Schwinger Model.

Let us see now how all these ideas work in cases where confinement *is* proven and the theoretical picture is clear. This is the case in certain two-dimensional and three-dimensional gauge theories.

We have seen that a sufficient condition for confinement is the growth of the interquark potential at large distances. In 3+1 dimensions, this behavior is highly nontrivial and can be brought about only by collective nonperturbative effects: the perturbative Coulomb potential falls off as $\propto$ $1/R$. But in 1+1 dimensions, the classical Coulomb potential *rises* linearly at large distances and confinement is obtained without tears just at the perturbative level.

Consider first the pure 2D photodynamics probed by external heavy charges. The string tension

$$\sigma = \lim_{R \to \infty} [V_{\bar{Q}Q}(R)/R] = V_{\bar{Q}Q}(R)/R$$

(as the theory is free, the linear law for the potential holds at *all* distances) is in this case just $\sigma = e^2/2$, where e is the charge of the source carrying the dimension of mass. This means that $W(C) \propto \exp\{-e^2 S/2\}$.

It is instructive to obtain this result in a slightly different way. We have

$$W(C) = \left\langle \exp\left\{ ie \oint_C A_\mu(x)dx_\mu \right\} \right\rangle = \left\langle \exp\left\{ ie \int_D E(x)d^2x \right\} \right\rangle$$

$$= \exp\left\{ -\frac{e^2}{2} \int_{D\times D} d^2x d^2y \, \langle E(x)E(y)\rangle \right\} = \exp\left\{ -\frac{e^2}{2}S \right\}. \quad (19.8)$$

In this derivation we used successively: *(i)* the Stokes theorem (D is the domain with the boundary C), *(ii)* the Gaussian nature of correlators in free theory, and *(iii)* a specific form of the correlator $\langle E(x)E(y)\rangle = \delta(x-y)$ in pure 2D photodynamics. One can also write

$$W(C) = \langle e^{i\Phi}\rangle = e^{-\langle\Phi^2\rangle/2}, \quad (19.9)$$

where Φ is the flux of electric field through the contour. The area law $\langle\Phi^2\rangle \propto S$ can be interpreted as "random walks in flux space" and is due to the stochastic nature of fluxes. The same stochastic flux picture should hold in pure 4D Yang–Mills theory. This serves as an "explanation" of confinement and would qualify as its proof if we could *derive* that non-Abelian fluxes in the pure Yang–Mills theory are stochastic.

The Schwinger model is not just 2D photodynamics, it also involves a dynamical Dirac fermion carrying the electric charge e. If the fermion is massless, the Euclidean path integrals

$$\int \prod_{x\mu} dA_\mu(x) \prod_x d\bar\psi(x)d\psi(x)O[A,\psi,\bar\psi] \exp\left\{ \int d^2x \left[-\frac{E^2}{2} + i\bar\psi\slashed{D}\psi \right] \right\}$$

$$(19.10)$$

are Gaussian, and the theory is therefore exactly solvable. The Gaussian character of the theory makes the derivation (19.8) valid until the very last step. The electric field correlator is, however, modified now and acquires the form

$$\langle E(x)E(y)\rangle^{\text{SM}} = \delta(x-y) - \frac{\mu^2}{2\pi}K_0(\mu|x-y|), \quad (19.11)$$

where μ is the Schwinger boson mass (14.35). The second term is negative and the coefficient is exactly such that $\int_{-\infty}^{\infty} d^2x\langle E(x)E(0)\rangle^{\text{SM}} = 0$. That means that $\int_{D\times D} d^2x d^2y \, \langle E(x)E(y)\rangle$ is not proportional to the large area S as before, but acquires contributions only from the regions close to the boundary. Actually, with the exact expression (19.11) in hand, we can calculate $W(C)$ analytically. For large smooth contours, one obtains

$$W(C) = \exp\left\{-\frac{\pi\mu P}{4}\right\}, \qquad (19.12)$$

where P is the perimeter of the contour. The appearance of the *perimeter law* and the absence of an area law implies the screening of heavy sources by dynamical light fermions discussed heuristically above for QCD.

Weak confinement is still there, however. As the theory is solvable, its spectrum is also exactly known. It involves only neutral bosons with mass (14.35) which can be interpreted as "massive photons" but also as some "meson" bound $\bar{f}f$ states. Charged fermion states are absent in the spectrum, and this means confinement.

Problem 1. Derive Eq. (19.11).

Solution. One has to calculate first the fermionic path integral in Eq. (19.10), which amounts to calculating the fermion loop in an external gauge field. The only nonvanishing contribution to the latter comes from the diangle anomalous graph calculated in Lecture 14. Using the result (14.34) derived there, the integral (19.10) acquires the form

$$\int \prod_{x\mu} dA_\mu(x) \exp\left\{-\int d^2x \left[\frac{E^2}{2} + \frac{\mu^2}{2}A_\mu\left(\delta_{\mu\nu} - \frac{\partial_\mu\partial_\nu}{\partial^2}\right)A_\nu\right]\right\}$$

$$= \int \prod_{x\mu} dA_\mu(x) \exp\left\{-\int d^2x \left[\frac{E^2}{2} - \frac{\mu^2}{2}E\frac{1}{\partial^2}E\right]\right\} \qquad (19.13)$$

(the identity $\delta_{\mu\nu}\partial^2 - \partial_\mu\partial_\nu = \epsilon_{\mu\beta}\epsilon_{\nu\gamma}\partial_\beta\partial_\gamma$ was used). Going to momentum space and coming back, we obtain

$$\langle E(x)E(y)\rangle = \frac{1}{4\pi^2}\int e^{i(k\cdot x - k\cdot y)}\frac{k^2}{k^2 + \mu^2}d^2k, \qquad (19.14)$$

which coincides with Eq. (19.11).

Problem 2. Derive Eq. (19.12).

Solution. We have got to calculate the double integral $\int_{D\times D}$ in (19.8). The task is basically reduced to calculating the integral

$$I(a) = \int_{-a}^{\infty} dy_1 \int_{-\infty}^{\infty} dy_2 \langle E(0)E(y)\rangle, \qquad (19.15)$$

where a is the distance of the point x from the boundary. Using the Fourier representation (19.14), we obtain

$$
I(a) = 1 - \frac{\mu^2}{2\pi} \int_{-a}^{\infty} dy_1 \int_{-\infty}^{\infty} dk_1 \frac{e^{-ik_1 y_1}}{k_1^2 + \mu^2} = 1 - \frac{\mu}{2} \int_{-a}^{\infty} e^{-\mu|y_1|} dy_1
$$

$$
= \frac{1}{2} e^{-\mu a} . \qquad (19.16)
$$

Integrating it further over a, multiplying by $e^2 P/2 = \pi \mu^2 P/2$, and exponentiating, we arrive at Eq. (19.12).

19.3 Polyakov Model.

Let us now go over to the (2+1)-dimensional world. The perturbative Coulomb potential rises as $\propto g^2 \ln R$ (g is the gauge coupling constant carrying the dimension $[m^{1/2}]$), which immediately leads to confinement of static heavy sources in pure 3D photodynamics.

In interacting non-Abelian theories the situation is less clear: 3D Feynman graphs involve power-like infrared divergences, and perturbation theory does not work at large distances. A common *conjecture* is that the interquark potential grows linearly at large distances, $V_{\bar{Q}Q}(R) \propto g^4 R$, as it presumably does in four dimensions as well. Pure 3D Yang–Mills theory is strongly coupled and cannot, so far, be treated analytically. There is a related model, however, where an analytic calculation is possible.

Consider a theory involving 3D gauge fields with the gauge group $SU(2)$ and a triplet of scalar fields $\Phi^a(x)$. The Lagrangian of the model reads

$$
\mathcal{L} = -\frac{1}{2g^2} \mathrm{Tr}\{G_{\mu\nu} G^{\mu\nu}\} + \frac{1}{2} (\mathcal{D}_\mu \Phi^a)^2 - \frac{\lambda}{4} (\Phi^a \Phi^a - v^2)^2 , \quad (19.17)
$$

where λ, v^2, and g^2 have all the dimensions of mass. We assume that the conditions

$$
v \gg g , \qquad v\sqrt{\lambda} \gg g^2 \qquad (19.18)
$$

are satisfied. In this case, the Higgs field acquires a large expectation value $\langle \Phi^a \rangle = v n^a$ (n^a is a unit color vector which can be brought to the form $n^a = \delta^{a3}$ by a gauge rotation), which gives mass to the color components A_μ^{1+i2} and A_μ^{1-i2} of the gauge field. The component A_μ^3 remains massless, and one sometimes says that the $SU(2)$ gauge symmetry of the Lagrangian (19.17) is broken spontaneously down to $U(1)$.[§]

Thus, the spectrum of the theory seems to involve a massless photon, a massive charged $W^\pm$ boson and a massive Higgs boson. If the conditions (19.18) are satisfied, the masses of the $W^\pm$ and Higgs bosons can be calculated perturbatively:

$$m_W = gv \gg g^2 \,, \qquad m_H = \sqrt{2\lambda}v \gg g^2 \,, \qquad (19.19)$$

and corrections are small.

This simple picture does not hold water, however. It *would* be correct in a (3+1)-dimensional theory with the Lagrangian (19.17) called *Georgi–Glashow model*. Actually, the Georgi–Glashow model enjoys a version of weak confinement: the physical states do not carry non-Abelian color quantum numbers, but only electric charges with respect to the "unbroken" $U(1)$.

In the 3D case the situation is more complicated and more interesting. To begin with, the charged $W^\pm$ bosons can be expected to be confined due to the logarithmic growth of the Coulomb potential. A further observation, due to Polyakov, is that the potential actually grows linearly, while the photons acquire mass.

The reasoning involves several steps. The first remark is that the Lagrangian (19.17) admits Euclidean classical solutions. Their explicit form depends on the gauge choice. A convenient symmetric form is

$$A_\mu^a = \frac{\epsilon_{a\mu\nu}x_\nu}{r^2}[1 - F(r)] \,,$$

$$\Phi^a = q\frac{x_a}{r}v[1 - H(r)] \,, \qquad (19.20)$$

where $q = \pm 1$ and the functions $F(r)$, $H(r)$ can be found numerically. They have the properties *(i)* $F(0) = H(0) = 1$ (so that the solution is regular at zero) and *(ii)* $F(r)$ and $H(r)$ fall off exponentially at infinity.

In the 4D Georgi–Glashow model, the solution (19.20) describes a static topologically stable soliton (the famous *'t Hooft–Polyakov monopole*). In

---

[§]These words are rather misleading. As we already mentioned in Lecture 12, gauge symmetry is not actually a symmetry and cannot be broken. A better terminology has never been established, however.

the 3D world, it is an Euclidean instanton solution. Its action is

$$S_{\text{monopole}} = \frac{4\pi v}{g} f\left(\frac{\lambda}{g^2}\right), \tag{19.21}$$

where $f(0) = 1$ and $f(\infty) \approx 1.79$.

Why is it called a monopole? This becomes clear if one performs a singular gauge transformation which brings the scalar field to the standard form $\Phi^a = v\delta^{a3}$ at infinity. Then only the component A_μ^3 of the gauge field survives at large distances, and it just coincides with the vector potential of the Dirac monopole with all its paraphernalia including the Dirac string. The magnetic field is

$$B_\mu = -\frac{\Phi^a}{2v}\epsilon_{\mu\nu\alpha}G_{\nu\alpha}^a \overset{r\to\infty}{\sim} -\epsilon_{\mu\nu\alpha}\partial_\nu(A_\alpha^3)^{\text{Abel. gauge}} = \frac{qx_\mu}{r^3}. \tag{19.22}$$

There are monopoles (with charge $q = 1$) and antimonopoles (with $q = -1$).¶

The monopoles $\equiv$ instantons contribute to the partition function of the theory. As the action (19.21) is large (if $v \gg g$), the corresponding functional integrals can be evaluated by semi-classical methods. The characteristic density of monopoles, $\zeta \propto \exp\{-S_{\text{monopole}}\}$ (the pre-exponential factor can also be found exactly), is small and the characteristic distance between them is large. This very much resembles the instanton gas model discussed in Lecture 5 and Lecture 15. In this case it should be called, however, *plasma* rather than gas. The issue is that the action of a two-monopole configuration is not just twice the individual monopole action, but involves also the long-distance Coulomb interaction term $\propto qq'/r$, where r is the distance between the monopoles and q, q' are their magnetic charges. The partition function of the model acquires the form

$$Z = \sum_{n_+,n_-} \frac{(\zeta)^{n_++n_-}}{n_+!n_-!} \prod_{a=1}^{n_++n_-} \sum_{q_a=\pm 1} \int dr_a \exp\left\{-\frac{4\pi}{g^2}\sum_{a<b}\frac{q_aq_b}{r_{ab}}\right\}, \tag{19.23}$$

where $n_\pm$ are the numbers of monopoles and antimonopoles in a finite three-dimensional box introduced as usual for infrared regularization. The factor $4\pi/g^2$ appears when comparing the normalization $\frac{1}{2g^2}\int B^2 d^3x$ of

¶We address the reader not satisfied with this patter to the books [52; 53] specially devoted to classical solutions.

the action following from Eq. (19.17) with the conventional normalization of the energy in classical electrodynamics $E = \frac{1}{8\pi} \int B^2 d^3x$.

The partition function (19.23) coincides exactly with the thermal partition function of the ordinary nonrelativistic plasma (of infinitely heavy particles) under the identification $g^2/4\pi \equiv T$. An important property of a plasma is that an external field is screened. The Debye screening radius is given by the formula [54]

$$r_D^{-2} = \frac{4\pi}{T} \sum_i e_i^2 n_i \,, \tag{19.24}$$

where n_i is the concentration of the charge carriers of type i and e_i is their charge.|| r_D^{-1} can be interpreted as a photon mass. In our case,

$$m_\gamma^2 = \frac{32\pi^2 \zeta}{g^2} \,. \tag{19.25}$$

We are ready now to determine the string tension in the Polyakov model. Consider a smooth flat contour C whose size is much larger than the Debye screening length and evaluate the mean square magnetic flux penetrating the contour. This flux is brought about by monopoles. A monopole sitting inside the contour provides the flux $2\pi q$. For a monopole at a distance z from the contour plane, the flux is, roughly speaking, $\Phi_{mon}(z) \sim 2\pi q e^{-m_\gamma |z|}$. The full mean square flux through the contour is

$$\langle \Phi^2 \rangle_C \approx (2\pi)^2 (2\zeta) S \int_{-\infty}^{\infty} e^{-2m_\gamma |z|} dz = \frac{8\pi^2 \zeta S}{m_\gamma} \,. \tag{19.26}$$

Substituting this into Eq. (19.9), we obtain

$$\sigma = C \frac{4\pi^2 \zeta}{m_\gamma} = C\pi g \sqrt{\frac{\zeta}{2}} \,, \tag{19.27}$$

where $C \sim 1$ can in principle be fixed if evaluating the screened monopole flux $\Phi_{mon}(z)$ more accurately, taking into account also nonlinear effects brought about by a large value of the monopole charge.

Probably, 3-dimensional Yang–Mills theory exhibits strong confinement by a similar mechanism, but this has not been proven yet.

|| This formula works up to a numerical factor also in the ultrarelativistic case. For the quark–gluon plasma, $n \sim T^3$, and $m_D^2 \sim g^2 T^2$ in agreement with Eqs. (18.14), (18.15).

19.4 Dual Superconductivity.

Let us go back to four dimensions. According to a hypothesis of 't Hooft and Mandelstam, strong confinement in pure 4D Yang–Mills theory is realized by the *dual superconductor* mechanism. To understand it, let us first recall conventional superconductivity. It is conveniently described by the Ginsburg–Landau effective Lagrangian, alias the Abelian Higgs model:

$$\mathcal{L} = -\frac{1}{4e^2}F_{\mu\nu}F^{\mu\nu} + |\mathcal{D}_\mu\phi|^2 - \frac{\lambda}{2}(\bar{\phi}\phi - v^2)^2 \ . \tag{19.28}$$

The condensate of the scalar charged field $\langle\phi\rangle = v$ gives mass to the photon, $m_\gamma^2 = 2e^2v^2$, which means screening of electric fields.

The model (19.28) admits Abrikosov–Nielsen–Olesen magnetic strings. Like monopoles in the Georgi–Glashow model, ANO strings are topologically stable solutions. The essential difference, however, is that they are not localized, but are extended objects with axial geometry. The solution can be written as

$$\phi(x) = ve^{i\alpha}A(r) \ ,$$
$$A_i(x) = -\frac{\epsilon_{ij}^\perp x_j}{r^2}B(r) \ , \tag{19.29}$$

where r is the distance from the axis, α is the azimuthal angle, and $A(r)$, $B(r)$ are profile functions interpolating between 0 at $r = 0$ and 1 at $r = \infty$. Magnetic field is directed along the string. It dies away exponentially at large distances from the axis, so that the total flux is 2π. The ANO string carries energy; the energy density per unit length is

$$\sigma_{\text{ANO}} = 2\pi v^2 f\left(\frac{\lambda}{e^2}\right) , \tag{19.30}$$

where $f(1) = 1$ and $f(x) \sim \ln x$ at large x.**

It is evident now that superconductivity is associated with confinement of magnetic charges. The energy of an individual magnetic charge immersed in a superconductor is infinite due to the infinite energy of the ANO string associated with it. Suppose we have a pair made of a monopole and an antimonopole carrying minimal admissible magnetic charges $q = \pm 1/2$ (this

---

**The solution (19.29) describes the string with minimal flux. For $\lambda < e^2$, there exist also stable axially symmetric solutions with higher fluxes n. They are characterized by the asymptotics $\phi(x) \sim ve^{in\alpha}$ at large distances.

corresponds to the total outgoing magnetic flux $4\pi q = \pm 2\pi$) and separated by a large distance R from each other. An ANO string will be streched between the static monopole sources, and the energy of such a system grows with distance as $\sigma_{\text{ANO}} R$.

Our goal is, however, to confine *electric* color charges. The dual super-conductivity mechanism assumes that a condensate of an effective *monopole* field is formed so that magnetic charges are screened and electric charges are confined. At first sight, this idea may seem somewhat exotic, but this is exactly a situation where a first impression is dangerous to rely upon. Remarkably, there exists a 4D gauge model which is known to involve con-finement and where confinement is brought about by the dual Meißner ef-fect: $\mathcal{N} = 2$ supersymmetric Yang–Mills theory. Its dynamics was studied recently by Seiberg and Witten.

We are able to say just a few words about it here and refer the reader to the original papers [55] for details. The simplest version of the model involves the gauge field, a Dirac fermion in the adjoint representation of the gauge group, and a complex adjoint scalar ϕ^a. The model involves so-called *vacuum valleys*. This means that the potential is flat with respect to some gauge-invariant *moduli variables*. For an $SU(2)$ gauge group, there is only one such variable $u = \phi^a \phi^a$. This is the situation at the classical level and, due to supersymmetry, also when quantum perturbative and/or nonpertur-bative effects are taken into account. Thus, *any* vacuum expectation value $\langle \phi^a \rangle$ is admissible and, after gauge freedom is factored out, we are left not with one, but rather with a continuous family of the theories characterized by a complex parameter $u = \langle \phi^a \rangle^2$.

A nonzero vacuum expectation value $\langle \phi^a \rangle \neq 0$ brings about "breaking" of the $SU(2)$ gauge symmetry (with all the reservations made) by the Higgs mechanism. Like QCD, the $\mathcal{N} = 2$ super-Yang–Mills theory is asymptot-ically free and involves a fundamental scale Λ_{SYM} where nonperturbative effects come into play. If $|u| \gg \Lambda_{\text{SYM}}^2$, we are in the perturbative weak coupling regime, and the dynamics is very simple. Supersymmetry requires that the massless photon is upgraded to the full $\mathcal{N} = 2$ gauge supermulti-plet. Its scalar component corresponds to free motion along the bottom of the vacuum valley. Besides that, the spectrum involves massive particles classified according to their electric and also magnetic charges (the model is akin to the Georgi–Glashow model and involves monopoles). The masses are

$$M_{Z_e, Z_m} = |Z_e a(u) + Z_m a_D(u)| , \qquad (19.31)$$

where $a(u)$ and $a_D(u)$ are known functions of the moduli parameter u. The state $(Z_e, Z_m) = (1, 0)$ describes a W boson and the state $(Z_e, Z_m) = (0, 1)$ describes a monopole. There are also many different *dyons* which carry both electric and magnetic charge.

A remarkable fact is that the monopoles become exactly massless at some point $u \sim \Lambda_{SYM}^2$. The effective low energy Lagrangian at the vicinity of this point involves besides the photon and its superpartners (actually, it turns out that the proper degree of freedom here is not the photon field A_μ, but rather a *dual photon* field B_μ such that $\partial_\mu B_\nu - \partial_\nu B_\mu = \tilde{F}_{\mu\nu}$) also the monopole field and its superpartners.

Now, it is possible to slightly modify the theory by adding a small mass term which breaks the $\mathcal{N} = 2$ supersymmetry down to $\mathcal{N} = 1$, so that the monopole field condenses, the dual photon acquires a mass, the magnetic charges are screened, and the electric charges are confined.

Pure Yang–Mills theory also involves certain field configurations which are akin to monopoles and may play their role. The 't Hooft–Mandelstam mechanism means that these quasi-monopole degrees of freedom are important and that the associated effective field condenses. There are nevertheless a number of differences distinguishing the known mechanism of confinement in the Seiberg–Witten model from a hypothetical dual superconductor mechanism of confinement in QCD. The most essential distinction is that the former goes actually in two stages. First, we have the Higgs effect which means weak confinement of non-Abelian charges, and it occurs for any value of the vacuum moduli u. For some particular values of u,[tt] and with adding an extra mass term, Abelian $U(1)$ charges are screened too. On the other hand, in pure Yang–Mills theory everything should be confined at once by the same mechanism.

Most of the models discussed in this lecture enjoy confinement in its strong form. Weak confinement is more elusive: an analysis of the Wilson loop does not give anything, and we do not have any other convenient physical criteria. The only known to us example where one *is* sure that

---

[tt]We discussed above only the "monopole point" u_{mon}, but there is also the point u_{dyon} where the dyon state with $(Z_e, Z_m) = (1, 1)$ becomes massless. Physics of these two points is identical.

the theory is confining in the weak, but not in strong sense is the exactly solvable Schwinger model.

We conclude by recalling that there is a very close, not yet fully understood connection between weak confinement and chiral symmetry breaking. In all known cases they come together, and 't Hooft's self-consistency condition discussed at the end of Lecture 14 tells us quite definitely that in many cases confinement *implies* spontaneous chiral symmetry breaking.

Problem 3. As we have seen, the greater the dimension of space-time, the more "freedom" charged particles have, meaning that it is more difficult to confine them. In two dimensions, confinement is trivial. Three-dimensional theories are also confining, but linear confinement is proven only for the weakly coupled Polyakov model. In four dimensions, not all theories are confining, and confinement is proven only for certain exotic supersymmetric models. There is no confinement for $d \geq 5$.

Maybe, this is the reason for which our world is four-dimensional? Write an essay on this subject.

Solution. To be given in one of the following editions of this book.

Appendix

Unitary groups

We are not going to plunge deeply into group theory here and will describe (with the physical standards of rigor!) only some properties of the $U(N)$ and $SU(N)$ groups which we use in the main text.

Let $\mathbb{C}^N$ be the N-dimensional complex Euclidean vector space. The inner product is defined as

$$\langle x, y \rangle = \sum_{i=1}^{N} \bar{\xi}^i \eta_i , \qquad (A.1)$$

where ξ_i and η_i are the (complex) coordinates of the vectors x, y in some orthonormal basis $\{e_i\}$. The coordinates of the vectors $\{e_i'\}$ constituting some other orthonormal basis define the elements of a unitary matrix:

$$\begin{cases} e_i' = \sum_k U_i^k e_k \\ \langle e_i', e_j' \rangle = \delta_{ij} \end{cases} \longrightarrow \bar{U}_i^k U_j^k = \delta_j^i \text{ or } U^\dagger U = \mathbb{1} \qquad (A.2)$$

It follows that $|\det \|U\| |^2 = 1$ and hence

$$\det \|U\| = e^{i\alpha} \quad \text{with real } \alpha . \qquad (A.3)$$

The set of all N-dimensional unitary matrices forms a group: *(i)* a product of two unitary matrices is unitary and *(ii)* a unitary matrix U is invertible due to (A.3) and U^{-1} is unitary. This group has a continuous infinite number of elements and belongs to the class of *Lie* groups. It is denoted by $U(N)$.

The group $U(N)$ is not simple, but is a direct product of a smaller group $SU(N)$ involving only the unitary matrices whose determinant is equal to unity (such matrices are called *unimodular*) and the group $U(1)$ involving complex numbers $e^{i\phi}$ lying on the unit circle. This just means that any unitary matrix g can be expressed as $g = \tilde{g}e^{i\phi} \equiv (\tilde{g}, \phi)$ with unimodular $\tilde{g}$, and the product of two such matrices is $gh \equiv (\tilde{g}, \phi)(\tilde{h}, \chi) = (\tilde{g}\tilde{h}, \phi + \chi)$, where $\tilde{g}\tilde{h}$ is unimodular.

Any unitary matrix can be brought to diagonal form

$$U = \text{diag}(e^{i\phi_1}, \ldots, e^{i\phi_N}) \tag{A.4}$$

by *conjugation* $U \to hUh^\dagger$, where $h \in U(N)$. For unimodular matrices (and only these will be discussed in the following), the relation

$$\sum_{i=1}^{N} \phi_i = 0 \quad \text{modulo } 2\pi \tag{A.5}$$

holds. The special matrices $U = e^{2\pi i/N}\text{diag}(1, \ldots, 1)$ form a discrete subgroup Z_N, which is called the *center* of the $SU(N)$ group. The elements of the center commute with all elements of $SU(N)$.

A general complex $N \times N$ matrix depends on $2N^2$ real parameters, but the constraint (A.2) together with the condition $\det \|U\| = 1$ brings the number of parameters down to $N^2 - 1$. Therefore, the $SU(N)$ group represents an $(N^2 - 1)$-dimensional manifold. One can provide it with a "metric structure", i.e. define a distance between two points on the manifold. A *natural* definition for the distance between infinitesimally close unimodular matrices U and $U + dU$ is

$$ds^2 = C\text{Tr}\{dU^\dagger dU\}, \tag{A.6}$$

with an arbitrary positive real number C. This metric is invariant under the action of the group, i.e. the distance between the matrices U, $U + dU$, the matrices gU, $g(U + dU)$, and the matrices Uh, $(U + dU)h$, with $g, h \in SU(N)$, is the same. The metric (A.6) induces a group-invariant measure on the group manifold called *Haar measure*. If diagonalizing the matrices as in (A.4), it can be expressed, following Weyl, as

$$\mathcal{D}U = \prod_{i=1}^{N} \frac{d\phi_i}{2\pi} \, 2\pi\delta\left(\sum_{i=1}^{N} \phi_i\right) |P|^2, \tag{A.7}$$

where

$$P = \prod_{k<l}^{N} \left(e^{i\phi_k} - e^{i\phi_l}\right). \tag{A.8}$$

The coefficient in (A.7) is related to the coefficent C in (A.6) and is chosen such that the total volume $\int DU$ of the group manifold is normalized to 1.

Let us dwell on the latter point. Note that the integration in Eq. (A.7) goes over so-called *Weyl alcove* defined such that any U can be conjugated to the diagonal form (A.4), with ϕ_i lying within the alcove, in a unique way. The explicit structure of the Weyl alcove is

$$\phi_N = \frac{2\pi}{N} \sum_{j=1}^{N-1} js_j \,,$$

$$\phi_i = \phi_N - 2\pi \sum_{j=i}^{N-1} s_j \,, \qquad i = 1,\ldots,N-1\,, \qquad \text{(A.9)}$$

where the *Kac variables* s_j vary within the range

$$s_j \geq 0\,, \qquad \sum_{j=1}^{N-1} s_j \leq 1\,. \qquad \text{(A.10)}$$

The angles (A.9) are ordered $-\pi(N-1) \leq \phi_1 \leq \ldots \leq \phi_N \leq \pi(N-1)$. It is not difficult to see that

$$\prod_{i=1}^{N-1} d\phi_i = \frac{(2\pi)^{N-1}}{N} \prod_{i=1}^{N-1} ds_i \,. \qquad \text{(A.11)}$$

Substituting it into Eq. (A.7) and integrating over the region (A.10), bearing in mind that

$$|P|^2 = N! + \text{something that vanishes after integration}$$

(see **Problem 1** in Lecture 15), we obtain $\int DU = 1$.

For $N = 2$, the group manifold is just the sphere S^3 * and the invariant measure (A.7) is reduced to

$$DU_{N=2} = \frac{2}{\pi} d\Theta \sin^2 \Theta \,, \qquad \text{(A.13)}$$

with $\Theta \equiv \phi_1 = -\phi_2$ being the polar angle on S^3.

---

*It is best seen if using the parametrization

$$g \in SU(2) = -in_\mu \sigma_\mu, \qquad \mu = 1,\ldots,4 \qquad \text{(A.12)}$$

with $n_\mu^2 = 1$ and $\sigma_\mu = (i,\boldsymbol{\sigma})$.

A set of N-dimensional columns $\{x_i\}$ with $x_i \in \mathbb{C}$ which are transformed under the action of the group as $x_i \to U_i^j x_j$ is called the *fundamental representation space*. Physicists simply say that such a column belongs to the fundamental representation of $SU(N)$. There is also the antifundamental representation formed by the rows $\{\bar{x}^i\}$ and transformed as $\bar{x}^i \to \bar{x}^j (U^\dagger)_j^i$.

Another important representation is constructed on the basis of the group manifold $\mathcal{G}$ itself. Let us define the action of the element of the group U on $\mathcal{G}$ as the congugation $U(g) = UgU^\dagger$ ($g \in \mathcal{G}$). The important advantage of this definition [compared to possible definitions $U(g) = Ug$ or $U(g) = gU$] is that the small neighborhood of the group unity $\mathbb{1}$ is invariant under such congugation. Therefore, the action of U can be restricted to the flat Euclidean space representing a hyperplane tangent to the group manifold at the point $g = \mathbb{1}$. This tangent space is called the *adjoint* representation space.

Let us discuss it in some more details. Any element of $SU(N)$ in the neighborhood of $\mathbb{1}$ can be uniquely represented as $\exp\{iA\}$, where $A \ll 1$ (i.e. all elements of A are small) is a *Hermitian* matrix $A = A^\dagger$ with zero trace. These Hermitian matrices form an (N^2-1)-dimensional real vector space $\mathfrak{G}$.[†] The action of the group on $\mathfrak{G}$ induced by its action on $\mathcal{G}$ is $A \to UAU^\dagger$.

The vector space $\mathfrak{G}$ is endowed with the norm induced by the metric (A.6):

$$\|A\|^2 = C\mathrm{Tr}\{A^2\} . \tag{A.14}$$

Correspondingly, the inner product is $\langle A, B \rangle = C\mathrm{Tr}\{AB\}$. Let $\{t^a\}$ be an orthogonal basis in $\mathfrak{G}$ normalized as

$$\mathrm{Tr}\{t^a t^b\} = \frac{1}{2}\delta^{ab} . \tag{A.15}$$

The t^a are called *generators* of the group $SU(N)$. Indeed, t^a are intimately related to the small neighborhood of $\mathbb{1}$, i.e. to infinitesimal group transformations which "generate" the whole group. In spite of the fact that $\{t^a\}$ constitutes a basis in the adjoint representation space, they are called generators of the group in the *fundamental* representation. The reason for that is that the $N \times N$ matrices t^a determine an infinitesimal transfor-

---

[†]You may call it, say, $\mathcal{A}$ if you like, but gothic letters are indispensable if you want to learn this subject *really* well ...

mation of the column $\{x_i\}$ belonging to the fundamental representations:
$\delta x_i = i\alpha_a(t^a)^j_i x_j$.

The standard choice for $SU(2)$ is $t^a = \sigma^a/2$ (σ^a being the Pauli matrices) and, for $SU(3)$, it is $t^a = \lambda^a/2$, where

$$
\lambda^{1,2,3} = \begin{pmatrix} \sigma^{1,2,3} & 0 \\ 0 & 0 \end{pmatrix}, \qquad
\lambda^4 = \begin{pmatrix} 0 & 0 & 1 \\ 0 & 0 & 0 \\ 1 & 0 & 0 \end{pmatrix},
$$

$$
\lambda^5 = \begin{pmatrix} 0 & 0 & -i \\ 0 & 0 & 0 \\ i & 0 & 0 \end{pmatrix}, \qquad
\lambda^6 = \begin{pmatrix} 0 & 0 & 0 \\ 0 & 0 & 1 \\ 0 & 1 & 0 \end{pmatrix},
$$

$$
\lambda^7 = \begin{pmatrix} 0 & 0 & 0 \\ 0 & 0 & -i \\ 0 & i & 0 \end{pmatrix}, \qquad
\lambda^8 = \frac{1}{\sqrt{3}}\begin{pmatrix} 1 & 0 & 0 \\ 0 & 1 & 0 \\ 0 & 0 & -2 \end{pmatrix} \tag{A.16}
$$

are the *Gell-Mann matrices*.

If allowing complex coefficients, the set $\left\{\frac{1}{\sqrt{2N}}\mathbb{1},\ t^a\right\}$ constitute a complete orthogonal basis for all $N \times N$ complex matrices. Consider the tensor $\delta^i_m \delta^l_k$ as a set of matrices $(F^i_k)^l_m$ and expand it over this basis. We obtain the *Fierz identity*:

$$
\delta^i_m \delta^l_k = \frac{1}{N}\delta^i_k \delta^l_m + 2(t^a)^i_k (t^a)^l_m . \tag{A.17}
$$

Multiplying it by δ^k_l, we arrive at

$$
t^a t^a = \frac{N^2-1}{2N}\mathbb{1} \equiv c_F \mathbb{1}, \tag{A.18}
$$

i.e. the matrix $t^a t^a$ (called *quadratic Casimir operator* in the fundamental representation) is proportional to $\mathbb{1}$. Multiplying (A.17) by $(t^b)^k_l$, we obtain another useful relation,

$$
t^a t^b t^a = -\frac{1}{2N}t^b = \left(c_F - \frac{N}{2}\right)t^b . \tag{A.19}
$$

The elements of $\mathfrak{G}$ can also be understood as traceless tensors A^j_i with the "fundamental" lower and "antifundamental" upper indices. Besides fundamental, antifundamental, and adjoint representations of $SU(N)$, there are also other representations which can all be written as tensors $T^{j_1\ldots j_m}_{i_1\ldots i_n}$ with some number of upper and lower indices. The law of transformation is

$$T^{j_1\cdots j_m}_{i_1\cdots i_n} \ \to\ U^{k_1}_{i_1}\cdots U^{k_n}_{i_n}T^{l_1\cdots l_m}_{k_1\cdots k_n}(U^\dagger)^{j_1}_{l_1}\cdots(U^\dagger)^{j_m}_{l_m}\ . \qquad (A.20)$$

Generically, a representation $\{T^{j_1\cdots j_m}_{i_1\cdots i_n}\}$ is *reducible*. This means that the corresponding vector space can be written as a direct sum of a certain number of smaller vector spaces which are invariant under the action of the group. In a more colloquial language, the set of components of $\{T^{j_1\cdots j_m}_{i_1\cdots i_n}\}$ can be organized into subsets so that the components of each such subset are transformed into each other.

Consider a generic tensor T^j_i as an example. It gives a reducible representation of dimension N^2 of the group $SU(N)$. Now, one can present

$$T^j_i \ = \ \tilde{T}^j_i + \frac{1}{N}\delta^j_i T^k_k$$

with traceless $\tilde{T}^j_i$ and notice that the trace T^k_k is invariant under the action of the group and is never mixed with the components of $\tilde{T}^j_i$. The latter form the irreducible (N^2-1)-dimensional adjoint representation of $SU(N)$:

fundamental $\otimes$ **antifundamental** $=$ **adjoint** $\oplus$ **trivial** .

For $SU(3)$, this means $\mathbf{3}\otimes\bar{\mathbf{3}} = \mathbf{8}\oplus\mathbf{1}$.

In physical applications, the formula

$$\mathbf{3}\otimes\mathbf{3}\otimes\mathbf{3} \ = \ \mathbf{1}\oplus\mathbf{8}\oplus\mathbf{8}\oplus\mathbf{10} \qquad (A.21)$$

for the tensor product of 3 fundamental $SU(3)$ representations is used. It means that a generic 27-dimensional vector space $\{T_{ijk}\}$ is decomposed into a direct sum of 4 vector spaces whose dimensions are written in (A.21). To derive it, notice first that the trivial representation appears sometimes in disguise. Thus, the completely antisymmetric tensors $\epsilon_{i_1\ldots i_N}$ and $\epsilon^{j_1\cdots j_N}$ preserve their form under the action of the group $SU(N)$ and correspond actually to its trivial representation. Now, the singlet on the right side of Eq. (A.21) is $\epsilon^{ijk}T_{ijk}$, the decuplet is the tensor $T_{\{ijk\}}$ symmetrized over its indices and the two octets are

$$\epsilon^{ijk}T_{ijl} - \frac{1}{3}\delta^k_l\epsilon^{ijp}T_{ijp} \ \text{ and } \ \epsilon^{ijk}T_{lij} - \frac{1}{3}\delta^k_l\epsilon^{ijp}T_{pij} \qquad (A.22)$$

[the components of the third such possible tensor are expressed as linear combinations of (A.22)].

Let us return to the adjoint representation.

The commutator $[A, B]$ of two Hermitian matrices is an anti-Hermitian matrix with zero trace. That means that $\mathfrak{G}$ is not just a vector space, but a *Lie algebra* where an additional antisymmetric binary operation $(A, B) \rightarrow i[A, B]$ is defined.

Consider the commutator $[t^a, t^b]$ and expand it over the basis $\{t^a\}$. The coefficients are purely imaginary:

$$[t^a, t^b] = i f^{abc} t^c . \tag{A.23}$$

f^{abc} are called the *structure constants* of the group. Writing down $f^{abc} = -2i \text{Tr}\{[t^a, t^b] t^c\}$ and using the cyclic property of the trace, it is easy to see that the f^{abc} are antisymmetric under exchange of any pair of indices. For the group $SU(2)$, $f^{abc} = \epsilon^{abc}$. The structure constants satisfy an important identity,

$$f^{abe} f^{ecd} + f^{bce} f^{ead} + f^{cae} f^{ebd} = 0 , \tag{A.24}$$

which follows from the Jacobi identity for commutators,

$$[[t^a, t^b], t^c] + [[t^b, t^c], t^a] + [[t^c, t^a], t^b] = 0 .$$

Another important relation is

$$f^{abc} f^{abd} = N \delta^{cd} , \tag{A.25}$$

which follows from the definition (A.23) and from the identity

$$-\text{Tr}\{[t^a, t^c][t^a, t^d]\} = c_F \delta^{cd} - 2\text{Tr}\{t^a t^c t^a t^d\} = \frac{N}{2} \delta^{cd} , \tag{A.26}$$

where the relations (A.18), (A.19) were also used.

We can also consider the anticommutator $\{t^a, t^b\}_+$ and define

$$d^{abc} = 2\text{Tr}\{t^c \{t^a, t^b\}_+\} . \tag{A.27}$$

d^{abc} is symmetric under exchange of any pair of indices. For $SU(2)$, d^{abc} vanishes. The identity

$$d^{abc} d^{abd} = \frac{N^2 - 4}{N} \delta^{cd} \tag{A.28}$$

holds.

A set of $(N^2 - 1)$-dimensional matrices

$$(T^a)^{bc} = -i f^{abc} \tag{A.29}$$

are called the generators of the group in adjoint representation. Indeed, the coefficients A^a of an element $A^a t^a \in \mathfrak{G}$ are transformed under an infinitesimal group transformation as

$$\delta A^a = 2\mathrm{Tr}\left\{ t^a e^{i\alpha^b t^b} A^c t^c e^{-i\alpha^b t^b} \right\} - A^a = 2i\alpha^b \mathrm{Tr}\{t^a[t^b, t^c]\} A^c + o(\alpha)$$
$$\sim -f^{abc}\alpha^b A^c = i\alpha^b (T^b)^{ac} A^c . \qquad (A.30)$$

The Jacobi identity (A.24) can then be rewritten as

$$[T^a, T^b] = i f^{abc} T^c . \qquad (A.31)$$

As for the relation (A.25), it can be interpreted as

$$\mathrm{Tr}\{T^a T^b\} = N\delta^{ab} \quad \text{and also as} \quad T^a T^a = N\mathbb{1} , \qquad (A.32)$$

where $\mathbb{1}$ is the unit matrix of dimension $N^2 - 1$. We see that the generators T^a form an orthonormal basis in a vector space which is isomorphic to $\mathfrak{G}$ and has the same algebraic structure. The Casimir operator $T^a T^a$ is also proportional to the unit matrix with the coefficient $c_A = N$.

Similarly, one can define the generators in any representation R. They are given by a set of $N^2 - 1$ matrices T^a whose dimension coincides with the dimension of the representation $\dim(R)$. Their defining property is

$$\delta R = i\alpha^a T^a R \quad \text{for} \quad g = 1 + i\alpha^a t^a + o(\alpha) .$$

It implies that the generators satisfy the same commutation relations (A.31). To see that, act on R by a group commutator $ghg^{-1}h^{-1}$ of two different infinitesimal group transformations g, h and use the definition (A.23). In addition, the properties

$$\mathrm{Tr}\{T^a T^b\} = d_R \delta^{ab} , \qquad T^a T^a = c_R \mathbb{1} \qquad (A.33)$$

hold, where the coefficient c_R is the quadratic Casimir invariant in the corresponding representation and d_R is called *Dynkin index* of the representation R. Multiplying the first relation in Eq. (A.33) by δ^{ab} and comparing it with the trace of the second relation, we obtain

$$d_R = c_R \frac{\dim(R)}{\dim(\mathcal{G})} . \qquad (A.34)$$

This holds not only for a unitary, but for an arbitrary Lie group.

Bibliography

[1] Feynman, R. (1967) *The Character of Physical Laws*, MIT Press.

[2] Peskin, M.E. and Schroeder, D.V. (1995) *An Introduction to Quantum Field Theory*, Addison–Wesley.

[3] Weinberg, S. (1995 (v. 1); 1996 (v. 2); 2000 (v.3)) *The quantum theory of fields*, Cambridge Univ. Press.

[4] Faddeev, L.D. and Slavnov, A.A. (1980) *Gauge fields: Introduction to Quantum Theory*, Benjamin/Cummings.

[5] Voloshin, M.B. and Ter-Martirosyan, K.A. (1984) *Teorija Kalibrovochnyh Vzaimodejstvij Elementarnyh Chastic*, Energoatomizdat, Moscow.

[6] Yndurain, F.J. (1999) *The Theory of Quark and Gluon Interactions* , 3-rd ed., Springer-Verlag.

[7] Muta, T. (1998) *Foundations of Quantum Chromodynamics*, 2-nd ed., World Scientific, Singapore.

[8] Berestetsky, V.B., Lifshitz, E.M., and Pitaevsky, L.P. (1982) *Quantum Electrodynamics*, Pergamon, Oxford.

[9] Bjorken, J.D. and Drell, S.D. (1965) *Relativistic Quantum Mechanics*, McGraw-Hill, New York.

[10] Landau, L.D. and Lifshitz, E.M. (1975) *The Classical Theory of Fields*, Pergamon, Oxford.

[11] Shifman M.A. (2001) In: *Handbook on QCD, Boris Ioffe Festschritt* (Shifman M.A., ed), World Scientific, Singapore, v. 1, 126.

[12] Biro, T.S., Matinian, S.G., and Muller, B. (1994) *Chaos and gauge field theory*, World Scientific, Singapore.

[13] Brown, L.S. (1992) *Quantum field theory*, Cambridge Univ. Press.

[14] Berezin, V. (1966) *Method of Second Quantization.*, Academic Press.

[15] Ogievetsky, V.I. and Sokatchev, E.S. (1980) *Sov. J. of Nucl. Phys.* **32**, 589.

[16] Actor, A. (1979) *Rev. Mod. Phys.* **51**, 461.

[17] Kugo, T. and Ojima, I. (1979) *Prog. Theor. Phys. Suppl.* **66**, 1.

[18] Celmaster, W. and Gonsalves, R. (1979) *Phys. Rev.* **D20**, 1420; Pascual, P.

and Tarrach, R. (1980) *Nucl. Phys.* **B174**, 123.

[19] Feynman, R. (1972) *Statistical Mechanics, Ch. 8* Addison–Wesley.

[20] Shifman, M. A. and Vainshtein, A.I. (1986) *Sov. J. Nucl. Phys.* **44**, 321; (1986) *Nucl. Phys.* **B277**, 456; Arkani-Hamed, N. and Murayama, H. (1997) hep-ph/9707133.

[21] Ovsyannikov, L.V. (1956) *Dokl. Akad. Nauk* **109**, 1112 (a good review of the results of this original paper can be found in [Bogolyubov, N.N. and Shirkov, D.V. (1980) *Introduction to the theory of quantized fields*, Wiley.]); Callan, C. (1970) *Phys. Rev.* **D2**, 1542; Symanzik, K. (1970) *Commun. Math. Phys.* **18**, 227.

[22] Doria, R., Frenkel, J., and Taylor, J.C. (1980) *Nucl. Phys.* **B168**, 93.

[23] Frenkel, J. and Taylor, J.C. (1976) *Nucl. Phys.* **B116**, 185.

[24] Guzenko, S. Ya. (1963) *ZhETF.* **44**, 1093.

[25] Smilga, A.V. (1979) *Nucl. Phys.* **B161**, 449.

[26] Vysotsky, M.I. and Smilga, A.V. (1979) *Nucl. Phys.* **B150**, 173.

[27] Smilga, A.V. (1979) *Phys. Lett.* **B83**, 357.

[28] Ioffe, B.L., Khoze, V., and Lipatov, L.N. (1984) *Hard Processes*, North Holland, Amsterdam.

[29] Moyal, I.E. (1949) *Proc. Cambr. Phil. Soc.* **45**, 99.

[30] Stalin, I.V. (1939) In: *Proceedings of the XVIII Party Congress*, Moscow.

[31] Smilga, A.V. (1987) *Witten Index* In: *Proceedings of the XXII LINP Winter School*, Leningrad.

[32] see e.g. Leutwyler, H. (1992) In: *Perspectives in the Standard Model*, Proc. 1991 Theor. Adv. Study, Institute in Elementary Particle Physics, World Scientific, Singapore, 97.

[33] see e.g. Dubrovin, B.A., Fomenko, A.T., and Novikov, S.P. (1984) *Modern Geometry — Methods and Applications*, Springer-Verlag, Part 2, 13.

[34] Neuberger, H. (1998) *Phys. Lett.* **B417**, 141; Hasenfratz, P., Laliena, V., and Niedermayer, (1998) *F. Phys. Lett.* **B427**, 125; Lüscher, M. (1998) *Phys. Lett.* **B428**, 342.

[35] Zyablyuk, K. (2000) *JHEP.* **0006**, 025

[36] Nussinov, S. and Lampert, M.A. (1999) hep-ph/9911532.

[37] Smilga, A. and Stern, J. (1993) *J. Phys. Lett.* **B318**, 531 .

[38] Osborn, J., Toublan, D., and Verbaarschot, J. (1999) *J. Nucl. Phys.* **B540**, 317.

[39] Verbaarschot, J. (1994) *Acta Phys. Polon.* **B25**, 133.

[40] Dolgov, A.D. and Zakharov, V.I. (1971) *Nucl. Phys.* **B27**, 525.

[41] Smilga, A. (1999) *Phys. Rev.* **D59**, 114021.

[42] Bagan, E., Lavelle, M., and McMullan, D. (2000) *Ann. Phys.* **282**, 471; 503.

[43] Bhanot, G., Demeterfi, K., and Klebanov, I. (1993) *Phys. Rev.* **D48**, 4980.

[44] Shifman, M. (ed.) (1992) *Vacuum structure and QCD sum rules*: reprints, North Holland, Amsterdam.

[45] Smilga, A.V. (1997) *Phys. Repts.* **291**, 1.

[46] Lifshitz, E.M. and Pitaevsky, L.P. (1982) *Physical Kinetics*, Pergamon Press,

Oxford.

[47] Drukarev, E.G. and Levin, E.M. (1988) *JETP Lett.* **48**, 338; Hatsuda, T., Hogaasen, H., and Prakash, M. (1991) *Phys. Rev. Lett.* **66**, 2851.

[48] Bailin, D. and Love, A. (1984) *Phys. Repts.* **107**, 325.

[49] Rapp, R., Schäfer, T., Shuryak, E.A., and Velkovsky, M. (1998) *Phys. Rev. Lett.* **81**, 53; Afford, M., Rajagopal, K., and Wilczek, F. (1998) *Phys. Lett.* **B422**, 247.

[50] Son, D.T. (1999) *Phys. Rev.* **D59**, 094019.

[51] Gross, D.J., Klebanov, I.R., Matytsin, A.V., and Smilga, A.V. (1996) *Nucl. Phys.* **B461**, 109.

[52] Rajaraman, R. (1982) *Solitons and Instantons in Quantum Field Theory*, North-Holland, Amsterdam.

[53] Rubakov, V.A. (1999) *Klassicheskie Kalibrovochnye Polja*, Editorial URSS, Moscow.

[54] Landau, L.D. and Lifshitz, E.M. (1980) *Statistical Physics, Part I*, §78, Pergamon Press, Oxford.

[55] Seiberg, N. and Witten, E. (1994) *Nucl. Phys.* **B426**, 19; **B431**, 484.

Index

The index is complementary to the table of contents. We have tried, whenever possible, to avoid "double counting".